1B15

In memory of

Wes Parrish

FPA   Book # 71

Automatic Sprinkler
& Standpipe Systems

by JOHN L. BRYAN

NATIONAL FIRE PROTECTION

ASSOCIATION NFPA®

470 Atlantic Avenue, Boston, MA 02210

ABOUT THE AUTHOR

Dr. John L. Bryan has thirty years of paid and volunteer fire department experience, with a current membership in the College Park, Maryland Volunteer Fire Department. He is a member of the National Fire Protection Association, The Fire Council of Underwriters Laboratories, the National Professional Qualifications Board for the Fire Service, and the Society of Fire Protection Engineers. He is an Associate Member of the International Association of Fire Chiefs and the American Society for Testing and Materials. He has served as a professional and educational consultant to various industrial and governmental agencies, and has been Professor and Chairman of the Fire Protection Curriculum in the College of Engineering at the University of Maryland since 1956.

NFPA No. TXT-1
Standard Book No. 87765-073-X
Library of Congress No. 76-5252
Printed in U.S.A.

ACKNOWLEDGMENTS

Grateful acknowledgment is due to the following for use of their material in this text:

J. C. Abbott; A. D. T. Security Systems; Robert J. Alban; American Insurance Association; American Mutual Insurance Alliance; Wayne E. Ault; Automatic Sprinkler Corporation; Charles F. Averill; Charles W. Bahme; R. Baldwin; David M. Banwarth; J. S. Barritt; Gordon M. Betz (Western Electric Co., Inc.); Horatio Bond; Ernest H. Bratlie, Jr.; G. Bray; David E. Breen; John K. Brouchard; Stanley G. Budzynski; Carrol E. Burtner; Building Officials and Code Administrators International; W. G. Butturini; Layard E. Campbell; Raymond J. Casey; Peter Joseph Chicarello, Jr.; Bert M. Cohn; College Park Volunteer Fire Department, College Park, Maryland; Richard L. P. Custer; Thomas M. Czarnecki; Gorham Dana; Dick A. Decker; J. R. DeMonbrun; Department of the Navy, United States Department of Defense; John B. Dietz; Seddon T. Duke; Lester A. Eggleston; Frank J. Fabin; Factory Insurance Association; Factory Mutual System; John M. Foehl; E. W. Fowler; J. R. Gaskill; Glen Echo Fire Department, Glen Echo, Maryland; Globe Automatic Sprinkler Company; Robert S. Godfrey; Alan I. Gomberg; G. J. Grabowski; Grinnell Company, Inc.; James M. Hammack; Gregory A. Harrison; William G. Hayne; William R. Herrera; A. J. M. Heseldon; Harry E. Hickey; P. L. Hinkley; James L. Houser; H. Hoyle; Insurance Services Office; International Fire Service Training Association, Oklahoma State University; Dan W. Jacobsen; Rolf Jensen; Donald J. Keigher; Warren Y. Kimball; Hugh B. Kirkman; Jacob B. Klevan; Alvin Koog; Boris Laiming; Lloyd A. Layman; Stephen C. Leahy (College Park Volunteer Fire Department, College Park, Maryland, and Environmental Safety Services Division, University of Maryland); William L. Livingston; Los Angeles City Fire Department, Los Angeles, California; John E. Luley; Don McClanahan; L. M. McLaughlin; J. W. McCormick; Michael W. Magee; Hugh M. Maguire; Gerald E. Marks; H. W. Marryatt; Duane McSmith; L. E. Medlock; Philip H. Merdinyan; Joseph T. Merry; William D. Milne; J. L. Murrow; P. Nash; National Aeronautics and Space Administration; National Automatic Sprinkler and Fire Control Association, Inc.; National Board of Fire Underwriters; National Fire Protection Association; T. Francis O'Connor; M. J. O'Dogherty; Richard M. Patton; Peerless Pump (FMC Corp.); Marshall E. Peterson; W. Robert Powers; E. N. Proudfoot; Raisler Sprinkler Company; J. Randall; Edward J. Reilly; Reliable Automatic Sprinkler Company, Inc.; Jack Rhodes; Richard E. Rice; Benjamin Richards; Richard E. Ritz; Rockwood Sprinkler Corporation; A. K. Rosenhan; Earl J. Schiffhauer; Chester W. Schirmer; Seco Manufacturing, Inc.; O. J. Seeds; Donald C. Shaw; Raymond E. Shea; Bernard J. Shelley; David Shpilberg; Carl M. Siefried; Orville M. Slye, Jr.; Paul D. Smith; S-R Products, Inc.; Star Sprinkler Corporation; Richard E. Stevens; J. John Stratta; P. H. Thomas; Norman J. Thompson; Richard P. Thornberry; Emile W. J. Troup; TRW Mission Manufacturing Company;

Underwriters Laboratories, Inc.; Union Carbide Corporation; United States Atomic Energy Commission; University of Maryland, Fire Protection Curriculum; Viking Corporation; T. E. Waterman; John M. Watts, Jr.; W. E. Weathersby; William A. Webb; Arthur Widawsky; Joel Woods (Maryland Fire and Rescue Institute); Worthington Corporation; L. Murray Young; R. A. Young.

PREFACE

The intent and objective of this book is to provide the student with information relative to the basic concepts and principles involved with the design, installation, and function of standpipe and automatic sprinkler systems. It is difficult to recognize any single concept which has so effectively altered the industrial fire experience in the United States as the automatic sprinkler system. It is hoped the students involved in courses of study or engaged in solitary self-improvement will have their curiosity aroused and their interests awakened to pursue the readings in the bibliographies provided at the end of each chapter. The history and development of the concept of the automatic sprinkler system is in reality the story of the development of the concept of professional responsibility for fire protection and life safety in the United States involving the creation of the National Fire Protection Association, and the development of public and personal awareness. It is recommended the student devote some effort and energy to a review of the pioneering concepts and principles in the development of the modern automatic sprinkler system.

It is earnestly hoped the students and readers of this text will develop an understanding of the basic principles and concepts involved in both standpipe and automatic sprinkler systems, and thus be able to provide the new and unique ideas necessary for the development and improvement of these basic fire protection and suppression systems. There presently appears to be a significant and broad interest in the development and improvement of the automatic sprinkler system. These improvement efforts have provided a system which activates with a faster response, and also provides a greater area coverage with equivalent water. Therefore, the application of the automatic sprinkler system in a greater number of occupancies and situations for the life safety of the individual occupant is being considered and implemented.

This text was not written as an inspection, maintenance, or installation manual, and the reader desiring such detailed information should consult the bibliographies and contact the equipment manufacturers, who provide detailed manuals for their specific systems.

I wish to thank the manufacturers and distributors of the various devices equipment, and systems utilized on both standpipe and automatic sprinkler systems, and the evaluation and certification organizations for their permission and generous contribution of illustrations, drawings, and information. I have attempted to identify the data and illustrations provided by both individuals and organizations, and without their assistance this publication would not have been possible.

I want to thank all of the many individuals who contributed to the content and context of this book, in particular the members of the various segments of the fire protection profession who generously allowed the utilization of information and illustrations from their articles or publications. To the present and past students of the Fire Protection Curriculum, I wish to acknowledge their contributions of stimulation, inspiration, understanding, and the respect for knowledge which they have transmitted in their interaction in the learning environment I have been priviledged to share with them at the University of Maryland.

To Mrs. Eloise McBrier I extend the most sincere appreciation and thanks for her understanding, efficient, and outstanding efforts in the typing, compilation, and preparation of this manuscript. I accept full responsibility for the limitations and omissions this book may contain. The recognition and appreciation of the contributions of both organizations and individuals is not intended to imply any sharing of this responsibility.

John L. Bryan

CONTENTS

Automatic Sprinkler

& Standpipe Systems

Chapter **1**

Standpipe and Hose Systems

STANDPIPE SYSTEMS

Standpipe and hose systems provide a means for the manual application of water to fires in buildings. Although standpipe systems are required in buildings of large area and in buildings more than four stories high so that fire departments can place hose lines in service with a minimum of delay, they do not take the place of automatic extinguishing systems. Horizontal standpipes are provided in warehouse, manufacturing, and shopping mall facilities. Vertical standpipes are found in most buildings more than four stories high and are essential in high-rise buildings beyond the reach of fire department aerial ladder equipment.

While standpipes with hose systems have been placed in buildings for the use of occupants, many fire departments prefer they be of the type designed for fire department use without the attached hose for occupant use. Some building codes require standpipes relative to the occupancy of the building; however, the primary factors to be considered are the area of the building for a horizontal standpipe system, and the height of the building for a vertical standpipe system.

Classification of Standpipe Systems

The National Fire Protection Association's NFPA 14, *Standard for the Installation of Standpipe and Hose Systems* (hereinafter referred to in this text as the *NFPA Standpipe and Hose Systems Standard*), contains a classification of standpipes based on the purpose or intended usage of the system.[1]

1

The Class I System is designed for use by fire department personnel or other persons (such as industrial fire brigade or security personnel) who are trained in fire fighting procedures involving $2\frac{1}{2}$-inch hose streams at high pressures.

The Class II System is designed for use primarily by building occupants who, presumably, have no specialized training. Class II Systems are usually limited to hose sizes of $1\frac{1}{2}$ inches and smaller. Some fire protection personnel prefer to restrict the installation of Class II Systems to industrial or military occupancies where personnel systematically receive fire suppression training. The provision of standpipe hose and nozzles for occupancies with limited numbers of personnel, or for occupancies populated extensively by the general public (including office buildings), would appear to create the risk of personnel injury from the operation of the hose lines. Therefore, the assumed benefits of fire control or suppression from the operation of the hose on a Class II Standpipe System are neither assured nor dependable. It would seem that the provision of a standpipe system with $1\frac{1}{2}$-inch hose for occupant use is usually favored by individuals who do not have: (1) an understanding of, or appreciation for, the hazards of manual fire fighting procedures, or (2) an awareness of the hazards to untrained personnel who do not have adequate personnel protective equipment, including breathing apparatus.

The Class III Standpipe System is designed for use by either the fire department or the building occupants. Thus, a Class III System will have both $2\frac{1}{2}$-inch and $1\frac{1}{2}$-inch or smaller outlets, and will usually have $1\frac{1}{2}$-inch or smaller hose and

Table 1.1.   Summary of National Fire Protection Association
Standpipe Standards*

TYPE	INTENDED USE	SIZE HOSE AND DISTRIBUTION	MINIMUM SIZE PIPE	MINIMUM WATER SUPPLY
Class I	Heavy Streams Fire Department Trained Personnel Advanced Stages of Fire	$2\frac{1}{2}$-in. connections All portions of each story or section within 30 ft of nozzle with 100 ft of hose	4 in. up to 100 ft 6 in. above 100 ft (275 ft maximum unless pressure regulated.)	500 gpm 1st standpipe 250 gpm each additional (2,500 gpm maximum) 30 minute duration 65 psi at top outlet with 500 gpm flow
Class II	Small streams Building occupants Incipient Fire	$1\frac{1}{2}$-in. connections (Distribution same as Class I)	2 in. up to 50 ft $2\frac{1}{2}$ in. above 50 ft	100 gpm per building 30 minute duration 65 psi at top outlet with 100 gpm flowing
Class III	Both of above	Same as Class I with added $1\frac{1}{2}$-in. outlets or $1\frac{1}{2}$-in. adapters and $1\frac{1}{2}$-in. hose.	Same as Class I	Same as Class I

*From NFPA 14, *Standard for the Installation of Standpipe and Hose Systems.*

nozzle attached. A Class I System can be converted to a Class III System by attaching a 2½-inch to 1½-inch reducer and 1½-inch hose and nozzle, provided the reducer and hose can be readily disconnected. Table 1.1 presents a summary of the hose, pipe size, and minimum water supply requirements for the three classes of standpipe and hose systems, as classified by the *NFPA Standpipe and Hose Systems Standard.* Class I, II, and III Standpipe Systems are shown in Figures 1.1, 1.2, and 1.3.

Fig. 1.1. Class I Standpipe System. (Glen Echo Fire Department, Glen Echo, Maryland)

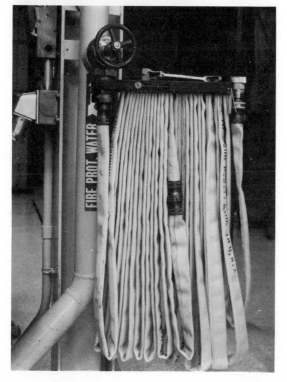

Fig. 1.2. Class II Standpipe System. Note size of standpipe and 1½-inch linen hose for building occupant use. (National Aeronautics and Space Administration)

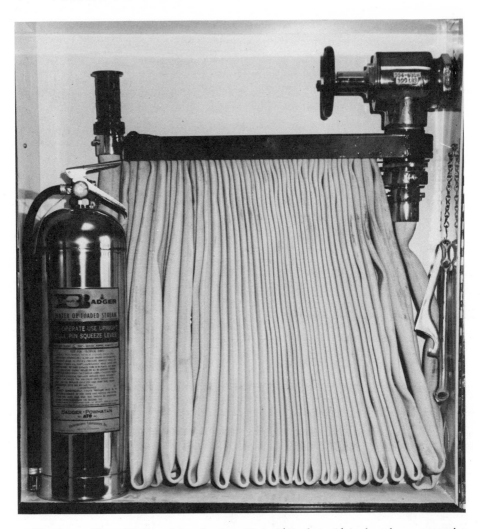

Fig. 1.3.   Class III Standpipe System. Note 2½-inch to 1½-inch reducer on outlet to provide 1½-inch synthetic fiber and rubber single-jacketed hose for building occupant use. (National Aeronautics and Space Administration)

Types of Standpipe Systems

Standpipe systems are also classified by type according to the design of the system relative to the water supply features of the system. The *NFPA Standpipe and Hose Systems Standard* defines the various types of standpipe systems as follows:*

   1. Wet standpipe system having supply valve open and water pressure maintained at all times.

---

*From NFPA 14, *Standard for the Installation of Standpipe and Hose Systems*, 1974 ed., NFPA, Boston, p. 4.

2. Standpipe system so arranged through the use of approved devices as to admit water to the system automatically by opening a hose valve.

3. Standpipe system arranged to admit water to the system through manual operation of approved remote control devices located at each hose station.

4. Dry standpipe having no permanent water supply.

In addition to the preceding types of systems, systems that are similar to wet standpipe systems can be found in some areas; however, in such systems the water supply valve which keeps the system full of water involves a limited 1-inch or ¾-inch connection. These systems are sometimes called "primed" systems, and operating pressure for their use with hose lines is not provided until the public fire department connects to the fire department connection, or until a fire pump is started. The reported advantages of the primed system are the reduction in the delivery time of the water as compared with the dry standpipe, and the reduction in corrosive effects on the inside of the standpipe. The dry standpipe may be attached to the outside of buildings, often adjacent to the exterior fire escape, with the system readily accessible for use by the public fire department. A standpipe system is primarily designed to save time for fire department personnel when placing hose streams in service on the upper floors of buildings. Figure 1.4 shows a 3-inch hose being connected from the public water system and the fire department pumper to the fire department connection for the standpipe system.

Fig. 1.4. Connection of 3-inch hose from fire department pumper and public water system to standpipe system. Note that both caps have been removed from fire department connection to facilitate connection of second hose line if required later. (Stephen C. Leahy, College Park Volunteer Fire Department, College Park, Maryland)

WATER SUPPLIES FOR STANDPIPES

The water supply to a standpipe will vary with the design of the particular standpipe system and according to the location of the structure which contains the standpipe. Acceptable water supply sources include connections to public or private water mains, pressure tanks, gravity tanks, and connections to fire pumps.

Some standpipe systems are designed so that when a discharge valve at the hose station is opened, the water from public or private water mains is automatically admitted into the system. Such systems may utilize a principle of operation similar to that of dry pipe sprinkler systems. Thus, the standpipe normally contains air pressure; the opening of the discharge valve causes a pressure drop which, in turn, allows water to flow into the standpipe system. The systems that are designed to operate with remote control devices often utilize a principle of operation similar to the deluge valve, with a remote manual release located adjacent to each standpipe discharge valve or hose station. The remote control devices may be electrical, pneumatic, or hydraulic in design and operation. These standpipe systems, which allow for the water to flow into the system when the discharge valve is opened, are usually found in freezing climates in occupancies where heat is not provided (including warehouses, piers, and similar structures).

The typical sources of water for standpipe systems are similar to the water sources for sprinkler systems. The most prevalent source of water is a direct connection to a public or private water main system. Although an adequate supply of water may be available in the water mains of high-rise buildings, the pressure of the water may be inadequate if the height of a building exceeds 10 stories (43.4 psi are required to overcome the pressure loss in 100 feet of height). Thus, the use of fire pumps is often required on standpipe systems; in high-rise buildings, series of fire pumps are usually arranged to supply vertical zones of the building. The *NFPA Standpipe and Hose Systems Standard* recommends that the maximum height per zone be 275 feet; this involves approximately 120 psi head pressure loss on the system. Also, the *NFPA Standpipe and Hose Systems Standard* recommends a water supply capability at the topmost outlet of the standpipe system to provide 500 gpm at 65 psi for Class I and III Systems, and 100 gpm at 65 psi for Class II Systems (see Table 1.1). At least one water supply source should be automatic and should be capable of supplying the initial hose streams on Class II and III Systems until the auxiliary or secondary sources of supply can be utilized. Many standpipe systems utilize the siamese fire department connection as the auxiliary or secondary source of water supply, and with the dry standpipe the fire department connection is the only source of water.

Fire Pumps

In order to supply the pressure needed to assure an adequate volume of water at an efficient pressure, a stationary pump with an automatic starting controller can be connected to the standpipe. With a wet standpipe normally filled with water from the connection to the public water main, the automatic fire pump is usually arranged with the controller on a pressure drop or water flow in the standpipe. Thus, while a minor pressure drop or flow will not start the fire pump, a pressure drop that indicates that a standpipe hose line valve is being opened will. Figure 1.5 shows an automatic fire pump controller, as installed for a wet standpipe system. Automatic starting controllers for fire pumps are expensive and require extensive maintenance and periodic testing. Fire pumps that are automatically operated preferably are driven by electric motors due to the flexibility of automatic controller operation. However, steam turbine or internal combustion engines using gasoline, diesel fuel, or natural gas can be arranged for automatic start operation. Gasoline engines are not recommended by many authorities, or in NFPA 20, *Standard for the Installation of Centrifugal Fire Pumps* (hereinafter referred to in this text as the *NFPA Centrifugal Pump Standard*), due to the fire hazard inside buildings.[2] Where electric motor drive is used, standby power generation may be required.

Fig. 1.5. An automatic fire pump controller. (Stephen C. Leahy, Environmental Safety Services Division, University of Maryland)

Manually controlled fire pumps can also be used to supply standpipe systems. The manually controlled pumps can be arranged on standpipe systems in combination with a pressure tank or a gravity tank. The pressure tank is often arranged to operate on the pressure drop in the standpipe system caused by the operation of a system discharge valve. The pressure tank operation then provides visual and audible signals on the fire alarm system which, in turn, alerts the responsible personnel to manually start the fire pump. This type of fire pump operation is often found in industrial and manufacturing occupancies where the fire pump is located in a power plant, boiler house, or other location having personnel in 24-hour attendance.

The manual fire pump operation is often found where the industry utilizes a steam operated turbine or reciprocating fire pump, since these pumps are often located at the source of steam, the power plant, or boiler station. Steam fire pumps will usually be found in occupancies where steam is required in the operation of the industry.

Figure 1.6 shows a diagram of a steam reciprocating stationary fire pump. Although these pumps have not been manufactured since the 1920s, they can still be found in industrial plants.

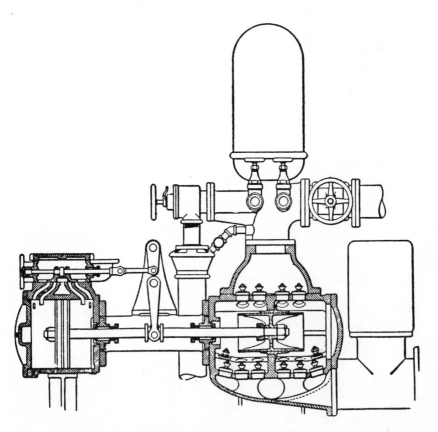

Fig. 1.6.   Steam reciprocating fire pump.

Standpipe risers and branch lines may be equipped with water flow alarm devices of the same type used in the wet pipe automatic sprinkler system. Such devices are useful in large systems to indicate the area of the building or the complex in which the standpipe system hose station on a Class II or III System has been activated. Figure 1.7 shows an electric motor driven horizontal centrifugal fire pump which is arranged for automatic start with a water pressure drop from the opening of a standpipe control valve. Figure 1.8 is a diagrammatic view of a horizontal split-case fire pump. Note that fire pumps are usually of single stage construction. A vertical submersible turbine pump is preferred when the water source is a well or reservoir that requires a lift (see Figure 1.9).

Fig. 1.7. Electric motor driven horizontal centrifugal fire pump. (Stephen C. Leahy, Environmental Safety Services Division, University of Maryland)

Fire pumps are also utilized to supply standpipes from aboveground reservoirs or suction tanks, as well as from public or private water mains. Fire pumps always have discharge headers or series of outlets located adjacent to them, often on the outside of the building. Some of these connections may be identified as wall hydrants. Fire department personnel should be familiar with the locations of fire pumps since these wall hydrants or header outlets may be valuable in a fire situation involving adjacent buildings or an exposure to the fire pump locations. Figure 1.10 shows an exterior wall hydrant located adjacent to an interior fire pump location.

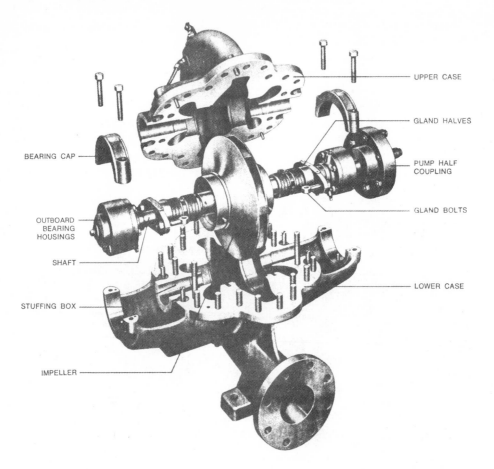

Fig. 1.8. Diagrammatic view of a horizontal split-case centrifugal fire pump. (Peerless Pump, FMC Corp.)

Pressure maintenance pumps, sometimes referred to as jockey or makeup pumps, are often found on sprinkler systems and standpipe systems. These pumps are mostly of the centrifugal type, although vane and positive displacement type pumps may be found on some systems. Usually provided with electric motor drive and having limited gpm capacity, these pumps operate on the initial pressure drop from the system and prevent the fire pump from cycling on and off due to pressure losses from surges on the system or leakage through valves.

The jockey or pressure maintenance pump first operates with the initial pressure drop; once the flow of water from the standpipe system or the sprinklers exceeds the capacity of the jockey pump, the continued drop in pressure activates the fire pump.

Figure 1.11 shows a jockey pump installed on a system supplying both wet standpipes and wet pipe sprinkler systems.

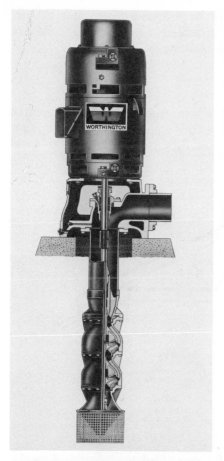

Fig. 1.10. Wall hydrant. (Stephen C. Leahy, Environmental Safety Services Division, University of Maryland)

Fig. 1.9. Vertical submersible turbine fire pump. (Worthington Corp.)

Fig. 1.11. Pressure maintenance pump (jockey pump). (Stephen C. Leahy, Environmental Safety Services Division, University of Maryland)

Stationary fire pumps installed in buildings to assure adequate water pressure and volume to the standpipe systems are an effective and valuable assistance factor to the fire department, especially in high-rise buildings.

Gravity Tanks

Gravity tanks are installed on the roofs of some buildings and are supported on towers in many industrial, manufacturing, and warehouse areas. The prime advantage of the gravity tank is its head pressure capability. Due to this capa-

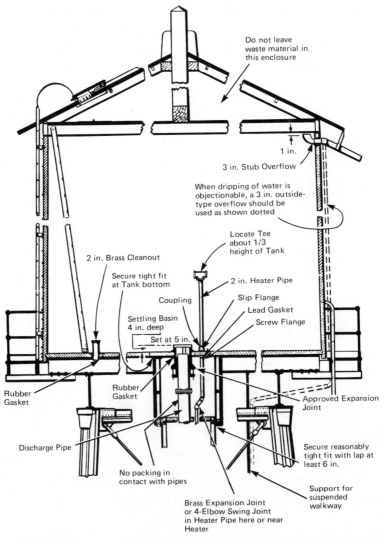

Fig. 1.12. Wood gravity tank. (From NFPA 22, *Standard for Water Tanks for Private Fire Protection*)[3]

bility, the gravity tank is able to provide a constant source of pressure to the standpipe or sprinkler system. For every foot the water is elevated above a gravity tank's discharge outlet, the pressure is increased by 0.434 pounds. However, gravity tanks require extensive and continuing maintenance; in many geographical areas, they also require protection against freezing.

Separate tanks or discharge pipes are recommended to prevent the gravity tank from being emptied of water when the tank supplies both operational and fire protection systems. Gravity tanks have collapsed from ice loads created by overflows when being filled, and from lack of maintenance. Gravity tanks are of various sizes and are manufactured from both wood and steel, with steel being the prevalent material utilized in modern tank construction. Figures 1.12, 1.13, and 1.14 illustrate the general features of these types of gravity tanks.

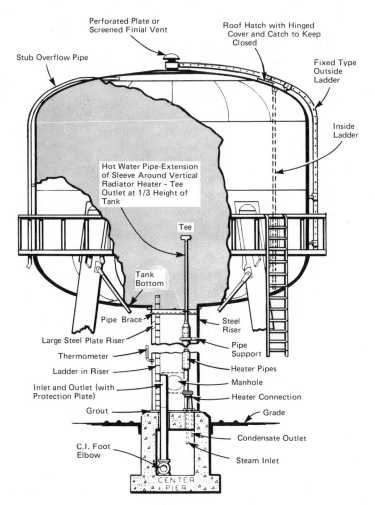

Fig. 1.13. Steel gravity tank of double ellipsoidal design. (From NFPA 22, *Standard for Water Tanks for Private Fire Protection*)[3]

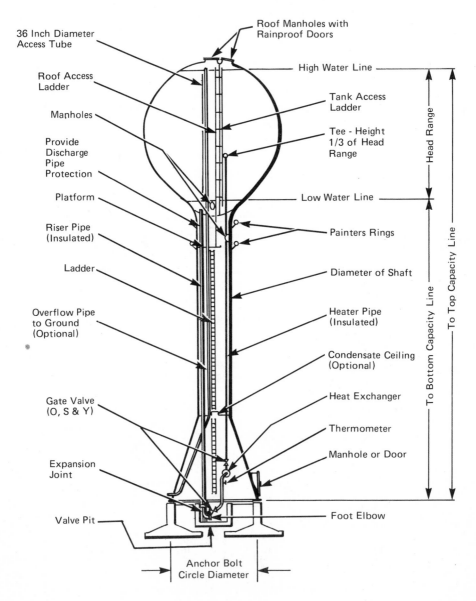

36 Inch Diameter
Access Tube

Roof Manholes with
Rainproof Doors

Roof Access
Ladder

High Water Line

Manholes

Tank Access
Ladder

Provide
Discharge
Pipe
Protection

Tee - Height
1/3 of Head
Range

Platform

Low Water Line

Riser Pipe
(Insulated)

Painters Rings

Ladder

Diameter of Shaft

Overflow Pipe
to Ground
(Optional)

Heater Pipe
(Insulated)

Condensate Ceiling
(Optional)

Gate Valve
(O, S & Y)

Heat Exchanger

Thermometer

Expansion
Joint

Manhole or Door

Valve Pit

Foot Elbow

Anchor Bolt
Circle Diameter

Head Range

To Top Capacity Line

To Bottom Capacity Line

Fig. 1.14. Steel gravity tank of steel pedestal design. (From NFPA 22, *Standard for Water Tanks for Private Fire Protection*)[3]

Pressure Tanks

Pressure tanks are enclosed water tanks of limited size, with air pressure maintained within the tank to provide the velocity energy for the discharge of the water from the tank. Thus, pressure tanks operate on a principle similar to that of pressurized air-water extinguishers. It is preferable to locate a pressure tank on a building's upper floor when building construction is sufficiently advanced to support the structural load of a tank filled with water. However, with facilities involving light construction such as health care facilities or nursing homes where public water mains are not available, the tanks should be located in basements or outside the structure. Pressure tanks provide a means of assuring immediate water discharge from standpipes in tall structures; thus, they are sometimes used in combination with fire pumps or gravity tanks.

Normally, pressure tanks are kept approximately two-thirds full of water and charged with a minimum air pressure of 75 psi. When gravity tanks are connected to a supply riser for sprinklers or standpipes with a pressure tank, the connection should be made 40 feet or more below the bottom of the gravity tank with the check valve for the gravity tank at the connection in order to avoid air lock. Air lock occurs when the gravity tank and the pressure tank are connected to a common riser, and the gravity water pressure from the gravity tank is less than the remaining air and water pressure in the riser at the gravity tank check valve. Thus, for new systems being installed, air lock can be avoided by providing sufficient gravity pressure at the gravity tank check valve; this can be accomplished by making the connection a minimum of 40 feet below the bottom of the gravity tank. In existing installations, a possible solution to air lock is to reduce the air pressure maintained on the pressure tank to the safe minimum to allow discharge of the water and also to increase the water level in the pressure tank to about 80 percent. A reduction in air pressure to approximately 60 psi with increased water should provide a minimum residual air pressure after the pressure tank has been completely discharged. Figure 1.15 illustrates the fittings and connections to a standard pressure tank.

Fire Department Connection

The fire department connection is mandatory on all standpipe systems, and is recommended on sprinkler systems. The fire department connection provides the only means of supplying water to the dry standpipe system (see Figure 1.4). The only valve allowed in the connecting pipe between the standpipe riser and the fire department connection is a check valve which is installed to allow water to flow into the system from the fire department's siamese connection. Because no type of control valve is allowed on the pipe, the fire department connection should always be available to supply water to the standpipe system. Figure 1.16 illustrates the arrangement of the fire department connection as recommended by the NFPA.

The fire department connection should always be inspected by fire companies when conducting in-service inspections, especially when such con-

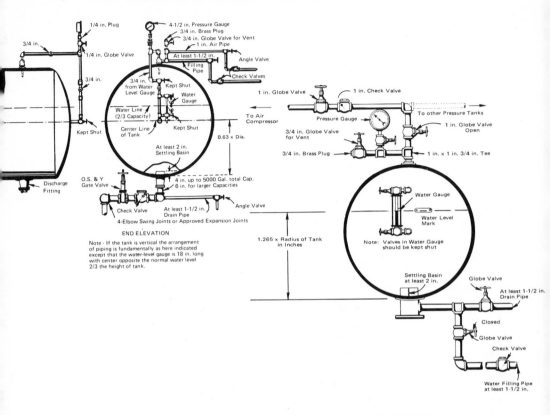

Fig. 1.15.   *Pressure tank fittings and connections.* (From NFPA 22, *Standard for Water Tanks for Private Fire Protection*)[3]

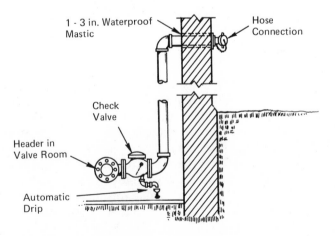

Fig. 1.16.   *Fire department connection.* (From NFPA 14, *Standard for the Installation of Standpipe and Hose Systems*)

nections are not protected with caps, and before attaching hose lines. These connections often become depositories for all types of trash and obstructions, including empty bottles. The Glen Echo Fire Department, Glen Echo, Maryland, has an effective inspection program whereby members of the department locate and inspect all fire department connections to both standpipes and sprinkler systems in their response area. With the cooperation and permission of property owners, shrubs that obstructed the connections were moved and reflective paint markings were made around the connections to make them more visible from a distance, especially at night. Figure 1.17 shows a fire department connection with the location marked on the side of the building, as painted by the Glen Echo Fire Department.

Fig. 1.17. Marking of fire department connections with reflective paint. Note the connections for both the standpipe and automatic sprinkler system. (Joel Woods, Maryland Fire and Rescue Institute, and Glen Echo Fire Department, Glen Echo, Maryland)

STANDPIPE TESTING PROCEDURES

The importance of the fire department connection for the effective utilization of the standpipe by the fire department is obvious. In 1967 a program for the periodic testing of standpipes was initiated in Los Angeles. The program, which required property owners to have the standpipe tests conducted by a private concern, made provision for fire department notification prior to the tests so that fire department representatives could witness the procedures. Many problems were corrected as a result of the program, including problems caused by the improper installation of the spring check valves at the fire department inlet of the siamese connection.

Another serious problem encountered was the installation of fire department connections with nonstandard threads that did not match the fire department's connections. In a survey of the program, Lyons[4] reported that in 125 buildings in Los Angeles the standpipe testing program produced the following results:*

    1. In 25 percent of the buildings it was impossible to deliver water to the dry standpipes.
    2. 45 percent of the buildings failed to pass the test requirements.
    3. 70 percent of the buildings had Pacific Coast threads.
    4. 14 percent of the buildings had faulty pipes.
    5. 12 percent of the clapper-type valves leaked excessively.
    6. 8 percent of the standpipes were never completely connected (either at the laterals to the outlets or at the top section of the standpipe).

A regular inspection and testing program relative to fire department connections for both sprinkler or standpipe systems, and complete testing of the standpipe is a necessity for every fire department. Serious consequences for both building occupants and fire department personnel may result from depending on standpipe systems which are not known to be in serviceable condition. The types of debris that were found in standpipes during the Los Angeles testing program are shown in Figure 1.18.

The critical importance of fire department connections indicates the need for careful testing, inspection, and marking of these connections under the supervision of the fire department, and the need for the testing and inspection of the complete standpipe system.

Faulty fire department connections found during the Los Angeles testing program, which effectively prevented the discharge of water into the standpipes, are shown in Figure 1.19.

For the testing of new standpipe systems, the *NFPA Standpipe and Hose Systems Standard* recommends the hydrostatic testing of the complete new system at a minimum pressure of 200 psi for 2 hours, or at 50 psi in excess of the normal pressure on the system when the normal exceeds 150 psi. When the

---

*From Lyons, Paul R., "Dry Standpipe Survey in Los Angeles," *Fire Journal*, Vol. 63, No. 3, May 1969, pp. 65–66.

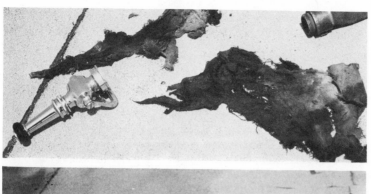

Fig. 1.18. *Debris consisting of rags and wood found in standpipes during the Los Angeles standpipe testing program.* (Los Angeles City Fire Department)

Fig. 1.19. *Faulty fire department connection installation found during Los Angeles standpipe testing program.* (Los Angeles City Fire Department)

standpipe system is installed in walls or interior portions of the building, the hydrostatic testing should be conducted prior to concealment of the piping. It should be noted that the hydrostatic test requirements for the standpipe systems are similar to the requirements established by the NFPA for automatic sprinkler systems.

For any system being restored to service, the *NFPA Standpipe and Hose Systems Standard* recommends an air pressure test at a pressure under 25 psi prior to admitting water into the system. Dry standpipe systems are recommended for hydrostatic testing at intervals of not less than five years.

The survey of the Los Angeles standpipe testing program indicated the following test requirements were adopted by the Los Angeles Fire Department:*

1. Notify the Chief at least one working day in advance of performing and testing.
2. Air-test at 25 psi to determine if the system leaks.
3. Hydrostatic-test the system with water 50 psi greater than the head pressure.
4. Flow-test the system by flowing 100 gpm through the standpipe to the roof outlet. A separate flow shall be conducted through each inlet. Install a test gage at the inlet to measure the inlet pressure. The maximum allowable pressure loss within the system shall not be more than 15 psi. The friction loss shall be determined by subtracting static pressure (head) and outlet pressure from the inlet pressure while there is a flow of 100 gpm at the roof.
5. Operate each system outlet valve to determine its reliability.
6. Fire hose connections must be equipped with American (National) Standard Hose Coupling screw threads as specified in NFPA 194, *Standard for Screw Threads and Gaskets for Fire Hose Connections*. Test with "Go" and "No Go" gages, as specified in NFPA 194, or with fittings approved by the Chief.
7. When it has been determined that the fire protection equipment is operable, the owner or agent shall certify that condition to the Chief, using the form "Fire Protection Equipment Certification."

In 1968 the city of Pasadena, California, initiated a testing program for dry standpipe systems. In an article titled "Dry Standpipes: Try Testing Them!"[5] Fire Marshal Don McClanahan states the following testing requirements were utilized in the Pasadena program:**

1. A hydrostatic pressure test of 150 psi, plus 5 pounds pressure for every story above ground.
2. A 20-minute period during which test pressure was to be held.
3. At the time of testing, certification in writing by the testing agency to the fire department that the entire dry standpipe system had been serviced, maintained, and repaired, and that the standpipe was completely operable in all respects.

The Pasadena program included 73 standpipes in buildings that varied in height from 4 to 9 stories. The most numerous failures involved the fire department connections and resulted with standpipes and connections being enclosed in the walls of the buildings. The second most numerous failures involved the valves and joints in the standpipes. In this program approximately 12 to 16 percent of the systems that were tested experienced complete failure, resulting in the standpipes being inoperable. The Pasadena standpipe testing program required that every standpipe be tested within a maximum 5-year interval. Figure 1.20 shows the type of metal tags that were attached to the fire department connections to indicate the dates the connections were tested.

*From Lyons, Paul R., "Dry Standpipe Survey in Los Angeles," *Fire Journal,* Vol. 63, No. 3, May 1969, pp. 65–66.
**From McClanahan, Don, "Dry Standpipes: Try Testing Them!" *Fire Journal,* Vol. 62, No. 3, May 1968, p. 62.

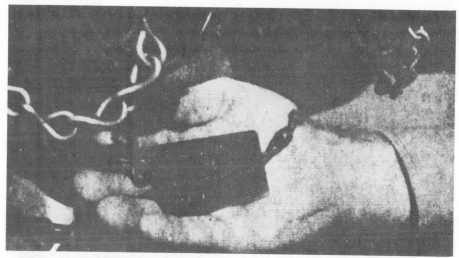

Fig. 1.20.   Metal tag attached to standpipes to indicate date of testing. (From "Dry Standpipes: Try Testing Them!" by Don McClanahan)

STANDPIPE INSPECTION PROCEDURES

Dry standpipe systems should be inspected at least annually. Wet pipe systems with connections to fire pumps, pressure tanks, or gravity tanks should be inspected more frequently, preferably on a quarterly or semiannual basis. The inspection procedure should proceed from the water supply connections or facilities through the system. The water supply control valves should be inspected to make certain that they are in the open position; preferably, they should be sealed in the open position. Figure 1.21 shows a fire company member checking the position of a P.I.V. (Post Indicator Valve) on the connection from the public water main to the standpipe system. Every standpipe discharge valve should be checked to ascertain that it is sufficiently operable to both initiate and terminate the flow of water through hose lines. It is not unusual in some locations to find valves jammed or valve control wheels missing. The threads on the discharge outlets of the standpipe should always be checked for nonstandard or damaged threads which prevent the attachment of fire department hose lines to the outlet. Figure 1.22 shows a fire officer inspecting the discharge valves of a wet standpipe system. On a wet standpipe system, it is important to drain the system prior to inspecting the operation of the discharge valves.

When the standpipe system is supplied from a gravity or pressure tank, these water supply facilities should also be inspected. The air pressure and water level in the pressure tank are critical and should be carefully noted. The usual condition is a minimum of 75 psi air pressure and the water level at the two-thirds capacity. The water level in the gravity tank is critical, especially when the tank is also utilized to supply operational water requirements. The performance of the heating system for the gravity tank is critical in freezing climates;

therefore, the water temperature must be checked during winter months. The water control valves for the water supply facilities (consisting of pressure tanks, gravity tanks, or water mains) must also be checked and sealed in the open position.

Where the standpipe system is also supplied with a fire pump, the pump should be checked for proper starting and operation at least weekly, according to the *NFPA Centrifugal Fire Pump Standard*. Some industrial facilities operate the fire pump at the time of the weekly drill session for the employee fire brigade.

The fire pump will usually be tested at rated discharge capacity and pressure at least once a year by the industrial or insurance authorities. Fire department personnel should arrange to be present at such tests in order to obtain an under-

Fig. 1.21. Inspector checking the position of a P.I.V. (Post Indicator Valve) on the connection from the public water main. (College Park Volunteer Fire Department, College Park, Maryland)

Fig. 1.22. Fire department officer checking the condition of the discharge valve from the standpipe system. (Stephen C. Leahy, College Park Volunteer Fire Department, College Park, Maryland)

standing of the operating procedures for the pump, the piping system, and the limitations of the particular installation relative to the capabilities and performance of the fire pump. The fire department connection should always be checked for visibility, accessibility, and operational features including proper connection threads, check valves, and freedom from debris that could interfere with the flow of water into the standpipe system. Some fire departments conduct annual tests of the standpipe systems, and the *NFPA Standpipe and Hose Systems Standard* requires the testing of dry standpipes at least once every five years.

Because they are provided with hose at the standpipe discharge outlets, Class II and Class III Standpipe Systems require extensive inspection procedures. The hose should be thoroughly inspected for damage and deterioration. The hose on standpipe racks or reels should be removed annually, inspected, conditioned, and then returned to the system. Threads on the couplings and gaskets should be inspected at the time of testing. In many installations, the unlined linen hose formerly used for standpipe purposes has been replaced with single-jacketed hose that has an outer covering of synthetic fiber or plastic. Since difficulties are usually experienced due to lack of care for standpipe hose, most fire departments utilize their own hose for standpipe operations. A suggested inspection form for use in the inspection of standpipe systems is illustrated in Table 1.2, as recommended by the NFPA.[6]

FIRE DEPARTMENT USE OF STANDPIPES

As stated earlier in this chapter, the standpipe system is primarily intended as a procedure for the transmittal of water to various areas of a building where the water is used for hose line manual fire fighting. Essentially, the advantage of

Table 1.2.   Suggested Inspection Form for Properties Having Standpipe Systems*

Name of Plant..

Fire Inspection District No.........................

Location...Fire Alarm Code No................................

Description of Property...No. of Stories.................

Plant Official Responsible for Standpipe System...

Phone..

Type of Standpipe Systems:

    First Aid Standpipes for Occupants..........................No. of Risers.............................

    Size of Risers.............................No. of Hose Outlets.............................Location of

    Valves Controlling Standpipes...

    ....

    Fire Dept. Standpipes..................No. of Risers.................Size of Risers.................

    No. of Hose Outlets..Location of Valves Controlling Stand

    pipes...

    ....

Water Supplies to Standpipes:

Gravity Tank..Pressure Tank.............................

    Volume in Storage...

    Fire Pumps: Manual or Automatic Operation..

    Capacity...Location..

    ....

    Public Water Main.......................................Static Pressure...

    Private Water Main......................................Static Pressure.......................................

    ....GPM Available at.....................Residual Pressure at.....................Elevation

Fire Department Connections:

    Location of Connections..

    Standpipe Supplied:

        First Aid..Fire Dept.......................................

    Company Assigned to Pump into Standpipe System on First Alarm..........................

    ....

    Hydrant to be Used (normal procedure)...

    Supplemental Pumper Supply Available...

Hose Provided at Standpipe Outlets:

    Type....................................... Length.......................................

    Size.. Condition......................................

Any Special Hose Threads or Adapters Provided?...

Type of Nozzle Provided: Straight Stream............................Spray.......................................

Are Standpipes Inter-connected?...

Standpipe Waterflow Alarm Devices...

Remarks: Include any pertinent special information that may affect fire department opera-

        tions..

    ....

Inspected by.................................Approved by.................................Date...............

---

*From NFPA 13E, *Recommendations for Fire Department Operations in Properties Protected by Sprinkler and Standpipe Systems.*

the standpipe system is the time gained in not having to provide hose lines for extensive distances, or up many flights of stairs or fire escapes. While a fire department pumper is being connected to a standpipe system, personnel will be proceeding to the fire floor either by elevator (under manual control with the fire switch procedures utilized in many cities) or by stairways. Some fire departments have developed useful procedures for transporting the hose, nozzles, and other required equipment to the fire area. Figure 1.23 shows a wheeled cart arrangement for the movement of the standpipe hose pack and equipment to the fire area, and Figure 1.24 shows a portable standpipe hose pack.

Since it is often necessary for fire department personnel to climb many floors to get to a fire area, the problem of the bulk and weight of equipment is of major concern. To help reduce the weight of the standpipe hose pack, some fire departments have utilized lightweight, single-jacketed hose in 100-foot lengths.

Efficient communications between fire department personnel operating the pumper and the personnel directing and operating the hose streams in the fire area is important. Many fire departments rely on portable radio communica-

Fig. 1.23. Wheeled standpipe hose pack assembly. Joel Woods, Fire Service Extension, Maryland Fire and Rescue Institute, and Rockville Fire Department, Rockville, Maryland)

Fig. 1.24.   Portable standpipe hose pack. (Joel Woods, Maryland Fire and Rescue Institute, and Glen Echo Fire Department, Glen Echo, Maryland)

tions; however, in order to assure positive communications between fire area and ground floor, some cities require the installation of a telephone communication system in the stairway of buildings (especially high-rise buildings) adjacent to the standpipe discharge outlets.

Pressure-reducing valves are installed on standpipes in some buildings to prevent the pressure at the standpipe discharge valve outlet from exceeding the safe operating pressure for manual hose line procedures. These valves are essential in Class II or III Systems which might possibly be used by building occupants. The provision of standpipes for the use of building occupants (other than in industrial occupancies with effectively trained fire brigades) would seem to be of little value. The concept of the effective utilization of standpipe hose lines by building occupants would seem to be based on naive assumptions concerning the behavior of untrained personnel, and on a lack of understanding of the fire threat to individuals not having protective equipment.

To effectively utilize the standpipe systems provided in buildings in their immediate response area, it is necessary for fire department personnel to inspect the systems and to practice their standpipe operating procedure on the systems. Prefire planning with standpipe familiarization as to the locations of water supply control valves, fire pumps, gravity or pressure tanks, and fire department connections are essential. Prior to a fire occurrence it is important to know the arrangement of the fire department connection relative to its attachment to one or all of the standpipes in a building.

COMBINED SPRINKLER SYSTEMS AND STANDPIPES

In 1971 the NFPA formally adopted provisions for the combination of standpipe and sprinkler systems. The provisions for these systems were adopted to promote the installation of automatic sprinkler systems in high-rise buildings. Since standpipes are generally required in buildings higher than four stories, it would seem the provision allowing the use of standpipe systems as risers for sprinkler system would most likely result in buildings having both automatic sprinkler and manual hose line protection. Hammack[7] indicated the following basic requirements for standpipe systems when utilized as risers for sprinkler systems:*

- The riser and hose valves shall be located in a fire-resistive stair enclosure.
- Sprinklers shall be under control of separate floor control valves located in the fire-resistive stair enclosure.
- The minimum size of the riser shall be 6 inches.
- The water supply shall be adequate for sprinklers and standpipes combined.

The requirements for the standpipe system are maintained relative to the location of the standpipe outlet to provide that all areas of the building are within 30 feet of the end of 100 feet of hose, as indicated in Table 1.1. Thus, the concept of the combined sprinkler system and standpipes has increased the element of flexibility in the design of sprinkler systems, and has helped bring about innovations including the use of new materials. These innovations are examined in detail in Chapter 5, "Basic Design of Automatic Sprinkler Systems," relative to the design of automatic sprinkler systems.

---

*From Hammack, James M., "Combined Sprinkler System and Standpipes (Some Random Thoughts)," *Fire Journal*, Vol. 65, No. 5, Sept. 1971, p. 68.

SUMMARY

Standpipe systems are essential aids to the efficient and effective use of hose streams in buildings of extensive area or in buildings of a height that cannot be serviced by local fire department ladders. The provision of standpipes with attached hose lines for occupant use (Class II and Class III Systems) in occupancies other than those having trained fire brigades would seem to create the possibility of extremely hazardous conditions for building occupants. The effectiveness of a standpipe system is directly determined by the understanding, training, and procedures of the fire department personnel using the system. Thus, every fire department must have a complete and thorough program of inspection, testing, and evaluation for all standpipes in its responsible area.

SI Units

The following conversion factors are given as a convenience in converting to SI Units the English units used in this chapter:

1 foot	= 0.305 m
1 inch	= 25.400 mm
1 pound (force)	= 4.448 N
1 psi	= 6.895 kPa

ACTIVITIES

1. Write your own definition of a standpipe and hose system. Also describe the purpose of these systems.
2. Standpipe systems may be grouped into three general classes of service for intended use in the extinguishment of fire: Class I, Class II, and Class III. Write a brief description of each class of system.
3. List at least three occupancies in which the use of horizontal standpipes is particularly effective, and at least two occupancies in which the use of vertical standpipes is particularly effective. Explain why.
4. Discuss the inspection and testing of standpipe systems, including the following:
   (a) List four major reasons for the inspection and testing of standpipe systems.
   (b) By whom do you think such inspections and testing should be accomplished? Why?
   (c) List five procedures of a standpipe system to be checked in an inspection.
5. Discuss the concept of "air lock." Then write a statement explaining how air lock can be prevented in a standpipe system.
6. Describe the principles of operation for the type of standpipe system known as a "primed system." What are some of the advantages of a primed system?

7. What is the purpose of a stationary fire pump with an automatic controller? Explain its arrangement and operation when connected to a wet standpipe normally filled with water from the connection to the public water main.
8. Pressure tanks operate on a principle similar to that of air-water extinguishers. Write an explanation of that principle, including in your explanation a description of how pressure tanks are utilized with standpipe systems.
9. Explain why Class II and Class III Standpipe Systems require particularly extensive inspection procedures. Make an outline of the items that should be inspected.
10. List the factors that you feel best determine the effectiveness of a standpipe system.

SUGGESTED READINGS

Lyons, Paul R., "Dry Standpipe Survey in Los Angeles," *Fire Journal*, Vol. 63, No. 3, May 1969, pp. 62–63.

NFPA 13E, *Recommendations for Fire Department Operations in Properties Protected by Sprinkler and Standpipe Systems.* 1973, NFPA, Boston.

NFPA 14, *Standard for the Installation of Standpipe and Hose Systems,* 1975, NFPA, Boston.

NFPA 22, *Standard for Water Tanks for Private Protection,* 1975, NFPA, Boston.

"Recommended Good Practice for Maintaining and Testing National Standard Reciprocating Steam Fire Pumps," Factory Insurance Association, Hartford, Conn., Feb. 15, 1958.

Safety Code for Inspection, Maintenance, and Protection of Standpipe and Inside Hose Systems, Fire Equipment Manufacturers Association, Inc., Pittsburgh, Pa., Oct. 1, 1957.

BIBLIOGRAPHY

[1]NFPA 14, *Standard for the Installation of Standpipe and Hose Systems,* 1975, NFPA, Boston.

[2]NFPA 20, *Standard for the Installation of Centrifugal Fire Pumps,* 1975, NFPA, Boston.

[3]NFPA 22, *Standard for Water Tanks for Private Protection,* 1974 ed., NFPA, Boston.

[4]Lyons, Paul R., "Dry Standpipe Survey in Los Angeles," *Fire Journal,* Vol. 63, No. 3, May 1969, pp. 63–66.

[5]McClanahan, Don, "Dry Standpipes: Try Testing Them!" *Fire Journal,* Vol. 62, No. 3, May 1968, pp. 62–63.

[6]NFPA 13E, *Recommendations for Fire Department Operations in Properties Protected by Sprinkler and Standpipe Systems,* 1973, NFPA, Boston.

[7]Hammack, James M., "Combined Sprinkler System and Standpipes (Some Random Thoughts)," *Fire Journal,* Vol. 65, No. 5, Sept. 1971, pp. 68–69.

Chapter 2

Fire Department Procedures for Automatic Sprinkler Systems

SPRINKLER SYSTEM PERFORMANCE

NFPA statistics indicate that automatic sprinkler systems established an efficiency record of approximately 96.2 percent satisfactory performance in the United States during the 45 years between 1925 and 1970.[1] This performance record was derived from detailed records of 81,425 fires that took place during that period.

The principal sources of performance reports on sprinkler systems have been the insurance companies and the Insurance Services Office (ISO). Thus, with the increasing application of insurance policies with an exclusion of coverage for small losses, there has been a selective reduction in the reporting of cases of successful sprinkler performance since these fire situations result in minor losses. Also, in most areas of the United States, uninsured losses are not included in these cases.

In those fires in which automatic sprinkler systems were unsuccessful, the principal reasons were: (1) closed water control valves, (2) obstructions to sprinkler distribution, and (3) only partial protection of occupancies by sprinkler systems. Primarily, the major cause of unsatisfactory performance of automatic sprinkler systems has been the result of human action: such action involves the closing of water supply control valves before the fire occurs, or before the fire is completely extinguished. Fire department, security, and industrial personnel have all been involved in the premature closing of water control valves.

Table 2.1 presents the NFPA performance summary for the operation of sprinkler systems in selected occupancies relative to satisfactory and unsatisfactory performance. The classification of the unsatisfactory sprinkler performance into the thirteen primary modes of sprinkler failure is presented in Table 2.2.

Table 2.1.   Sprinkler Performance Summary for
Selected Occupancies*

OCCUPANCIES	TOTAL NO. OF FIRES	TOTAL UNSATIS- FACTORY	TOTAL SATIS- FACTORY	TOTAL SATIS- FACTORY PER- CENT
Residential.............................	1,073	48	1,025	95.5
Assembly	1,551	52	1,499	96.6
Educational............................	241	20	221	91.7
Institutional	305	12	293	96.1
Office......................................	494	13	481	97.4
Mercantile	6,237	176	6,061	97.2
Industrial				
Beverages, essential oils	543	64	479	88.2
Chemicals	4,147	198	3,949	95.2
Fiber products	539	25	514	95.3
Food products	2,484	133	2,351	94.6
Glass products.....................	519	23	496	95.6
Leather, leather products	2,864	114	2,750	96.0
Metal, metal products...........	9,807	305	9,502	96.9
Mineral products..................	394	19	375	95.2
Paper, paper products...........	7,147	234	6,913	96.7
Rubber, rubber products	1,489	61	1,428	95.9
Textiles—manufacturing	16,119	291	15,828	98.2
Textiles—processing	6,527	127	6,400	98.1
Wood products....................	5,353	492	4,861	90.8
Miscellaneous industries	9,013	265	8,748	97.1
Total (Industrial)	66,945	2,351	64,594	96.5
Storage Occupancies	4,160	375	3,785	91.0
Other Occupancies	419	87	332	79.2
Total (All Occupancies)	81,425	3,134	78,291	96.2

As previously indicated, the most significant single cause of unsatisfactory performance is the water shutoff category. Therefore, in their inspection and prefire planning procedures, fire department personnel should always check to be certain that the water control valves to the automatic sprinkler systems are open. Table 2.3 is a detailed analysis of 3,134 cases of unsatisfactory sprinkler performance for the 9 principal occupancies in Table 2.1, and for the 13 modes of unsatisfactory performance in Table 2.2. The analysis in Table 2.3 includes 47 causative conditions.

Some evidence shows that when fire departments are automatically notified

*From "Automatic Sprinkler Performance Tables, 1970 Edition," NFPA.

of automatic sprinkler system operation and systems have been inspected regularly, performance records will be significantly improved. Horatio Bond, in an article titled "Sprinklers—Australia and New Zealand," reported in Australia, the satisfactory performance record for automatic sprinkler systems is 99.7 percent.[2] This record is based on the reports of fires involving sprinkler systems (since 1886), with a total sampling of 5,734 fires. The three

Table 2.2. Classification of Unsatisfactory Sprinkler Performance Modes*

OCCUPANCIES	WATER SHUT OFF	PARTIAL PROTECTION	INADEQUATE WATER SUPPLIES	SYSTEM FROZEN	SLOW OPERATION	DEFECTIVE DRY-PIPE VALVE	FAULTY BUILDING CONSTRUCTION	OBSTRUCTION TO DISTRIBUTION	HAZARD OF OCCUPANCY	EXPOSURE FIRE	INADEQUATE MAINTENANCE	ANTIQUATED SYSTEM	MISCELLANEOUS AND UNKNOWN
Residential	13	9	5	1	...	...	11	3	1	...	2	2	1
Assembly	23	10	3	...	1	...	9	1	...	1	4	...	...
Educational	4	8	1	...	...	...	5	...	...	...	1	1	...
Institutional	3	3	2	...	...	...	1	...	1	...	...	...	2
Office	4	2	1	...	...	1	2	...	1	...	1	1	...
Mercantile	83	11	4	4	4	5	35	11	12	1	4	1	1
Industrial													
Beverages, essential oils	17	4	9	...	...	1	2	1	18	3	3	5	1
Chemicals	33	11	19	...	3	3	1	13	95	2	12	1	5
Fiber products	6	...	4	1	...	2	...	5	4	...	2	1	...
Food products	43	11	8	1	2	1	7	9	29	4	12	1	5
Glass products	8	...	3	1	...	...	2	1	5	...	3	...	...
Leather, leather products	43	8	7	3	2	4	9	7	9	4	9	6	3
Metal, metal products	91	36	22	3	6	6	15	35	43	6	29	7	6
Mineral products	10	4	2	...	...	...	1	...	...	...	1	1	...
Paper, paper products	75	16	34	3	2	2	16	32	21	2	23	4	4
Rubber, rubber products	21	4	3	...	1	1	1	10	14	1	5	...	...
Textiles—manufacturing	109	15	32	3	5	3	11	27	18	1	50	9	8
Textiles—processing	52	6	11	...	5	1	8	13	15	2	7	1	6
Wood products	137	57	84	9	16	14	27	19	77	8	24	12	8
Miscellaneous industries	146	15	14	8	3	...	12	11	18	3	27	8	...
Total (Industrial)	791	187	252	32	45	38	112	183	366	36	207	56	46
Storage Occupancies	122	24	48	5	6	9	10	57	38	11	40	3	7
Other Occupancies	67	...	...	2	...	...	2	1	5	3	3	1	3
Total (All Occupancies)	1,110	254	311	44	56	53	187	256	424	52	262	65	60

principal factors causing the improved performance of the Australian sprinkler systems over automatic sprinkler systems in the United States were: (1) the sprinkler systems have better than average water supplies, (2) waterflow connections to fire departments are generally installed, and (3) weekly inspection service is usually a part of a contract for the installation of the automatic sprinkler system. The practice of requiring a connection between the sprinkler system and the fire department was initiated in Australia in 1909, and, to prevent a possible false alarm problem, the procedure for the weekly inspection of the sprinkler system was initiated.

*From "Automatic Sprinkler Performance Tables, 1970 Edition," NFPA.

Table 2.3. Detailed Analysis of Unsatisfactory Sprinkler Performance*

	RESIDENTIAL	ASSEMBLY	EDUCATIONAL	INSTITUTIONAL	OFFICE	MERCANTILE	INDUSTRIAL	STORAGE	MISCELLANEOUS OCCUPANCIES	TOTAL
Water to sprinklers shut off										
Valve defective or leaky	...	...	...	...	...	...	3	...	1	4
Unsupervised valve closed for undetermined reason	2	9	...	1	...	16	176	36	7	247
Premature shutoff	2	4	1	...	...	13	193	23	7	243
Alterations or repairs to system	3	5	...	...	1	14	179	28	10	240
To prevent freezing	1	1	1	2	1	24	127	18	26	201
Cold-weather valve closed out of season	...	1	...	...	...	7	24	4	...	36
To abet arson	1	...	2	...	...	5	33	6	2	49
Fear of water damage	...	...	...	...	2	...	18	1	...	21
Miscellaneous other reasons	4	3	...	...	...	4	38	6	14	69
Partial protection										
Originated in unsprinklered area	7	10	8	3	2	10	180	23	1	244
Spread to unsprinklered area	2	...	...	...	...	1	7	...	...	10
Inadequate water supply										
Insufficient water or low water pressure from public supply	2	2	...	...	...	2	114	19	...	139
Insufficient water or low water pressure from private supply	1	...	1	1	...	...	26	6	...	35
Insufficient water for both sprinklers and hose streams	1	1	...	...	1	...	53	13	...	69
Gravity tank empty	1	...	...	...	...	1	22	3	...	27
Pump failure or pump not started	...	...	...	1	...	...	14	...	...	15
Mains broken	...	...	...	...	...	1	11	1	...	13
Miscellaneous reasons	...	...	...	...	...	...	12	1	...	13
System frozen										
Pipes or valves frozen	1	...	...	...	...	4	32	5	2	44
Slow operation										
Excessive heads on dry-pipe valve	...	...	...	...	...	...	19	1	...	20
High-temperature sprinklers	...	1	...	...	...	...	9	...	...	10
Failure of quick-opening device	...	...	...	...	...	...	4	...	...	4
Heat-actuating devices inadequate or inoperative	...	...	...	...	...	...	3	...	...	3
Miscellaneous reasons	...	...	...	...	...	4	10	5	...	19
Defective dry-pipe valve										
Defective or improperly adjusted dry-pipe valve	...	...	...	...	1	5	38	9	...	53
Faulty building construction										
Concealed horizontal or vertical spaces lacking protection	11	9	4	1	2	34	94	5	1	161
Floor or roof collapse	...	...	1	...	...	1	12	6	1	21
Miscellaneous deficiencies	...	...	...	...	...	...	5	...	...	5
Obstruction to distribution										
Fires under benches, etc.	...	...	...	...	...	2	98	8	...	108
High piling of stock	...	...	...	...	...	3	60	41	...	104
Partitions erected	1	...	...	...	...	5	12	4	1	23
Miscellaneous reasons	2	1	...	...	...	1	13	4	...	21
Hazard of occupancy										
Hazard too severe for sprinkler equipment as installed	...	...	...	...	...	6	182	23	3	214
Explosion damaged system	1	...	...	...	1	6	161	13	2	184
Water overflowed containers of flammable liquids	...	...	...	1	...	...	15	...	...	16
Miscellaneous reasons	...	...	...	...	...	...	8	2	...	10
Exposure fire										
Exposure fire overpowered sprinkler system in exposed building	...	1	...	...	...	1	36	12	2	52
Inadequate maintenance										
Plugged sprinklers	...	2	...	...	...	1	39	8	...	50
Sprinklers dirty, corroded, or coated	...	...	...	...	...	1	41	4	2	48
Obstructed piping	2	2	1	...	...	2	117	25	1	150
Defective check valve	...	...	...	...	...	...	2	2	...	4
Miscellaneous reasons	...	...	...	...	1	...	9	...	...	10
Antiquated system										
Pipe sizes, sprinkler spacing substandard or old-standard	2	...	1	...	...	1	51	3	...	58
Valves substandard or old-standard	...	...	...	...	...	...	1	...	...	1
Sprinklers substandard or old-standard	...	...	...	...	1	...	3	...	1	5
Miscellaneous deficiences due to age	...	...	...	...	...	...	1	...	...	1
Miscellaneous and unknown										
Causes of unsatisfactory sprinkler performance unknown or cannot be otherwise classified	1	...	...	2	...	1	46	7	3	60
Total	48	52	20	12	13	176	2,351	375	87	3,134

*From "Automatic Sprinkler Performance Tables, 1970 Edition," NFPA.

Marryatt, in his book titled *Fire: Automatic Sprinkler Performance in Australia and New Zealand, 1886–1968*,[3] estimates the importance of fire department notification and weekly inspections as follows:*

> It soon became apparent that false alarms could become a serious prob-
> lem unless automatic sprinkler systems were maintained in first-class work-
> ing order at all times. This requirement led to the weekly inspection and
> testing of alarms by the sprinkler contractors and there is no doubt what-
> ever that the combination of regular inspection and testing of alarms and
> the examination of all operating parts of a system regularly, in association
> with the use of automatic alarms to the fire brigades, has contributed
> greatly to the exceptionally high standard of efficiency which the record of
> automatic sprinkler systems in Australia and New Zealand shows.

The objective of every automatic sprinkler system is to control fire occurrence and to notify the appropriate personnel or response agency. Therefore,
the Australian procedures have concentrated on the improvement of these
essential features of automatic sprinkler systems. The fire department procedures for the proper support of automatic sprinkler systems should be
initiated with the identification of the types and locations of automatic sprinkler
systems with which the department may be involved during a fire situation.

PREPLANNING OF AUTOMATIC SPRINKLER OPERATIONS

Procedures should be established within fire departments whereby necessary
information concerning locations and types of automatic sprinkler systems is
obtained and made available to department personnel. Of major importance to
the personnel of responding companies are diagrams of the locations of all
water control valves, fire department connections, sprinkler risers and sprinkler valves. The types of alarm devices provided on sprinkler systems should
also be noted.

The most reliable and effective alarm procedure is the automatic notification of the fire department when the automatic sprinkler system operates. However, many fire protection engineers, insurance authorities, and property owners continue to rely on local alarm devices and human action for notification of
response agencies. Such dependence would seem questionable since the use
of local alarm systems depends on human response, and such response can be
extremely unreliable and unpredictable. For example, Jacob B. Klevan conducted a study of the apparent effectiveness of water motor gongs.[4] The study
involved an awareness of water motor gong devices by the employees and
occupants of various buildings. Klevan found that while 35 percent of the
employees in sprinklered buildings could identify the water motor gong, only
20 percent could state its function. He also found that while 15 percent of the
occupants of various buildings.

Klevan found that while 35 percent of the employees in sprinklered buildings
could identify the water motor gong, only 20 percent could state its function.

*From Marryatt, H. W., *Fire: Automatic Sprinkler Performance in Australia and New Zealand, 1886–1968*,
Australian Fire Protection Association, Melbourne, Australia, 1971, p. 36.

He also found that while 15 percent of the occupants of sprinklered buildings in shopping malls could identify the water motor gong, only 10 percent could state its function. Therefore, it seems neither reasonable nor logical to provide buildings with automatic sprinkler systems that depend on people passing by to notify the fire department. To make such provisions involves two assumptions which may not be valid: (1) that someone will be available to hear the alarm, and (2) that whoever hears the alarm will notify the fire department. Figure 2.1 shows a typical water gong installation with a sign requesting anyone hearing the alarm to call the fire department or police.

Fire departments should be familiar with the location of the water motor gong or electric alarm on sprinklered buildings; this information should be provided to the communications or alarm center so that reports of alarm bells ringing in certain areas can be correlated with the location of sprinklered buildings. Many areas might well follow the experience of the Australian authorities by encouraging the connection of the automatic sprinkler systems to the fire department, and by requiring periodic inspections of the systems.

Prefire planning procedures of fire departments should be concentrated on supporting automatic sprinkler systems by ensuring adequate water supplies. Since two of the principal causes of unsatisfactory sprinkler performance (as identified in Table 2.3) are the turning off of the sprinkler water control valves and the lack of sufficient water to the automatic sprinkler system, fire department procedures should include consideration for providing automatic sprinkler systems with sufficient water capacity and pressure.

Prefire survey procedures should consider the location and condition of the water control valves to the automatic sprinkler system, and sectional or gate valves on the public or private water system. A complete diagram of the water

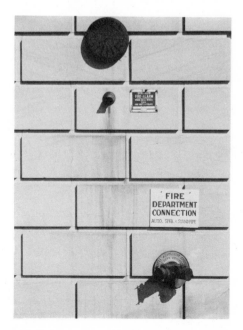

Fig. 2.1. Typical water motor gong installation. (Stephen C. Leahy, Environmental Safety Service Division, University of Maryland)

supply system to the sprinkler system should be constructed. The principal types of water supply for automatic sprinkler systems will be from public or private water mains, fire department connections, fire pumps from suction tanks or reservoirs, and gravity tanks or pressure tanks. Figure 2.2 illustrates a building with an automatic sprinkler system, with four of these types of water supplies available. These principal types of water supply systems were previously examined in detail, relative to standpipe systems, in Chapter 1, "Standpipe and Hose Systems."

An important aspect of the prefire survey activities of the fire department is the inspection of all the water control valves to make certain they are in the full open position. The principal types of water control valves are the O.S. & Y. (outside screw and yoke valve), the P.I.V. (post indicator valve), and the P.I.V.A. (post indicator valve assembly). Figure 2.3 shows an O.S. & Y. valve on a sprinkler riser as typically installed directly below the sprinkler valve. Figure 2.4 illustrates an O.S. & Y. valve with supervisory service which indicates, by both audible and visual signals, when the O.S. & Y. water control valve is opened or closed. Due to government requirements, such supervision to sound local alarm signals is required in many health care facilities.

Note the O.S. & Y. valve shown in Figure 2.3 is in the full open position, since the stem of the valve protrudes the full distance beyond the control wheel of the valve; the O.S. & Y. valve in Figure 2.4 is in the closed position, since the stem is flush with the control wheel. For years, insurance inspectors have used wire and lead seals to show that O.S. & Y. or P.I.V. valves were found in the open position. Recently, some fire departments have adopted a similar method relative to their prefire survey procedures for both standpipe and sprinkler systems.

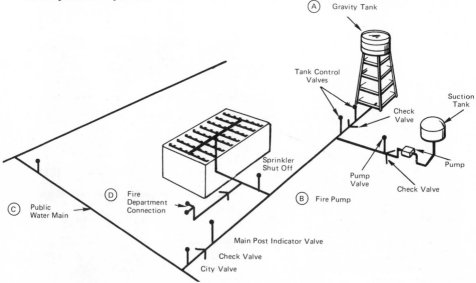

Fig. 2.2. *Various types of water supply to a sprinkler system.* (From NFPA 13E, *Recommendations for Fire Department Operations in Properties Protected by Sprinkler and Standpipe Systems)*

The Glen Echo Fire Department, Glen Echo, Maryland, provides a wire and lead seal marking on the water control valves of all the sprinkler and standpipe systems in their response area. Figure 2.5 shows a P.I.V. with the valve locked open following the prefire survey procedures on a P.I.V. control to an automatic sprinkler system. Figure 2.6 is a diagram of O.S. & Y. and P.I.V. valves, and Figure 2.7 shows a FM (Factory Mutual System) designed P.I.V.A. which indicates the actual position of the valve.

Table 2.4 shows an NFPA recommended inspection form for sprinklered properties, suggested for use by fire department personnel during the prefire survey procedures.[5] The prefire survey procedures and understanding of the sprinkler systems, the water supply systems, control valves, and fire department connections are greatly improved when written reports with maps or diagrams of the property and the sprinkler system have been prepared by the personnel conducting the survey. In order to provide a proper understanding of the sprinkler system on a property, maps or diagrams that include standard plan symbols relative to the essentials of the sprinkler system are necessary. Graph paper is helpful when making both the original sketch and final drawing

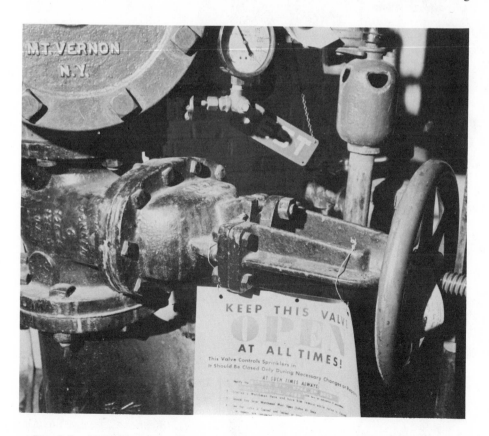

Fig. 2.3. Example of an O.S. & Y. control valve on an automatic sprinkler system riser. (Stephen C. Leahy, Environmental Safety Services Division, University of Maryland)

of a map or diagram. Figure 2.8, a typical prefire survey diagram of a sprinklered property, may be utilized with the inspection form shown in Table 2.4.

The primary purpose of the fire department's prefire survey activities is to ensure that when a fire occurs, a check will be made to make sure that the sprinkler system water supply control valves are fully opened, and, upon arrival at the scene of a fire, personnel will be assigned to prevent the valves from being closed prematurely. Secondly, the fire department must be familiar with the location of the fire department connections to supplement the water supply of the sprinkler system, and must make sure that these connections are accessible to the placement of hose lines. Figure 2.9 shows a fire department connection that has been made inaccessible by a fence.

It is necessary to determine the availability of water for fire department operations in order to keep the fire department from connecting pumping engines to hydrants on private or public mains; when such connections would reduce the flow of water to the sprinkler system. It is essential to determine

Fig. 2.4.   O.S. & Y. control valve with electrical supervision on the operation of the valve. (Stephen C. Leahy, Environmental Safety Services Division, University of Maryland)

Fig. 2.5.   P.I.V. locked in the open position following fire department survey. (Stephen C. Leahy, Environmental Safety Services Division, University of Maryland)

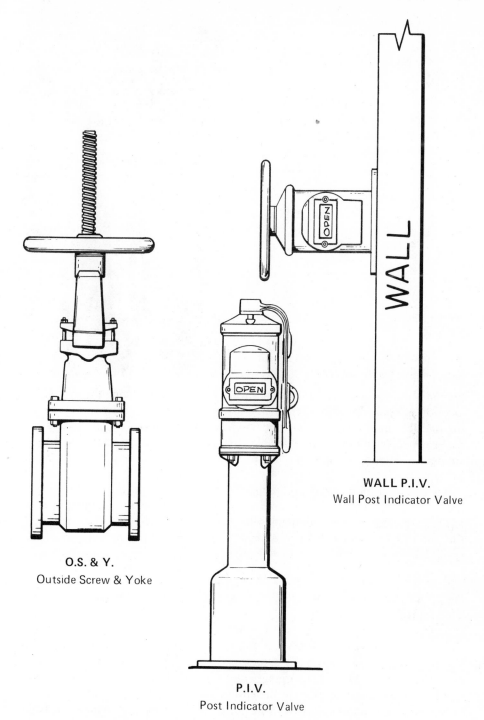

O.S. & Y.
Outside Screw & Yoke

WALL P.I.V.
Wall Post Indicator Valve

P.I.V.
Post Indicator Valve

Fig. 2.6. Diagram of water control valves for sprinkler systems. (Insurance Services Office)

Fig. 2.7. Factory Mutual designed P.I.V.A. which indicates position of valve in water main. (Michael W. Magee)

the amount of water available at an effective pressure in the public mains, the location of connections to the private mains, and the amount of water needed for the sprinkler system. The recommended procedures for determining the water capacity and pressure available from public or private mains will be examined in detail in Chapter 4, "Water Specifications for Automatic Sprinkler Systems."

An essential element in prefire planning is to provide for distribution of the information collected in the prefire survey and the diagram of the property, to the personnel of all fire companies due to respond. Effective dissemination of this information should help to achieve the coordination and support of the sprinkler system as planned.

OPERATIONS IN SUPPORT OF AUTOMATIC SPRINKLER SYSTEMS

At the time of a fire occurrence, fire department operations should be designed to supplement sprinkler systems since the automatic sprinkler heads over the fire should be discharging water into the fire area. Usually, the immediate procedure is to provide the sprinkler system, by means of the first arriving engines, with at least two 2½-inch hose lines or one 3-inch hose line from a good public hydrant with the engine connected to the hydrant. Such a procedure should allow an immediate additional supply of at least 500 gpm to the sprinkler system. With most systems, this supply to the sprinkler system should provide adequate capacity and pressure for approximately 25 sprinkler heads. The pumper supplying the fire department connection should be utilized solely for this purpose, and additional hose lines should not be taken from this engine unless absolutely necessary (see Figure 2.10).

Table 2.4.   Suggested Preplanning Survey Form for Sprinklered Properties*

Name of Plant...

                       Fire Inspection District No. ....

Location...Fire Alarm Code No. ....

Description of Property...

Plant Official Responsible for Sprinklers..

Phone....................................

Type of Automatic Sprinkler System (Wet, Dry, Deluge, etc.).................................
....

No. Sprinkler Heads............................ Size of Connection to Main.....................

Spare Sprinkler Heads Available.......................................Stored......................

Automatic Sprinkler Valves:

    Location........................... Type........................... Controlling.......................

    Cold Weather Valve..

Water Supplies to Sprinklers:

    Public Main....................., Gravity Tank....................Pressure Tank......................

    Private Fire Pump ...., Volume in Storage.................................

    Static Pressure...............psi. .... gpm   Available at ....Residual Pressure

    Post Indicator Valve Location..

Fire Department Connections:

    Location...

    Company Assigned to Pump Into Sprinkler on First Alarm...

    Hydrant to be Used (normal procedure)...

    Supplemental Pumper Supply Available...

Sprinkler Alarm Devices:

    Local Waterflow .... Master Alarm Box ....

    Central Station.................................. Other ....

    Supervised by...

Watchman Service:

Remarks: Include any pertinent special information that may affect fire department operations ....
....
....

Inspected by.........................Approved by.........................Date.....................

---

*From NFPA 13E, *Recommendations for Fire Department Operations in Properties Protected by Sprinkler and Standpipe Systems.*

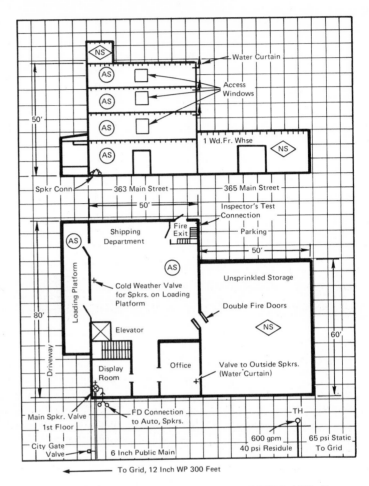

Fig. 2.8. Typical prefire survey diagram of a sprinklered property. (From NFPA 13E, *Recommendations for Fire Department Operations in Properties Protected by Sprinkler and Standpipe Systems*)

Fig. 2.9. Fire department connection with access blocked by fence. (Michael W. Magee)

Fig. 2.10. Fire department engine connected with hose lines to sprinkler system through the fire department connection. (Stephen C. Leahy, College Park Volunteer Fire Department, College Park, Maryland)

Warren Kimball, in his book titled *Fire Attack: Command Decisions and Company Operations*,[6] stated the primary function of the fire attack procedures in sprinklered buildings should consist of the following:*

> In any event, the primary responsibility of the fire department on responding to fires in sprinklered buildings is to see that the sprinklers have ample water supply and pressure, and that small hose streams are provided for mop-up. The fire department will also provide ventilation, overhaul, and salvage service.

As indicated in Chapter 1, "Standpipe and Fire Hose Systems," the operating condition and accessibility of the fire department connection is most critical

---

*From Kimball, Warren Y., *Fire Attack: Command Decisions and Company Operations*, National Fire Protection Association, Boston, 1966, p. 134.

since this connection is the fire department's primary means for supplementing the sprinkler system's water pressure and capacity. When a standpipe system has a damaged or inoperative fire department connection, the standpipe can still be utilized by attaching a double female connection at a first floor outlet on the interior of the building, attaching hose lines, and supplying water into the standpipe system.

However, with a sprinkler system, there is no readily available connection unless the system is a combination system with the automatic sprinklers supplied with water from the standpipe risers on each floor (see Chapter 1). The typical automatic sprinkler system utilizes a riser above the valve, and the usual emergency connection is the replacing of the main drain pipe on the sprinkler valve with a fire department hose coupling adapter. Figure 2.11 shows the proper fire department connection arrangement to a single sprinkler system riser. Note, the fire department connection is attached to the sprinkler riser above the sprinkler valve and the O.S. & Y. water control valve, thus allowing the fire department to supply water to the system even with the O.S. & Y. valve on the sprinkler riser turned off. When a single fire department connection has to supply multiple sprinkler risers and valves, it should be attached to the supply system main or manifold. Therefore, if one riser's O.S. & Y. valve is closed, the other risers can be supplied. Notice also in Figure 2.11 that the only connection between the siamese and the sprinkler riser is the check valve; this allows the fire department to attach hose lines into the siamese with the water main pressure on the sprinkler riser.

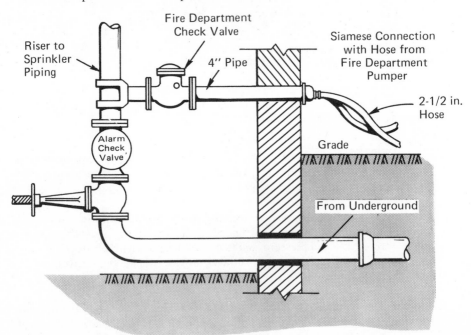

Fig. 2.11. Fire department connection to the sprinkler system riser. (From NFPA 13E, *Recommendations for Fire Department Operations in Properties Protected by Sprinkler and Standpipe Systems*)

The sprinkler valve on the wet pipe sprinkler system of the alarm check type shown in Figure 2.11 prevents the water that is being pumped into the sprinkler system from being distributed into the water main. When wet pipe sprinkler systems are installed without alarm check valves, a water flow indicator and a check valve will usually be installed to perform the function of the alarm check valve. Figure 2.12 is a diagram of the check valve as horizontally installed on sprinkler systems, standpipe systems, and water mains.

Harry E. Hickey, in an article titled "An Approach to Evaluating And Maintaining Sprinkler Performance," recommended the installation of a pressure gage at the fire department connection to indicate: (1) when the regular water supply ceases to provide the required volume-pressure relationship, and (2) when the fire department should provide additional volume or pressure through the fire department connection.[7] In addition, the differences in the pumper discharge gage and the fire department connection pressure gage can serve to indicate the volume of water being supplied to the sprinkler system.

NFPA 13E, *Recommendations for Fire Department Operations in Properties Protected by Sprinkler and Standpipe Systems,* hereinafter referred to in this text as the *NFPA Fire Department Operations in Protected Properties*

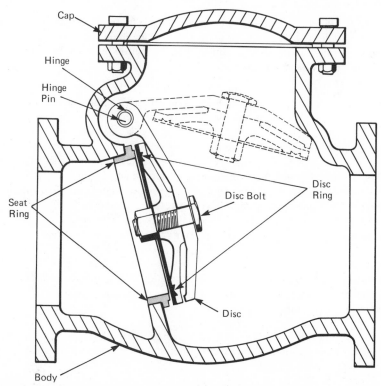

Fig. 2.12.   Check valve of the type found installed on sprinkler systems. (Insurance Services Office)

Standard, recommends the following good practice suggestions for the guidance of chiefs or officers in charge of fire department operations in sprinklered buildings:*

1. Immediately send a fire fighter to the proper control valve to:
    (a) Determine that the valve is fully open. (Valves are designed to show open or closed position.)
    (b) Open the valve if it has been closed. (One possible exception would be where valve is "tagged for repairs.")
    (c) Shut the valve only when ordered to do so by the officer in charge.
    (d) Remain at the valve so that in the event of rekindling or any detected extension of fire, the valve can be reopened immediately.

NOTE: The man assigned to the valve should take a light and portable radio so that no time will be lost in transmitting orders to open or close the valve. He should remain at the valve until orders are given for the companies to make up and return to quarters.

2. One of the first alarm pumper companies on arriving at the fire should immediately connect two lines to the proper sprinkler siamese connection and start pumping at about 150 psi. If there is more than one sprinkler system in the fire area, pumpers should be connected to provide adequate pressure and volume to each sprinkler system that may be in operation.

NOTE: Pumpers should not be connected to hydrants on private water systems unless such systems are designed to maintain the flow needed by fire department pumpers in addition to that required by sprinklers or other private fire protection facilities.

3. An officer should immediately be ordered to the fire area to determine the location and extent of fire. He will determine whether sprinklers have extinguished the fire or are holding the fire under control. He will check for any possible extension of fire both horizontally or vertically including concealed spaces. He should have a portable radio so that he can immediately report to the officer in charge. Where there are sectional or floor valves he should make certain these are open. In such cases when the fire is extinguished, sprinklers may be shut off at the sectional or floor valve without shutting off the entire system. The inspection records should indicate the location of such valves. In all cases, the officer in charge must use good judgment before he orders any sprinkler system valve to be closed.

4. As quickly as possible and before any sprinkler valve is closed, hose lines should be used to prevent any extension of fire and to complete extinguishment. In general, 1½-inch hose with combination spray nozzles can be used for this purpose.

5. The ladder company, squad company, or other unit performing "truck" duties should be ordered to ventilate the fire area as needed in

---

*From NFPA 13E, *Recommendations for Fire Department Operations in Properties Protected by Sprinkler and Standpipe Systems*, 1973, NFPA, Boston, pp. 13–19.

order that there be no delay in advancing hose lines to complete extinguishment.

6. Only when the fire is completely extinguished should sprinklers be
shut off. If it becomes necessary to enter the fire area to mop-up residual
fire, it is important to have charged hose lines in place and provide adequate ventilation.

Where only a few sprinkler heads are operating, sprinkler tongs or
tapered wooden wedges may be used to immediately stop the flow from
the opened heads.

7. As soon as manpower permits, salvage operations should be started.

8. When overhauling is completed so that there is no possible danger
of rekindling requiring use of sprinklers, the lines from the pumper to the
sprinkler system may be ordered disconnected.

The application of these suggestions is illustrated in Figure 2.13, which
shows the preferred fire department pumper location for supplying a sprinkler
system through the fire department connection and manual hose lines.

Kimball, in another article titled "Planning Fireground Action for Sprinklered Buildings," indicated that an analysis of the fire department operations
following every fire in a sprinklered building is essential in order to improve
the operational procedures of the fire department.[8] Therefore, he recommends
the development of a consolidated report to include the following information:*

Number of sprinkler heads operating.
Location of heads operating.
Result of sprinkler operation.
Reason for any unsatisfactory operation.
What fire company was assigned to connect to sprinkler connection?
What member of department was assigned to check control valve?
Who closed valve after fire?
Is sprinkler protection fully restored? By whom?
Did the private water supply to sprinklers operate satisfactorily?
Was water pumped into system?
Did fire department connect to sprinkler system? Why not?
Fire Company connected to sprinkler system. Company number.
Number of sprinkler heads replaced by fire company. Type of heads
    installed.

Fire Department Operations to Support Sprinklers in Specialized Occupancies

In an article titled "Fire Fighting Tactics for Racked Storage," J. S. Barritt
indicated that in high-racked storage buildings prefire planning by the fire
department is a necessity, and a map indicating the sprinkler control valves,
the fire department connections, and the arrangement of water supply facili-

---

*From Kimball, Warren Y., "Planning Fireground Action for Sprinklered Buildings," *The Sentinel*, Vol. 15,
No. 4, Factory Insurance Association, Apr. 1959, p. 6.

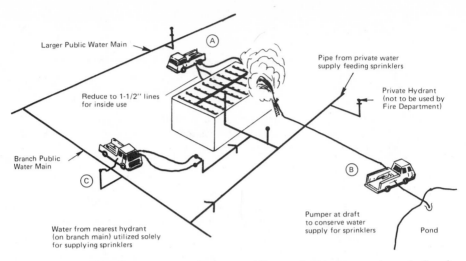

Larger Public Water Main

Pipe from private water supply feeding sprinklers

Reduce to 1-1/2″ lines for inside use

Private Hydrant (not to be used by Fire Department)

Branch Public Water Main

Water from nearest hydrant (on branch main) utilized solely for supplying sprinklers

Pumper at draft to conserve water supply for sprinklers

Pond

Fig. 2.13. Fire department pumpers supplying sprinkler system through fire department connection and manual hose lines. (From NFPA 13E)

ties is of immense value.[9] Barritt believes that it is essential to maintain and support the sprinkler system operation from both the roof and the stack sprinklers. Because both of these sprinkler systems may have to be turned off and on during the overhaul operations, personnel with radios should be located at the control valves to assure a successful extinguishment procedure. It is recommended that fire departments with high-racked storage buildings in their response areas should preplan their operations with both the insurance authorities and the operators of the facilities.

In an article titled "Fighting the Stored Plastics Fire," the Factory Mutual Engineering Division indicated that in fires involving the storage of plastics the sprinkler system should be supported and the structure closed as much as possible rather than vented, and the sprinklers should not be turned off, even if they have to remain in operation for hours.[10] In another article, "Two Plastics Fires—Two Disasters,"[11] the Factory Mutual Engineering Division recommended the following procedures for the support of sprinklers and the extinguishment operations in the "button-up" approach to fire extinguishment in plastic storage fires:*

- Evacuate the building of all personnel.
- Close up the building as tightly as possible to limit the air supply available to the fire.
- Connect a pumper engine to the fire department sprinkler connection and pump into the system.
- Keep sprinkler control valves wide open and sprinklers operating at all times, for an hour or more as necessary, until the fire has essentially been extinguished and can be manually attacked.
- During the final stages of sprinkler operation, but before manual

*From "Two Plastics Fires—Two Disasters," *F.M. Record*, Vol. 51, No. 6, Factory Mutual Systems, Nov.–Dec. 1974, p. 15.

attack is begun, attempt to mechanically exhaust smoke from the building if the equipment is available. This will not only facilitate manual extinguishment, but will also help to prevent the chance that combustible gases, built up inside the building from fire in an oxygen-starved atmosphere, could flash or explode when the building is opened.

● Before shutting off sprinklers, attempt to evaluate fire severity. This can be done either by a reconnaissance using life lines and air packs, or by breaking through the roof or a wall if the location of the seat of the fire can be fixed from questioning employees.

● Then shut off sprinklers, open up the building to vent smoke, and attack with hose streams. But keep a man posted at the sprinkler control valve at all times (with a two-way radio if available), ready to turn sprinklers back on if the fire threatens to flare up.

SUMMARY

Automatic sprinkler systems, properly installed in buildings and in operation when fires occur, are one of the greatest aids to fire departments. However, automatic sprinkler systems can only function effectively if there is sufficient water capacity and pressure. The major cause of unsatisfactory performance of sprinkler systems involves the closing of water supply control valves (the shutting off of water control valves prior to the fire, or before the fire is completely extinguished).

Upon arrival at a sprinklered building, one of the fire department's first actions should be to station a man at the operating sprinkler system water supply control valve to prevent the water from being turned off until the fire is extinguished. Another necessary action upon arrival is to supplement the sprinkler system's water supply with hose lines from adequate public water mains or natural water sources to the fire department connection on the sprinkler system. These essential and simultaneous procedures are dependent on information which is obtained during fire department preplanning, which concerns sprinkler system water supplies, control valves, and fire department connections. This information can best be obtained by the personnel conducting prefire surveys prior to the fire occurrence. Therefore, in order to provide the information for the operations necessary to properly control fires by supplementing automatic sprinkler systems, fire departments that are required to respond to sprinklered properties should have on-going preplanning and prefire survey programs.

SI Units

The following conversion factors are given as a convenience in converting to SI Units the English units used in this chapter:

1 inch = 25.400 mm
1 psi = 6.895 kPa
1 gpm = 3.785 litres/minute

ACTIVITIES

1. What are the three principal reasons for the unsatisfactory performance of automatic extinguisher systems, as established in the NFPA's 1970 "Automatic Sprinkler Performance Tables"? What is the major cause of unsatisfactory performance, and who might be responsible for such action? How might such action be avoided?

2. The NFPA's 1970 "Automatic Sprinkler Performance Tables" establish an efficiency record of approximately 96.2 percent satisfactory performance for automatic sprinkler systems in the United States for the years between 1925 and 1970. In Australia the satisfactory performance record for automatic sprinkler systems is 99.7 percent. Explain the variables that contribute to the improved performance of the Australian systems, describing why the Australian procedures may or may not be applied to systems in the United States.

3. What information concerning automatic sprinkler systems should be made available to responding company personnel during the prefire survey procedures? How is such information best presented, and who is best qualified to prepare such material? Explain why.

4. Using as a guide the form shown in Table 2.4, prepare, on a separate sheet of paper, your own prefire survey for a completely sprinklered building. Then, on another sheet of paper make a prefire survey diagram for your building. Compare your survey and diagram with those of your classmates. Obtain suggestions for improving your survey and diagram.

5. Present, in outline form, what you feel is a satisfactory operational plan for use by a fire department in support of an automatic sprinkler system at the time of a fire occurrence. Discuss your outline with others, describing in detail the procedures therein.

6. What might be the result if a fire department were to connect pumping engines to hydrants on private or public mains supplying the sprinkler system?

7. Describe the proper fire department connection arrangement to a single sprinkler system riser; also describe how an emergency connection is usually made.

8. List the information that you feel should be included in an analysis of fire department operations following fires in sprinklered buildings. Why is the inclusion of such information important?

9. Describe the "button-up" approach to fire extinguishment in plastic storage fires.

10. Who should be included in the prefire planning of fire department operations in response areas having high-racked storage buildings? Why?

SUGGESTED READINGS

"Automatic Sprinkler Performance Tables, 1970," *Fire Journal*, Vol. 64, No. 5, NFPA, Jul. 1970, pp. 35–39.

Casey, James F., "Standpipe, Sprinkler and Automatic Alarms Systems," Chapter 13, *The Fire Chief's Handbook*, 3rd ed., Donnelley, New York, 1967, pp. 339–366.

Clark, William E., "Sprinkler Operations," Chapter 9, *Fire Fighting: Principles and Practices*, Dun-Donnelley, New York, 1974, pp. 189–201.

"Fire Department Operations—Use of Automatic Sprinkler Systems," Special Interest Bulletin No. 270, 1970, Engineering and Safety Service, American Insurance Association, New York.

Fried, Emanuel, "Sprinkler Operations," *Fireground Tactics*, Chapter 12, Ginn, 1972, Chicago, pp. 125–134.

Marryatt, H. W., *Fire: Automatic Sprinkler Performance in Australia and New Zealand, 1886–1968*, Australian Fire Protection Association, Melbourne, Australia, 1971.

NFPA 13E, *Recommendations for Fire Department Operations in Properties Protected by Sprinkler and Standpipe Systems*, NFPA, 1973, Boston, pp. 13–19.

BIBLIOGRAPHY

[1]"Automatic Sprinkler Performance Tables, 1970," *Fire Journal*, Vol. 64, No. 4, NFPA, Jul. 1970, pp. 35–39.

[2]Bond, Horatio, "Sprinklers—Australia and New Zealand," *Fire Journal*, Vol. 66, No. 1, Jan. 1972, pp. 28–32.

[3]Marryatt, H. W., *Fire: Automatic Sprinkler Performance in Australia and New Zealand, 1886–1968*, Australian Fire Protection Association, Melbourne, Australia, 1971, p. 36.

[4]Klevan, Jacob B., "A Study of the Apparent Effectiveness of Water Motor Gongs in Prince George's County, Maryland," Undergraduate paper, Fire Protection Curriculum, University of Maryland, College Park.

[5]NFPA 13E, *Recommendations for Fire Department Operations in Properties Protected by Sprinkler and Standpipe Systems*, 1973, NFPA, Boston.

[6]Kimball, Warren Y., *Fire Attack: Command Decisions and Company Operations*, National Fire Protection Association, Boston, 1966, p. 134.

[7]Hickey, Harry E., "An Approach to Evaluating and Maintaining Sprinkler Performance," *Fire Technology*, Vol. 4, Nov. 1968, pp. 292–303.

[8]Kimball, Warren Y., "Planning Fireground Action for Sprinklered Buildings," *The Sentinel*, Vol. 15, No. 4, Factory Insurance Association, Apr. 1959, pp. 5–6.

[9]Barritt, J. S., "Fire Fighting Tactics for Racked Storage," *Fire Journal*, Vol. 67, No. 6, Nov. 1973, pp. 19–23.

[10]"Fighting the Stored Plastics Fire," *F.M. Record*, Vol. 51, No. 6, Factory Mutual System, Nov.–Dec. 1974, pp. 13.

[11]"Two Plastics Fires—Two Disasters," *F. M. Record*, Vol. 51, No. 6, Factory Mutual System, Nov.–Dec. 1974, pp. 14–15.

Chapter 3

Introduction to
Automatic Sprinkler Systems

DEFINITION AND PURPOSE OF AUTOMATIC SPRINKLER SYSTEMS

An operational definition for the basic concept of an automatic sprinkler system (based on the Chapter 1 discussions of standpipe systems and the Chapter 2 discussions of fire department operations in support of automatic sprinkler systems) is as follows:

> An automatic sprinkler system is a system of pipes, tubes, or conduits provided with heads or nozzles, that is automatically activated and (in some types) deactivated, utilizing the sensing of fire induced stimuli consisting of light, heat, visible or invisible combustion products, and pressure generation, to distribute water and water-based extinguishing agents in the fire area.

The various types of automatic sprinkler systems will be dealt with in detail in the following chapters of this text: Chapter 7, "The Wet Pipe Automatic Sprinkler System"; Chapter 8, "The Dry Pipe Automatic Sprinkler System"; Chapter 9, "Deluge and Preaction Automatic Sprinkler Systems"; and Chapter 10, "Specialized Automatic Sprinkler Systems" (including the dwelling systems and limited water supply systems).

HISTORY OF AUTOMATIC SPRINKLER SYSTEMS

The automatic sprinkler system is primarily a 19th century development, although the perfection of the various types of systems and heads beyond the dry pipe automatic sprinkler system occurred in the 20th century. In his book titled *Automatic Sprinkler Protection*, Gorham Dana reported the first recognized patent for a sprinkler system was issued in 1723 to a chemist named Ambrose Godfrey.[1] Godfrey's system consisted of a cask of fire extinguishing liquid, usually water, containing a pewter chamber of gunpowder. The chamber of gunpowder was connected with a system of fuses which were ignited by the flame of the fire, thus exploding the gunpowder and scattering the extinguishing liquid. According to Dana, the first automatic device for the application of water through a system of pipes was invented in 1806 by John Carey. Carey's system consisted of rows of perforated sprinklers connected to a system of pipes supplied with water from an elevated water tank. The main supply pipe from the tank contained a normally closed valve connected to a system of cords and weights. The system of cords and weights was arranged so the burning of a cord would release the valve by operation of a counterweight. Carey's system was not very practical because the cord tended to stretch, thus causing the valve to leak. Also, there were times when the valve stuck and failed to open properly.

In 1809 Sir William Congreve improved on Carey's concept by locating the valves outside the building to be protected. Congreve apparently incorporated the first concept of a fire department connection into a sprinkler system by means of his reported arrangements for an additional supply of water from water mains or hose connections from fire engines. In 1812 Congreve improved his system by replacing the cords with a fusible cement designed to operate at approximately 110°F. In the patent for his system, Congreve included an attachment that is considered to be the first practical alarm valve. The alarm in Congreve's attachment operated by the dropping of a weight.

In the middle of the 19th century additional improvements were reflected in the automatic sprinkler systems developed in England. In 1852 William Macbay patented a sprinkler system which consisted of a series of pipes that had outlets that were sealed either by caps of fusible gutta percha, or by a related low-temperature melting substance. In 1855 James Smith utilized cords or gutta percha as the initiating medium. In 1861 Lewis Roughton invented a system using fusible metal, and in 1863 Roger Dawson used perforated pipes with rose or fan-tail outlets.

In his book, Dana also reported the first automatic sprinkler system patented in the United States was developed in 1872 by Philip W. Pratt of Abington, Massachusetts. The Pratt system consisted of two revolving hollow arms containing rows of perforations. The arms were attached to a pipe containing a valve under water pressure, with a system of cords or fuses connected to the valve. When a fuse melted, the valve opened and allowed the water to flow into the arms and out of the perforations on the arms. The resulting velocity energy and pressure exerted a reactive force on the arms, rotating them and thereby increasing the water distribution throughout the area.

At about this same time, steam sprinkler systems were invented. In 1872 John Souther patented a system which consisted of perforated brass steam pipes. In Souther's system the steam was automatically turned on by the expansion of the brass pipe or the burning of fuse cords, while operation of the steam valve activated a steam whistle alarm device. In April of 1873 J. C. Meehan patented a steam system consisting of perforated steam pipes and a control valve operated by a charge of gunpowder. When the fuses were ignited by the flame contact, they ignited the gunpowder which, in turn, operated the steam valve, thus allowing the steam to flow into the pipe system. During this era, the concept of a perforated pipe system that used water as the extinguishing agent and could be turned on from a control valve for each area was widely adopted throughout New England.

Perforated Pipe Systems

Dana reported that around 1852 the first perforated pipe sprinkler systems were installed in the United States. The early perforated pipe sprinkler systems were installed primarily to protect the roofs of textile mill buildings, but were later installed in the picker, card, and spinning rooms of the mills. James B. Francis, a hydraulic engineer for the Locks and Canal Company of Lowell, Massachusetts, reportedly conducted the first experiments to determine the essential performance characteristics of the perforated pipe system. Francis determined the optimum size and location of the perforations, the proper size of the feed pipes or branch lines, and the most advantageous location for the pipes.

This system was installed across the mill in the center of each bay, with the perforated pipes close to the ceiling. The holes in the pipe were $\frac{1}{16}$ inch in diameter, and were placed 9 inches apart alternately on different sides of the pipe at a point slightly above the pipe's horizontal center. This design resulted in the discharge of water from the pipe toward the ceiling, thus covering the ceiling as well as the floor area.

To provide adequate residual pressure, Francis recommended a pipe schedule based on the principle that the area of the cross section at any point should be twice the area of the perforation orifices to be supplied from the pipe. This system, with 20 pounds pressure, would discharge enough water to cover the floor to a depth of $\frac{1}{14}$ inch in approximately 1 minute.

In 1880 the recommendations for the pipe schedule were modified by the Boston Manufacturers Mutual Insurance Company. This company advised property owners to be certain for systems supplied by reservoirs or tanks, the area of the orifices on the perforated pipe did not exceed 50 percent of the area of the supply pipe.

With systems supplied by pumps, the area of the perforated pipe orifices could be 66 percent of the area of the supply pipe. A system developed by William B. Whiting utilized holes that were $\frac{1}{12}$ inch in diameter, and 3 inches apart alternately on the top of the pipe at a point 30 degrees from the vertical on each side. Whiting recommended the installation of the pipe across the bays and under the beams; his perforated pipe system was installed extensively in textile mills throughout New England.

Dana also stated that during this era, in order to help prevent the orifices in perforated pipe from clogging due to corrosion, Frederick Grinnell, President of the Providence Steam and Gas Pipe Company, devised a countersunk orifice for pipe perforations. In 1875 Grinnell further helped improve resistance to corrosion by developing a noncorrosive orifice. Figure 3.1. shows a perforated pipe system in operation.

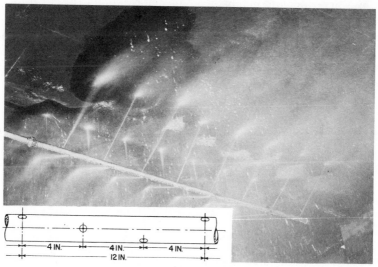

Fig. 3.1. Perforated pipe sprinkler system in operation. The inset shows the locations of perforations and the distances between them on a typical length of perforated pipe as was installed by the Providence Steam and Gas Pipe Company. (Grinnell Fire Protection Systems Company, Inc.)

The operation and design of the pipe arrangements for perforated pipe systems were similar. Perforated piping systems covered the entire room or area, the piping ran in parallel lines 10 feet or less apart, and ¾-inch pipe was the smallest size pipe used. Pipe branch lines were connected with feed pipe joined to the riser supplying the entire floor. Thus, the basic piping system used in today's automatic sprinkler systems originated with the early perforated pipe systems. However, with perforated pipe systems, each floor had separate control valves and risers. The valves were grouped and labeled by the section of the building controlled by each valve. The valves were often located in the building, in stair towers of the building, or outside the building.

In some complexes with several buildings, valves were placed in separate valve houses. The perforated pipe systems were sometimes connected to private hydrant systems when these systems were capable of supplying the needed volume and pressure.

The operation of the perforated pipe system was relatively simple. When fire was discovered, the valve for the area in which the fire was located was opened and water was discharged from the perforations in the pipe onto the ceiling. Thus, the entire area was covered by water projected off the ceiling. If the fire spread to another area or floor, another valve was opened. In some cases the opening of too many valves caused a lack of water pressure, which resulted in an inadequate amount of water being distributed for fire control.

The disadvantages of the perforated pipe systems were readily apparent. The primary problem, since water was not restricted to the area involved in fire, was the entire floor or area being subjected to water spray. Secondly, the system was dependent upon human action and reaction to discover the fire and to open the correct valve. Maintenance of the perforated pipe system was still another problem since orifices in the pipe became clogged from paint and corrosion. Testing the systems to determine if the orifices were blocked caused extensive water damage. Opening the valves accidentally, or opening incorrect valves during a fire occurrence, also caused water damage. However, because they helped establish an improved fire control record, perforated pipe systems were used extensively in textile mills throughout New England from approximately 1852 to 1885.

The Automatic Sprinkler Head

The development of a reliable automatic sprinkler system was dependent upon the development of the automatic sprinkler head. In *Automatic Sprinkler Systems*, Gorham Dana reported that the first automatic sprinkler head was invented by Major A. Stewart Harrison of London, although the device was neither patented nor applied in practice.[1] In 1874 Henry S. Parmelee of New Haven, Connecticut, patented the first practical automatic sprinkler head. Parmelee's sprinkler head consisted of a perforated head containing a valve which was held closed against water pressure by a heavy spring. The spring was held in place by two metal eyes manufactured from a low-fusing material. Figure 3.2 is a drawing of Parmelee's 1874 automatic sprinkler head.

In 1875 an entirely new type of automatic sprinkler head was developed by Parmelee. The new sprinkler head consisted of a brass cap soldered over a

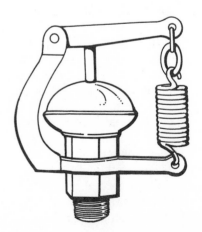

Fig. 3.2.   1874 Parmelee automatic sprinkler head. (From *Automatic Sprinkler Protection,* by Gorham Dana)

distributor. Although Parmelee's improved design was one of the simplest automatic sprinkler heads ever constructed, it was actually less sensitive than the original design because its solder joint was exposed to water on the inside of the head. Figure 3.3 shows the design and appearance of the 1875 Parmelee automatic sprinkler head—the first automatic sprinkler head extensively installed, and the head which made possible the concept of an automatic sprinkler system. In 1878 the Parmelee automatic sprinkler head was further

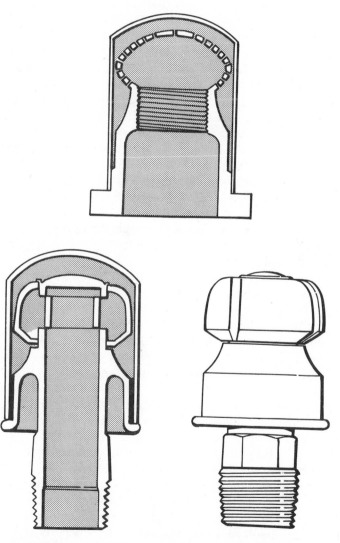

Fig. 3.3.   The 1875 (top) and 1878 (bottom) Parmelee automatic sprinkler heads. (From *Automatic Sprinkler Protection*, by Gorham Dana)

improved when the perforated head was replaced by a rotating slotted turbine. The rotating turbine provided improved distribution of water and less possibility of impairment by corrosion particles and sediment. The operation of the Parmelee automatic sprinkler head consisted of the fusing of the solder element when a temperature of approximately 160°F was reached. The water pressure then forced the cap off, and the revolving turbine distributed the water.

The Wet Pipe Automatic Sprinkler System

The design and installation of the Parmelee automatic sprinkler system involved some basic design principles that are still being utilized in today's automatic sprinkler system installations. The Parmelee automatic sprinkler system was installed with a single riser to feed all the floors, with the riser being large enough to supply the greatest number of heads on any one floor. Thus, the design of a separate riser for each floor (as installed with perforated pipe systems) was discarded because of the assumption that only one floor would be involved with fire at any one time, due to the early discovery of the fire with the automatic operation of the sprinkler head. This basic assumption in sprinkler system design is still being utilized today in essentially the same form.

The Parmelee automatic sprinkler system also contained an alarm valve which operated a bell or whistle when one or more sprinkler heads operated. The alarm valve consisted of a check valve installed in the main riser with a lever connected to the hinged end of the check valve clapper. The lever, which extended through a stuffing box, was connected by wire to a steam whistle or a mechanical gong. When water flowed through the check valve, the clapper would move the lever. The lever then moved the wire which operated the steam whistle or mechanical gong.

Until he developed his own automatic sprinkler in 1882, most of the Parmelee automatic sprinkler systems were installed by the Providence Steam and Gas Pipe Company under Frederick Grinnell. By 1881 Parmelee sprinkler systems were reported to have been installed in 214 plants, although eventually over 200,000 of the Parmelee automatic sprinkler heads were installed. Figure 3.4 shows the water discharge from the 1875 Parmelee automatic sprinkler head.

The initial development of and subsequent improvements to the automatic sprinkler head made possible the concept of wet pipe automatic sprinkler systems for heated buildings. In *Automatic Sprinkler Protection*, Gorham Dana stated that Underwriters Laboratories (UL) in 1917 approved a combination heating and sprinkler system. Dana reported the original sprinkler system installed by Parmelee in his New Haven piano factory in 1874 was also used for heating purposes during extremely cold weather. This was accomplished by letting steam into the sprinkler system's pipes; the sprinkler heads were installed upon inverted U pipes, forming a trap which remained full of water and prevented the steam from heating the sprinkler heads. The Parmelee tree system for piping layout was well suited for this purpose since the heat circulating in the main feed pipes would not create much circulation in the branch pipes on which the sprinklers were located. The system approved by the

Fig. 3.4. Water discharge from 1875 Parmelee automatic sprinkler head. (Grinnell Fire Protection Systems Company, Inc.)

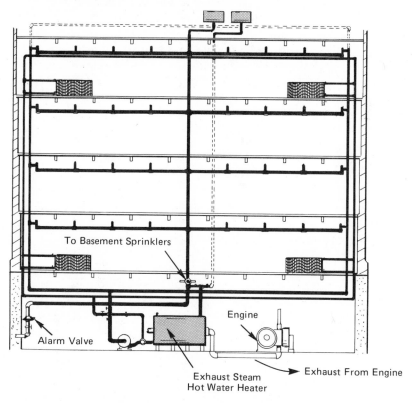

Fig. 3.5. Combined heating and sprinkler system. (From *Automatic Sprinkler Protection*, by Gorham Dana)

Underwriters Laboratories and the general design utilized in the early 1900s involved hot water in the sprinkler pipes, with the hot water being circulated either by pump or by means of a gravity system. These combination heating and sprinkler systems generally utilized enlarged branch lines from $1\frac{1}{2}$ to 2 in. in diameter, all connected together, with the branch lines then connected to auxiliary risers or return pipes for complete circulation of the hot water.

Thus, the concept of looped feed mains was initially utilized in the combination heating and sprinkler system, although these systems were not installed to any great extent. Marshall E. Peterson, in an article titled "Automatic Sprinkler Protection—Further Developments," estimated that only about 25 of these systems were ever installed, the last one having been installed before 1940.[2] Figure 3.5 shows a diagram of a combined heating and sprinkler system.

The Dry Pipe Automatic Sprinkler System

The dry pipe sprinkler system was made possible by the development of the dry pipe valve which was a result of the need to provide protection in unheated properties. Initial attempts to provide such protection involved the use of antifreeze solutions in the wet pipe systems. Dana indicated that in 1861 Osmund Williams received a patent for a nonfreezing chemical solution to be used in the pipes of sprinkler systems. In 1864 William Gilbert, Edwin Cooper, and G. R. Webster utilized a fusible cord that would melt at temperatures ranging from 90°F to 120°F. The cord was arranged to release a hammer on an alarm and open the water valve on the sprinkler supply pipe when the cord melted.

There were other attempts at designing dry pipe valves; however, the first dry pipe valve to be accepted for general use was the Grinnell bellows differential type dry valve patented in 1885. This dry pipe valve was an improvement of an earlier design patented in 1879. Figure 3.6 shows a drawing of the Grinnell

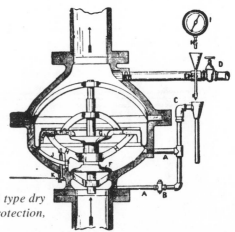

Fig. 3.6.   Grinnell Bellows differential type dry pipe valve. (From *Automatic Sprinkler Protection,* by Gorham Dana)

bellows differential type dry pipe valve. The improvement of the Grinnell bellows type dry pipe valve resulted in a valve that was probably the most generally used dry pipe valve prior to 1900. The Grinnell differential dry pipe valve No. 12, illustrated in Figure 3.7, was the predecessor of the modern differential dry pipe valve.

Many of the dry pipe valves were originally of the mechanical type, and were in general use until the mid-1920s. Very few mechanical valves were manufactured after this time. The principal problems with these valves were their susceptibility to corrosion and their need for precise adjustment.

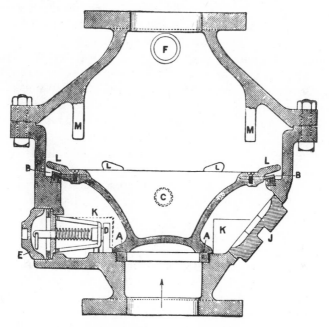

Fig. 3.7. Grinnell differential dry pipe valve No. 12.
(From *Automatic Sprinkler Protection*, by Gorham Dana)

Additional Automatic Sprinkler System Refinements

The period from the 1920s to the 1950s marked the development and perfection of automatic sprinkler systems. The deluge and the preaction sprinkler systems were initially developed in the early 1900s. However, in order to provide a supervisory alarm when the piping system was damaged or the supervisory air pressure was released from the system, the preaction system was further modified in 1933.

The automatic sprinkler head was continually developed during the period from 1900 to 1955. In his book titled *Fire Behavior and Sprinklers*, Norman J. Thompson indicated the Grinnell "Quartz bulb" type automatic sprinkler head was introduced in 1924, and the Globe "Saveall" head, which was activated by the melting of an organic compound, was introduced in 1931.[3] The

Grinnell "Duraspeed" head was developed in the 1930s, and today, because of the design and configuration of the heat collector arrangement on the actuating linkage, it is still one of the fastest operating sprinkler heads available.

The most extensive change in the automatic sprinkler head occurred as a result of experimental studies conducted by the Factory Mutual Laboratories from 1947 to 1950. These studies resulted in the development of the spray sprinkler head which, in 1955, became the standard automatic sprinkler head, while the previous standard sprinkler head became the old-type sprinkler head. This modification was basically a change in the design of the deflector on the sprinkler head to discharge the water downward in a more uniform, finely dispersed water spray. The details of the difference in the discharge water distribution patterns from the standard and the old style sprinkler heads will be discussed in Chapter 6, "The Automatic Sprinkler Head."

The ultra high-speed automatic sprinkler system with the explosive activated valve and the activation of the system from flame or pressure sensitive detectors was introduced in 1958 and 1959. These systems are examined in detail in Chapter 9, "Deluge and Preaction Automatic Sprinkler Systems."

The multicycle automatic sprinkler system, designed so the water control valve can be automatically turned on and off by heat sensors, was developed and approved in 1967. This system is examined in Chapter 10, "Specialized Automatic Sprinkler Systems." The on-off sprinkler head was approved by both the Factory Mutual and the Underwriters Laboratories in 1972. The on-off sprinkler head will also be examined in detail in Chapter 6.

Future Automatic Sprinkler System Developments

At present, there is much developmental interest in the improvement of the automatic sprinkler system. In an article titled "Automatic Sprinklers, The Past, The Present, and a Glimpse Toward the Future," Jack Rhodes indicated future developments relative to the automatic sprinkler system will involve improvements in both the hardware and the extinguishing medium.[4] One improvement cited by Rhodes is the NOVA (Normally Open Valve Assembly) valve which remains in the open position until manual effort is applied to put it in the closed position and keep it closed. Seemingly, such a valve might have value in correcting the most serious continuing cause of unsatisfactory sprinkler performance—the closing of water supply control valves before the fire occurs, or before the fire is completely extinguished. (As stated in Gorham Dana's book, *Automatic Sprinkler Protection,* according to the 1918 record of unsatisfactory sprinkler performance the largest single cause of sprinkler system failure to control fires was the closed water control valve. Also, see the "Sprinkler System Performance" section of Chapter 2, this text.)

Rhodes' article reported the Factory Mutual Laboratories have also been experimenting with the use of ablative water in sprinkler systems, and the modification of the size of the water droplets. He indicated the Factory Mutual studies have determined that high initial densities and large water droplets tend to penetrate the fire plume, achieving more effective fire extinguishment performance with more efficient extinguishing times. Thus, future developments will probably involve sprinkler heads with extra large orifices. Rhodes predicted the $\frac{17}{32}$-inch large orifice sprinkler head is becoming the standard sprinkler head for some specific situations, including plastic storage situations. Rhodes also described Factory Mutual's experimental interest in the development of a two-way automatic sprinkler head; essentially, a sprinkler head with two orifices. In such a sprinkler head, one orifice would be designed to produce a coarse water discharge spray to achieve good fire plume penetration, while the other orifice would produce a fine water spray for the efficient cooling of ceiling areas.

Development of the Sprinkler Standards

As previously indicated in this chapter, James Francis developed the initial standards for the installation of perforated pipe systems in the United States. A pipe schedule and procedures for the installation of the Parmelee sprinkler systems were developed by the Providence Steam and Gas Pipe Company under the direction of Frederick Grinnell, and the development of formal sprinkler rules was initiated by the insurance industry. In 1884 C. J. H. Woodbury of the Boston Manufacturers Company and F. E. Cabot of the Boston Board of Fire Underwriters conducted a study of the performance of automatic sprinkler systems. From this study of sprinkler performance, John Wormald of the Mutual Fire Insurance Corporation of Manchester, England, developed the first manual of rules for the installation of automatic sprinklers. In 1887 similar rules were prepared by the Factory Improvement Committee of the New England Fire Insurance Exchange.

In an article titled "The Development of Sprinkler Standards," E. W. Fowler reported that by 1895, in the Boston area alone, insurance organizations were using nine different standards for the size of pipe and sprinkler spacing.[5] Because of this situation, a committee was formed to develop standardized procedures for the installation of automatic sprinkler systems. The initial committee proceeded to the formation of a permanent organization on November 20, 1896, in New York City. Thus, a permanent organization was established that, at the time of its formation, had an existing standard consisting of recommendations for the installation of automatic sprinkler systems. This organization was formally titled the National Fire Protection Association (NFPA). Three years later, the standards for the installation of automatic sprinkler systems had been generally adopted in both the United States and Canada.

Fowler reported that sprinkler pipe schedules have been continually developed with extensive changes occurring in the sprinkler standards of 1896, 1905, 1940, and in the 1955 sprinkler standard in which the now standard spray sprinkler was adopted. The 1896 schedule adopted the $\frac{3}{4}$-inch pipe for the end sprinkler head on the branch line, as originally developed by Parmelee approximately 20 years earlier. In 1940 the smallest size pipe allowed was the 1-inch pipe for the end sprinkler head on the branch line. In 1955 changes were made on the number of sprinklers allowed on the 6-inch pipe, and a schedule was developed for an 8-inch pipe.

Fowler also reported in 1930 the sprinkler standard recognized for the first time a schedule for Light Hazard and Extra Hazard Occupancies. (Prior to 1930 the schedule was developed entirely for the Ordinary Hazard Occupancy.) In 1953 the standards were again modified for combined dry pipe and preaction systems. In 1955, with the introduction of the now standard sprinkler head, the rules for the spacing of the automatic sprinkler heads were modified. The area per sprinkler head in the Ordinary Hazard Occupancy was increased from 100 to 130 square feet. For the first time, the standards included provisions for the introduction of the hydraulically balanced design of automatic sprinkler systems, followed in 1966 by extensive instructions and procedures. The 1966 standard also included procedures for the utilization of the low differential dry pipe valve.

In 1972 the *NFPA Sprinkler System Standard* was modified to include a completely new section on the design and installation of automatic sprinkler systems in high-rise buildings. At this writing (1976) a new section to the 1975 standard has been proposed. The new section deals with the installation of automatic sprinkler systems in one- or two-family dwellings and mobile homes. Such a proposed addition to the *NFPA Sprinkler System Standard* could seemingly be interpreted as formal recognition of the value of automatic sprinkler systems as a factor in the prevention of loss of human life. Continuing research and developments in the design of automatic sprinkler systems are accelerating the frequency of changes in the sprinkler standards.

ECONOMIC VARIABLES OF AUTOMATIC SPRINKLER SYSTEMS

The economic factors relative to automatic sprinkler systems include the costs of the material and labor involved in installing the systems, including costs for maintenance, testing, and modification of the system. The charges made by many municipalities for the installation of water supply connections and meters on the supply main to the automatic sprinkler system is one of the variables in the continuing cost of the automatic sprinkler system. However, one of the principal economic benefits of an automatic sprinkler system is the amortization of the system's cost over a period of years, relative to the reduction of the insurance costs on both the building and the contents of the building.

Insurance Variables

The Rhodes article reported that in 1875 the Factory Mutual System was assessing charges of approximately $0.30 per $100.00 of insurance coverage when the Parmelee sprinkler system was introduced, and that by 1905, just thirty years later, the cost had been reduced to approximately $0.05 per

$100.00 for the identical class of risk equipped with automatic sprinkler protection.[4] He also reported during the decade from 1877 to 1888, the total number of fires involving the Factory Mutual unsprinklered properties was 759, with a total loss of $5,707,000, or approximately $7,500 per fire. The total number of fires in properties having automatic sprinklers, however, was 206, with a total loss of $224,480, or approximately $1,080 per fire. Thus, in the 10-year period, the average loss per fire in properties having automatic sprinklers was approximately 14 percent of the average loss per fire in the occupancies not having automatic sprinklers. It should be kept in mind the automatic sprinkler system originated in the textile mills of New England, and early system experimentation was developed at those mills with the active participation of the Factory Mutual Insurance System.

Obviously, the costs for the basic materials and the labor required to install an automatic sprinkler system will vary, depending upon the location in the United States and the time when the system was installed. Dana's book gives an indication of the costs of installing a sprinkler system in 1919, as follows:*

> The cost per head of equipping a building with sprinklers has gone up very rapidly in the last few years; while formerly an equipment could be put in for $3.00 to $4.00 per head, the cost today is nearer $10.00 a head. This is due to an increase both in stock and in labor, and is probably a temporary condition. The type of building has considerable effect on the price and under some conditions, the cost may go up to $15.00 or $20.00 a head. Extra sprinklers cost about $2.00 each.

In 1974 the costs of automatic sprinkler installations varied according to the type of building and the location of the building in the United States, as a result of varying material, freight, and labor rates. However, considering the many cost variations, *Building Construction Cost Data* reported an average cost of between $65.00 and $90.00 per head for a wet pipe sprinkler system in 1974 for mercantile, office, and health care occupancies.[6]

In an article titled "Fire Safety for High-Rise Housing for the Elderly," Chester W. Schirmer stated in 1973 a complete automatic sprinkler system could be installed in a high-rise residential building for the elderly, using the standpipes to supply the sprinkler system, at a cost of approximately 40 to 60 cents per square foot.[7]

Peterson's article ("Automatic Sprinkler Protection—Further Developments"[2]) provided an example from the late nineteenth century, specifically 1885, with a Gray dry pipe system installed in a cotton mill, relative to the cost of the system with the resulting reduction in insurance costs, as follows:**

> Before it had installed a Gray dry pipe system, it had paid $2,000 on fire insurance premiums; after the system was installed, the premiums were $1,200 per year. Over a five year period, the savings were as follows:

---

*From Dana, Gorham, *Automatic Sprinkler Protection*, 2nd ed., Wiley, New York, 1919, p. 83.
**From Peterson, Marshall E., "Automatic Sprinkler Protection—Further Developments," The Building Official and Code Administrator, Vol. 5, No. 12, Dec. 1971, p. 5.

Five years' savings in premiums at $800 per year	$4,000
Less cost of equipment	−$2,500
Maintenance	−$ 100
Net savings	$1,400
Add value of equipment	$1,000
Total savings ......,.....................................	$2,400

Five years' premium at old rates—$10,000
    (a net economy of 24 percent)

In an article titled "Convincing Consumers to Install Automatic Sprinklers," Raymond J. Casey explained the modern automatic sprinkler system, when installed throughout a building at the time of construction, can reduce the construction costs of the building as much as 60 percent.[8] Casey also explained a reduction of 40 to 95 percent can be achieved in the insurance premiums, and some of the tangible and indirect savings of automatic sprinkler systems involve the following assets:

- ● Sprinklers prevent costly downtime and assure continuity of business operations.
- ● Sprinklers protect the jobs of workers.
- ● Sprinklers safeguard business income.
- ● Sprinklers prevent the loss of customers who might go to another source of supply after a fire.
- ● Sprinklers permit greater design freedom and building flexibility.
- ● Sprinklers conserve water.

In "The Economics of a Fire Protection Program," a 1965 study concerning warehouses owned by an agency of the Federal Government, Carrol E. Burtner stated the average sprinkler systems' costs were $0.32 per square foot —calculated to be approximately 4.64 percent of the total building cost.[9] However, when the value of the contents of the warehouses were considered, the sprinkler systems' costs for the warehouses were less than 1 percent. Burtner estimated that a cost ranging from 2.0 to 2.5 percent of the value protected, usually building and contents, can be economically justified for fire protection expenditures. In addition, he reported the reduction in insurance rates, while varying greatly in some situations, may be as high as 75 percent. Burtner claims that such an insurance rate reduction is usually sufficient to amortize the cost of the automatic sprinkler system in a period from 6 to 10 years, considering the value of the contents to be high relative to the value of the building. He explained some sprinkler manufacturers estimate in order for the cost of the sprinkler system to be amortized within 10 years, the value of the contents should be at least three times the value of the building.

Burtner also conducted a survey of the fire protection costs of large industrial firms. From 37 responses, he discovered the average cost for installed systems (including but not exclusive of sprinklers) was $.202 per $100 of value, while the insurance rates on the sprinklered properties averaged $.097 per $100 of value.

In "Life Safety System vs. NFPA No. 13," Richard M. Patton claimed with his life safety system type of automatic sprinkler system, installation savings over the NFPA standard automatic sprinkler system can vary from 50 to 85 percent with existing buildings.[10] According to Patton, the difference in cost will usually vary from 10 to 20 percent in new buildings. The life safety system type of automatic sprinkler installation is examined in detail in Chapter 4, "Water Specifications for Automatic Sprinkler Systems," and in Chapter 5, "Basic Design of Automatic Sprinkler Systems."

In "Fire Protection Where There Was None," Edward J. Reilly described the installation of an automatic sprinkler system in an Illinois motel complex located outside a public water main service.[11] The system required a limited water supply system from a pressure tank. Reilly reported the value of the property protected by the system was approximately $3,000,000, and the reduction in the insurance premiums following installation of the sprinkler system was approximately 85 percent. With these values, total costs of the installation of the automatic sprinkler system would be amortized in less than four years.

In a 1965 article titled "The Economics of Automatic Sprinkler Systems," Seddon T. Duke explained one of the continuing factors that increases the cost of both the installation and the operation of an automatic sprinkler system is the practice (by both public and private water companies) of requiring a standby charge for the connection to the water system.[12] In some areas, the water company also required the installation of a meter and a meter vault. A fire flow check meter is necessary to ensure adequate water flow to the automatic sprinkler system. Because water is conserved since less of it is required to extinguish fires in sprinklered buildings than in nonsprinklered buildings, water companies can consider the installation of an automatic sprinkler system as an economy factor. Also, in most cities, less fire apparatus and fewer personnel are required to extingush fires in buildings equipped with automatic sprinklers than in nonsprinklered buildings. Thus, sprinklers help effect additional savings in the cost of this type of municipal service.

Governmental Variables

The prime governmental economic variables would seem to be the costs for service to private fire protection facilities. This service includes the automatic sprinkler system, with the assessment of standby charges, water meter charges, and water meter vault costs. The NFPA has consistently opposed the practice of assessing monetary charges on private water connections that are utilized solely for fire protection purposes. This opposition was stated in the NFPA's January 1968 issue of *Fire Journal*,[13] as follows:*

> For most sprinkler systems the water utility needs to make no charge
> at all except for the actual work of making the connection, and for the in-
> frequent repairs that might be necessary.

*From National Fire Protection Association, "Letters to the Association," *Fire Journal*, Vol. 61, No. 1, Jan. 1967, p. 87.

In 1960 the Factory Mutual Engineering Division conducted a survey of the practice of assessing charges of water connections to automatic sprinkler systems; the survey involved 1,250 private and public water companies serving approximately 3,100 communities in the United States and Canada.[14] The results of this survey, as summarized by Edward J. Reilly in *Private Fire Protection Serves a Public Purpose*, and as related to the number of private and public water systems which assess standby charges for sprinkler systems, indicated the charges were applied by 51 percent of the public and 88 percent of the private water systems in the United States.[15] In Canada, however, where all the respondents in the study involved public water systems, only 39 percent of the water systems assessed the standby charges. The basis for the charges, as summarized by Reilly, are presented in Table 3.1.

Table 3.1.   Basis of Standby Charges Assessed by Water Departments in the United States and Canada, 1960*

| | UNITED STATES | | CANADA |
BASIS FOR CHARGE	PUBLIC (1000)	PRIVATE (145)	PUBLIC (105)
Connecting Pipe Size	21%	27%	8%
Hydrants	3%	3%	5%
Sprinklers	1%	1%	5%
Combination	21%	57%	17%
Flat Rate	4%	—	1%
Floor Area	0.5%	—	2%
Other Basis	0.5%	—	1%
TOTALS	51%	88%	39%

Reilly also summarized the Factory Mutual survey data relative to the requirement for the installation of meters on the water supply line for private fire protection by the water company. Meters are typically required by a water company, with the justification the occupant may use the fire protection water connection for other domestic or industrial purposes. The utilization of standard meters on water mains supplying standpipes or sprinkler systems is not recommended, and only fire flow check meters should be installed. Fire flow check meters normally measure the flow of water up to a given flow capacity. When the flow of water is at the rated capacity for the operation of the sprinkler system or hose lines, a bypass opens allowing a large flow of water with small friction loss.

Reilly found that 28 percent of the public and 38 percent of the private water systems in the United States required the installation of meters on the water main supplying the automatic sprinkler system. In Canada, 18 percent of the

*From *Private Fire Protection Serves a Public Purpose*, by Edward J. Reilly.

105 public water systems required the meter to be installed on the water supply line to the sprinkler system. Some water companies required both standby charges and meters on water supply lines to private fire protection facilities.

In an undergraduate paper titled "A Study of Economic Constraining Factors Relative to the Cost of Required Meters, Meter Pits, and Standby Charges for Fire Protection Water Supply Connections," John K. Brouchard stated that in 1970–71 for the suburban Maryland area adjacent to Washington, D. C., the cost for a meter (dependent upon size) varied from $720 to $1,800.[16] In addition, the cost of a meter pit (required with each meter) varied from $2,400 to $5,000. Therefore, the total costs for the installation of a meter could vary from $3,120 to as high as $6,800.

In 1971 Brouchard also conducted a survey of 93 cities located throughout the United States with populations over 20,000. His survey was an attempt to replicate the information collected in the Factory Mutual survey of 1960, and to determine if any changes had occurred in the practice of assessing charges. Brouchard restricted his survey to public water departments in the United States; consequently, a comparison of the practices of the private water companies was not obtained. He found that 40 percent of the water companies required meters on the water supply lines to private fire protection systems. In addition, 75.6 percent of the companies requiring meters also required the installation of a meter pit. Brouchard also found that of the 93 public water departments surveyed, 65 percent required the standby charges for the private fire protection water supply connections. Table 3.2 presents the basis for the standby charges as determined by Brouchard's study. A comparison of these results with the information presented in Table 3.1 concerning Factory Mutual's 1960 study is interesting.

Table 3.2. Basis of Standby Charges
Assessed by Water Departments in
the United States, 1971*

BASIS FOR CHARGE	UNITED STATES PUBLIC (93)
Connecting Pipe Size	39%
Hydrants	3.5%
Sprinklers	1%
Combination	8.5%
Flat Rate	3.5%
Floor Area	1%
Meter Size	8.5%
TOTAL	65%

*From "A Study of Economic Constraining Factors Relative to the Cost of Required Meters, Meter Pits and Standby Charges for Fire Protection Water Supply Connections," by John K. Brouchard.

Brouchard found that 22 of the cities surveyed had changed their rate schedules in the period from 1960 to 1970: 77 percent increased the charges; 9 percent decreased the charges; 14 percent changed the schedule requirements without a change in the rates. Brouchard discovered that some of the standby charges were as high as $364 per year. When considered with the charges for the meters and meter pits, these financial charges imposed by the water department could be assumed to be an economic constraint to the installation of private fire protection facilities, including automatic sprinkler systems in some cities. Considering that 40 percent of Brouchard's sampling required meters on private water supply connections and 65 percent assessed standby charges, these practices appear to indicate a general trend.

In "How the City of Fresno Achieved Better Fire Protection," J. Randall reported the City of Fresno, California, is in the process of initiating two new programs that will create substantial economic incentives for the installation of automatic sprinkler systems.[17] These two programs are: (1) a revolving building incentive fund, and (2) a fire service demand charge.

The revolving building incentive fund, as proposed by the City of Fresno, would provide low interest loans for the installation of automatic fire extinguishing systems, primarily automatic sprinkler systems. The fund, once established and operating, would be expected to be self-supporting and revolving. Randall stated the revolving building fund has received the support of the construction industry. He believed the program will encourage industrial and commercial properties to build within the city, and that it will encourage property owners to annex existing buildings.

The fire service demand charge is a service charge similar to charges for water and sewer service, and as such may be applied to tax exempt property. The fire service demand charge is presently applied only to commercial and industrial buildings, and is established by measuring either the required fire flow of each building or the insurance premium of the building and the contents. Thus, the installation of automatic sprinkler systems will reduce the fire service demand charge since the fire flow and the insurance premiums are reduced in sprinklered properties.

Construction Variables

The installation of automatic sprinkler systems has another economic effect relative to the tradeoffs or allowances provided from features required in passive fire protection in the building when the active or dynamic fire protection (the sprinklers) is provided. In an article titled "Passive and Active Fire Protection, the Optimum Combination," R. Baldwin and P. H. Thomas defined the optimum level of fire protection as the fire protection that minimizes the sum of costs and the anticipated fire losses.[18] They have also correctly observed that building codes rely primarily on passive fire protection, fire resistance, flame spread, and compartmentalization requirements (including the height and area limitations), and tend to neglect adequate consideration of automatic sprinklers and other forms of fire suppression and protection systems.

Baldwin and Thomas reported the primary objective of passive fire protection is to protect the structural frame and integrity of the building in order to prevent the fire spread, thereby containing the fire to the compartment or floor of fire origin. Consequently, passive fire protection is principally of value when the fire becomes large enough to threaten the structural integrity of the building. The effect of providing increased fire resistance in the materials and assemblies in the building is to increase the threshold of sensitivity at which structural fire damage or significant fire loss may occur.

Baldwin and Thomas also reported the effect of automatic sprinklers is to reduce the frequency of fires above a given threshold, and the sensitivity of this threshold is modified by the spacing and the temperature rating of the sprinkler heads. The threshold of sensitivity for the sprinklers is generally expected to be less than the threshold of sensitivity for the fire resistance or other aspects of passive fire protection. They also explained that if the fire resistance or the sprinklers fail to operate or perform their design function, the results will be no different than if the measures were never provided in the building. Figure 3.8 is a conceptualized diagram of Baldwin and Thomas's concept of the role of active and passive fire protection measures.

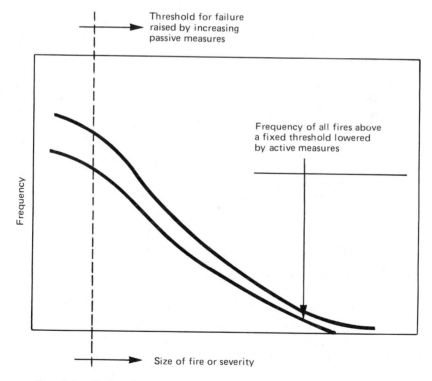

Fig. 3.8.   Role of active and passive fire protection measures. (From "Passive and Active Fire Protection—The Optimum Combination," by R. Baldwin and P. H. Thomas)

Richard E. Ritz, in an article on the fire resistivity of a sprinklered high-rise building, described the utilization of tradeoffs in the design of a high-rise office building completed in 1969.[19] These tradeoffs involved the following modifications of the passive fire protection in the building as a result of the installation of a complete automatic sprinkler system: modification of smoke shaft and entrance vestibules; extension of dead-end corridors; elimination of two wet standpipes with retention of two dry standpipes; extension of travel distance to exits; interior finish of Class C rating permitted in tenant spaces; and modification of the requirement for a fire resistant core in wood paneling and wired glass in tenant spaces.

Ritz reported that an economic analysis of the effects of these tradeoffs indicated the building could be completely sprinklered at no additional cost. Also, the annual savings from additional rental space and reduced insurance costs were approximately $50,000 per year which, projected over the 50-year estimated life of the building, totaled $2\frac{1}{2}$ million dollars. The 27-story building involved both high- and low-rise water supply systems with fire department connections, fire pump, and pressure tank. With the supply features, the cost of the complete automatic sprinkler system was approximately $0.70 per square foot.

In an article concerning the validity of tradeoffs for automatic sprinkler protection, Bert M. Cohn stated, using tradeoffs in the construction features of a building will reduce the requirements for passive protection, which is usually fire resistance.[20] However, as the dynamic fire protection, (specifically automatic sprinklers) is increased, these factors should be considered in relation to the probability of the failure of the automatic sprinkler system. Cohn also stated the reliability statistics for automatic sprinkler systems indicate that the NFPA estimates may be considered extremely conservative. Cohn related his studies suggest the automatic sprinkler system failure probably should be stated as approximately .004 for all fires, and about .006 for fires large enough to activate the system. He also stated in Robert W. Powers' article titled "Automatic Sprinkler Performance in High-Rise Buildings" (which deals with high-rise buildings in New York City), Powers reported a failure probability of .01.[21] Cohn explained (using the NFPA definition of a fire exceeding $250,000 in property damage) the probability of a large loss fire occurring in all structures is approximately .004, and in buildings with automatic sprinklers it is between .006 and .010.

Cohn questioned the validity of the tradeoffs allowed in some building codes for increased or unlimited area of openings in fire walls, in completely sprinklered buildings, and the substitution of automatic door closer requirements with automatic sprinkler systems installed in the corridors. He also questioned the practice of allowing the use of higher flame spread materials for interior finishes, including ceilings in buildings with automatic sprinklers. The practice of tradeoffs relative to passive and active fire protection varies from building code to building code. Obviously, the trend to provide some consideration for the fire suppression and fire protection capabilities of the automatic sprinkler system, relative to both the building and the occupants, will continue in the future. In a survey published in 1975, Gregory A. Harrison and James L. Houser presented the opinions of 187 professional individuals concerned with fire safety in buildings from the following occupational areas: academic, building official, general interest, fire service, federal government, insurance, and architects-engineers.[22] The results of the survey were based on the responses to a total of 38 questions, four of which dealt with fire suppression systems. These four questions were:*

1. Do you favor fire suppression systems in apartment buildings?
2. Do you favor fire suppression systems in office buildings?
3. Do you favor fire suppression systems in high-rise buildings?
4. Do you favor fire suppression systems in single-family residences?

The study population favored the provision of fire suppression systems in all but one of the occupancies, the exception being the single-family residence. While fire service personnel, government personnel, insurance personnel, and architects-engineers were generally in favor of fire suppression systems in high-rise buildings, they were generally against such systems in single-family residences. However, analysis of the injury and fatality statistics from fire has continually shown that the life loss and fire injury problem is a one- and two-family residence fire problem. The responses to these four questions from the study by Harrison and Houser are presented as Table 3.3.

Table 3.3.   Professional Opinions Relative to Fire Suppression Systems in Apartment, Office, High-Rise Buildings, and Single-Family Residences**

SAMPLE	APARTMENTS		OFFICES		HIGH-RISE		SINGLE-FAMILY RESIDENCE	
	YES	NO	YES	NO	YES	NO	YES	NO
Academic	5	2	4	3	6	1	2	5
Building Officials	11	3	12	3	20	1	5	5
General Interest	9	7	12	4	11	5	6	8
Fire Service	68	28	67	29	91	5	19	77
Government	0	2	0	2	2	0	0	2
Insurance	2	3	3	2	4	1	0	5
Architect-Engineer	9	8	11	6	17	0	1	16
TOTALS	104	53	109	49	157	13	33	118

*From Harrison, Gregory A., and Houser, James L., "A Survey for the Collection of Professional Opinion on Selected Fire Protection Engineering Topics," Technical Note 861, 1975, Center for Fire Research, National Bureau of Standards, Washington, D.C.
**Ibid.

Life Safety Variables

Human fatalities and injuries are both a moral, legal, and economic consideration relative to the installation of automatic sprinkler systems. The record of automatic sprinkler systems in terms of life safety is excellent, and has been established over a period of many years. The NFPA's *Fire Protection Handbook*[23] provides the following information in regard to life safety and automatic sprinkler systems:*

> The only fatalities in fully sprinklered properties reported to the NFPA were caused by explosions or flash fires; by ignition of the bedding or clothing of a person who was too young, too old, too intoxicated, or too handicapped in some other way to protect himself properly; by closure of water supply valves to the sprinkler system; or by hazards too severe for effective sprinkler performance in the protected property. Explosions in sprinklered properties have caused fatal injuries to occupants or have so damaged sprinkler piping as to render the systems virtually useless, with resultant loss of life. Severe flash fires have under unusual conditions traveled in advance of sprinkler operation, trapping victims before they had time to reach safety.
>
> In those isolated instances of fatalities to sleeping, handicapped, or intoxicated persons, ignition of clothing or bedding caused fatal burns or asphyxiation either because the small fire did not generate sufficient heat to fuse a sprinkler, or because the victim had suffered fatal injuries by the time the sprinkler operated. In these latter instances however, the sprinklers protected the lives of persons in adjoining areas.

Schirmer's article ("Fire Safety for High-Rise Housing for the Elderly")[7] reported the superior sprinkler performance record in Australia and New Zealand (discussed in Chapter 2) is even more outstanding when considered with the loss of life in sprinklered buildings. In the 82 years between 1886 and 1968, only 5 fatalities were related to fire situations in sprinklered buildings.

In "Misconceptions on Sprinklers and Life Safety," Richard E. Stevens discussed the types of objections that can arise when the installation of an automatic sprinkler system is considered.[24] Stevens categorizes these objections as follows:**

> ● There will be excessive water damage, either because of the fire or because of accidental physical damage to the sprinkler.
> ● The smoke generated will obscure exits and suffocate everyone.
> ● The water discharge will drown everyone.
> ● The steam generated will scald everyone.

Individuals citing water damage as an objection to the installation of an automatic sprinkler system fail to understand the operation of the sprinkler system. That is, only the sprinkler heads in the immediate fire area operate,

---

*From Tryon, George H., and McKinnon, Gordon P., eds., *Fire Protection Handbook*, 14th ed., National Fire Protection Association, Boston, 1976, p. 16–5.

**From Stevens, Richard E., "Misconceptions on Sprinklers and Life Safety," *Fire Journal*, Vol. 60, No. 4, Jul. 1966, pp. 28–29.

thereby releasing a lower application of water than will occur from fire department hose streams. When applied, department hose lines will commence at a later time than sprinkler system discharge, and will usually consist of a 1½-inch hose line at approximately 100 gallons per minute, or a 2½-inch hose line at 250 gallons per minute. It should be remembered that one sprinkler head will usually discharge approximately 20 gallons per minute, which is similar to the discharge from a single fire department ¾-inch hose line.

Water damage from accidental operation of the sprinkler head is unusual. Insurance companies generally charge approximately half of the fire insurance rate for the sprinkler leakage rate, since the loss experience is highly favorable. The principal cause of sprinkler leakage loss is the freezing of some portion of the sprinkler system. Such situations are not usually prevalent in heated buildings involved with occupants and thus, life safety considerations. The sprinkler heads utilized on systems are tested by both the Underwriters Laboratories (UL) and the Factory Mutual Laboratories (FM) to ensure that the head will not operate accidentally. Performance tests include a leakage examination, a water hammer test, a heating and cooling test, a strength of frame test, and a vibration test. In addition, the sprinkler piping is required to be hydrostatically tested to a minimum of 200 psi. Therefore, the automatic sprinkler system piping and heads are performance tested individually and again when assembled; however, such testing is not required on the domestic water and plumbing systems in buildings. The accidental operation of sprinkler heads is usually caused by freezing, mechanical damage, or placement of the heads in close proximity to heating devices or systems with heat conditions too high for the rating of the sprinkler head.

The quantity, opaqueness, and toxicity of the smoke produced in a given fire situation will change with the following variables: the fuel involved, the moisture content of the fuel, the air or oxygen ratio to the fuel, the intensity of the heat exposure to the fuel, and the duration of the heat exposure to the fuel. Extensive fire development is necessary in order to reach ceiling temperatures that are sufficient to operate the sprinkler heads on standard automatic sprinkler systems. In some cases sprinkler systems which will operate by the activation of smoke detectors have been installed in critical life safety areas and in concentrations of high property values. However, it should be kept in mind that conditions as observed in both test and actual fires usually involve less smoke generation and less spread of smoke with a sprinkler system.

Danger from drowning from automatic sprinkler operation is an objection usually cited by persons completely unfamiliar with the characteristics of automatic sprinkler systems. The discharge pattern from the standard sprinkler head at 4 feet below the deflector is approximately 16 feet in diameter, and the average water density in life safety occupancies is approximately 0.1 gpm/per square foot, given the usual sprinkler head discharge rate of 20 gpm. Stevens' article explains that this discharge density is approximately equal to 1 inch of rain an hour, and is about $\frac{1}{50}$ the density from the normal shower head with a rate of 4 gpm where the discharge density is measured with the stream approximately 1 foot in diameter.

The generation of steam in a fire situation is primarily determined by the velocity of the water particles, the size of the particles, the temperature of the atmosphere, and the convection fluid flow conditions created by the thermal column in the fire area. Since the sprinklers will usually be in operation before the fire department arrives, the fire temperature will usually be less severe, and, generally, less steam will be created with automatic sprinkler system operation as compared to fire department hose streams.

Also, untenable temperature conditions for human occupancy of an area will be exceeded before any appreciable steam generation will occur. Therefore, an individual unable to leave the immediate fire area will have suffered serious and possibly fatal injuries prior to the possibility of steam induced injuries. Steam condenses and cools very rapidly, and its ability to cause human injury is limited to the immediate area of fire involvement. The transmission of steam vapor tends to be induced and guided by the convection column which generally moves along the upper levels of compartments. Thus, in many situations the individuals within compartments at lower levels can avoid serious injury.

GENERAL MAINTENANCE AND INSPECTION PROCEDURES

The automatic sprinkler system, especially the wet pipe type, requires minimal maintenance. The more elaborate systems, such as the dry pipe, deluge, preaction and other specialized systems, require additional inspection procedures to assure efficient system operation. Every sprinkler system requires periodic inspections in order to detect human modifications to the system or changes in the physical environment it protects, either of which might prevent the proper functioning of the system. Maintenance procedures may also involve the conduct of repairs to the system caused by mechanical damage or modifications. The remainder of this chapter deals with those aspects of sprinkler systems that require regularly scheduled inspections.

Water Control Valves

Water control valves should be inspected every week. These valves should be identified with numbers or letters that indicate the portion of the system or building the valve controls. The valves should always be accessible, and, as recommended by the Factory Mutual Engineering Division since 1971, should be sealed or locked in the open position. A procedure should be adopted whereby the fire department is always notified when water control valves are shut down for repairs or modifications. Figure 3.9 shows the attachment of a tag to the water control valve; the tag indicates that persons closing the valve should notify the fire department. This procedure has been utilized in Montgomery County, Maryland, by the Glen Echo Fire Department. Efficient fire departments will send personnel to determine the extent and duration of the closed valve condition and to modify their prefire plans due to the shutdown of the sprinkler system. Figure 3.10 shows a riser with an O.S. & Y. valve obstructed by the storage of materials.

Fig. 3.9. Notification tag attached to O.S. & Y. valve. (Glen Echo Fire Department, Glen Echo, Maryland)

Fig. 3.10. Impaired access to an O.S. & Y. valve on the sprinkler system riser. (Michael W. Magee)

Conducting water flow tests is a valuable procedure in determining closed or partially closed water control valves or obstructions in the water supply mains. Water flow tests from the main drain lines and the inspector's test valve on the automatic sprinkler system are usually conducted by the owner's representative at the time of the insurance authorities' inspection. The recommended procedure for conducting a main drain test is explained in Chapter 4, "Water Specifications for Automatic Sprinkler Systems."

Fire department connections should be inspected regularly; a good practice is to inspect them weekly, simultaneously with the water control valve inspections conducted by the owner or occupants. The fire department should also inspect and check the fire department connections when they conduct their prefire planning surveys. They should make a positive check to see that hose lines can be attached to the fire department connection and that proper threads are provided. The caps should be in place on the intakes of the connection, the siamese clappers should be operating, the ball drip or drain should be operable, and the check valve should not be leaking. It is advisable for fire department personnel to remove both caps from the fire department connection even if they do not immediately attach hose lines to the connection. Water can become trapped between the clapper and the caps, exerting pressure on the cap on the unused intake and making later removal of the cap extremely difficult. Figure 3.11 shows a sign attached to a fire department connection; the location of the sign effectively hinders removal of the caps and the attachment of the hose lines.

Fig. 3.11. Fire department connection impaired by attachment of sign. (Michael W. Magee)

Fire Pumps

Fire pumps should be inspected and preferably operated weekly. The creation of a water flow sufficient to activate the automatic start controller will check the operation of both the automatic start fire pump and the automatic start controller. In systems that utilize a pressure maintenance or jockey pump, this pump operation should also be checked weekly. An annual full flow test on the fire pump is usually conducted by insurance authorities. Fire department personnel should be provided to witness full flow tests in order for such personnel to gain a better understanding of the system layout, connections, valves, and operation of the particular fire pump. In the past, many horizontal split-case pumps have been installed with a suction lift. Current practice is to install a vertical shaft turbine-type pump where a suction lift is necessary. In older installations where the fire pump is equipped with a suction priming tank, the water level should be checked weekly. If the fire pump is arranged to take suction from a water source, the suction pipe, foot valves, and trash screens must be checked regularly to avoid obstructions in the suction line.

Gravity and Pressure Tanks

Weekly checks should be made of gravity tanks, especially the water level in the tanks. Wooden tanks may suffer shrinkage and will leak if not maintained with the proper water level, while corrosion is prevalent in steel tanks. In freezing climates the heaters for the tanks must be checked, and the insulation and casing on the tank riser must be maintained. Freezing in the tank riser can block the flow of water; even a thin layer of ice in the tank can affect the flow of water. Whenever it is necessary to repair or paint the inside of gravity or storage tanks, a cover should be placed over the riser to prevent debris from falling into it. After the repairs are completed and the tank is refilled, the riser and underground mains should be thoroughly flushed, preferably through a hydrant, to remove all the small debris from the riser and to check the water flow.

Pressure tanks should have the water level and air pressure checked regularly, preferably weekly, when these conditions are not automatically supervised. During freezing weather the heat in the area of the pressure tank must be maintained.

Sprinkler Heads and Piping

The condition of sprinkler heads should be checked regularly, preferably weekly, by occupancy personnel. Sprinkler heads should not be painted, and materials, stock, shelving, or racks should be kept at least 18 inches below the sprinkler heads. Where bulk materials are involved, a 36-inch clearance from the sprinkler heads may be required. Locations with mechanical handling equipment and storage of materials on pallets should maintain the 36-inch clearance from the sprinkler heads. Extra sprinkler heads should be kept attached to or adjacent to the sprinkler riser and water control valve. To provide protection from damage, extra sprinkler heads should always be kept in a cabinet in an area where temperatures do not exceed 100°F. Sprinkler heads should be inspected to prevent corrosion and blockage from dust, dirt, paint residue, and bird or insect manifestations. The sprinkler piping should be kept in good condition, and should not be used to support materials, stock, or ladders. Any pipe hangers that are missing or damaged should be replaced. Should the sprinkler piping become impaired or partially obstructed with sediment or corrosion, regular sprinkler maintenance and installation companies should be contacted for flushing and cleaning operations.

Most of the major automatic sprinkler companies provide service contracts with provision for periodic surveys and complete maintenance checks. Such contracts, however, cannot be considered as adequate replacements for the weekly inspections that are necessary for every sprinkler system. Detailed inspection procedures for the various types of automatic sprinkler systems will be examined in later chapters.

Table 3.4 shows the NFPA recommended automatic sprinkler system inspection form.

Table 3.4.   NFPA Recommended Automatic Sprinkler System Inspection Form*

REPORT OF INSPECTION

Inspection Report
No........................

Conferred With
....

Inspection Contract
No.........................

Bureau File
No........................

REPORT TO.. BUILDING OR LOCATION.............................

STREET.. INSPECTOR...

CITY & STATE.. DATE..

	Yes	N.A.‡	No*
1. GENERAL			
a. Is the building occupied?...● ● ● ● ●............			
b. Is the occupancy same as previous inspection?..........................● ● ● ● ●............			
c. Are all systems in service?...● ● ● ● ●............			
d. Are all fire protection systems same as last inspection?..............			
e. Is building completely sprinklered?....................................● ● ● ● ●............			
f. Are all new additions and building changes properly protected?........			
g. Is all stock or storage properly below sprinkler piping?..............			
h. Was property free of fires since last inspection? (Explain any fire on separate sheet)			
i. In areas protected by wet system, does the building appear to be properly heated in all areas, including blind attics, perimeter areas and are all exterior openings protected against entrance of cold air?........			
2. CONTROL VALVES (See Section 16)			
a. Are all sprinkler system main control valves open?....................● ● ● ● ●............			
b. Are all other valves in proper position?..............................● ● ● ● ●............			
c. Are all control valves in good condition and sealed or supervised....● ● ● ● ●............			
3. WATER SUPPLIES (See Section 17)			
a. Was a water flow test made and results satisfactory?.................● ● ● ● ●............			

	Yes	N.A.‡	No*
4. TANKS, PUMPS, FIRE DEPT. CONNECTIONS			
a. Are fire pumps, gravity tanks, reservoirs and pressure tanks in good condition and properly maintained?....................................			
b. Are fire dept. connections in satisfactory condition, couplings free, caps in place and check valves tight?..................................			
5. WET SYSTEMS (See Section 13)			
a. Are cold weather valves open or closed as necessary?.................			
b. Have anti-freeze systems been tested and left in satisfactory condition?........			
c. Are alarm valves, water flow indicators and retards in satisfactory condition?.......			
6. DRY SYSTEMS (See Section 14)			
a. Is dry valve in service and in good condition?.......................			
b. Is air pressure and priming water level normal?......................			
c. Is air compressor in good condition?.................................			
d. Were low points drained during fall and winter inspections?..........			
e. Are Quick Opening Devices in service?................................			
f. Has piping been checked for stoppage within past 10 years?...........			
g. Has piping been checked for proper pitch within past 5 years?........			
h. Have dry valves been trip tested satisfactorily as required?.........			
i. Are dry valves adequately protected from freezing?...................			
j. Valve house and heater condition satisfactory?......................			
7. SPECIAL SYSTEMS (See Section 18)			
a. Were valves tested as required?......................................			
b. Were all heat responsibe systems tested and results satisfactory?....			
c. Were supervisory features tested and results satisfactory?...........			
8. ALARMS			
a. Water motor and gong test satisfactory?..............................			
b. Electric alarm test satisfactory?....................................			
c. Supervisory alarm service test satisfactory?.........................			

*From NFPA 13A, *Recommended Practice for the Care and Maintenance Of Sprinkler Systems.*

Table 3.4 (Continued)

9. SPRINKLERS—PIPING	Yes	N.A.‡	No*
a. Are all sprinklers in good condition, not obstructed, and free of corrosion or loading?..	● ● ● ● ●		
b. Are all sprinklers less than 50 years old?...	● ● ● ● ●		
c. Are extra sprinklers readily available?..	● ● ● ● ●		
d. Is condition of piping, drain valves, check valves, hangers, pressure gauges, open sprinklers, strainers satisfactory?....................................	● ● ● ● ●		
e. Are all sprinklers of proper temperature rating?..............................	● ● ● ● ●		
f. Are portable fire extinguishers in good condition?...........................			
g. Is hand hose on sprinkler system satisfactory?................................			

*Explain "No" answers in Item #19   ‡Not Applicable

10. Date Dry System Piping last checked for stoppage. ....
11. Date Dry System Piping last checked for proper pitch. ....
12. Date Dry Pipe Valve last trip tested. ....
13. Wet Systems: No?   Make and Model?..
14. Dry Systems: No?   Make and Model?..
15. Special System: No?   Type...
    Make and Model?..Condition?......................

16. CONTROL VALVES No? Type?	Open		Secured		Closed		Signs		Condition
	Yes	No	Yes	No	Yes	No	Yes	No	
City Connection Control Valve.									
Tank Control Valves.									
Pump Control Valves.									
Sectional Control Valves.									
System Control Valves.									

17.   WATER FLOW TEST

Water Pressure? ....CITY...............PSI   TANK...............PSI   FIRE PUMP...............PSI
Water Flow Test?..(If none made, Why?)...

Test Pipe Located	Size Test Pipe	Pressure Before	Flow Pressure	Pressure After	Test Pipe Located	Size Test Pipe	Pressure Before	Flow Pressure	Pressure After

18. Heat Responsive Devices: Type?   Type of Test?
    Valve No. ........A.....B.....C.....D.....E.....F.....   Valve No. ........A.....B.....C.....D.....E.....F.....
    Valve No. ........A.....B.....C.....D.....E.....F.....   Valve No. ........A.....B.....C.....D.....E.....F.....
    Valve No. ........A.....B.....C.....D.....E.....F.....   Valve No. ........A.....B.....C.....D.....E.....F.....
    Valve No. ........A.....B.....C.....D.....E.....F.....   Valve No. ........A.....B.....C.....D.....E.....F.....
    Auxiliary equipment: No?...........Type?.....................Location?.................Test Results?...........

19. Explanation of any "No" answers.
20. Recent changes in building occupancy or fire protection equipment.
21. Adjustments or corrections made.
22. Desirable Improvements.

DUPLICATE TO:...
STREET:...CITY & STATE:...

Maintenance Procedures

The repair and maintenance of a sprinkler system often requires the closing of water control valves. In all cases, the principal rule to be followed is to shut off only the minimum amount of the system necessary to conduct the repairs.

Table 3.5.   Freezing and Testing Times for Various
Pipe Sizes with the Pipe Freeze Method*

FREEZING TIME

PIPE SIZE (IN INCHES)	APPROXIMATE TIME TO FREEZE (IN MINUTES)
1	5
$1\frac{1}{4}$	10
$1\frac{1}{2}$	16
2	23
$2\frac{1}{2}$	30

Time intervals which should be allowed—before
testing for freezing—vary with pipe sizes.

TESTING TIME

PIPE SIZE (IN INCHES)	TIME UNTIL PIPE IS TESTED (IN MINUTES)
1	10
$1\frac{1}{4}$	15
$1\frac{1}{2}$	23
2	30
$2\frac{1}{2}$	37

Where provided, sectional cut-off valves should be utilized rather than the main system water control valve. Blanks may be inserted at flanges to reduce the area that must be turned off. In an article titled "Sprinkler Piping Put in Deep Freeze for Repairs," Gordon M. Betz presented a unique procedure for shutting off the water in wet pipe systems.[25] In Betz's procedure, blockage is accomplished by freezing a small section of the pipe at the desired pipe section, thus enabling removal of pipe sections beyond the ice blockage point. As his freeze solution, Betz used alcohol that had been cooled by dry ice. He constructed a freeze container to fit pipe sizes varying from 1 to $2\frac{1}{2}$ in.; the freeze container served to contain the 190-proof anhydrol alcohol cooled to $-100°F$.

The freeze method is more efficient on horizontal pipe; it has the advantage of enabling modifications to the sprinkler piping with a minimum of sprinkler heads out of service. Table 3.5 illustrates the freezing and testing times established by Betz, and Figure 3.12 illustrates the pipe freeze container in the open position. Figure 3.13 shows it in use on a horizontal run of pipe. The pipe freeze method developed by Betz appears to have many advantages for sprinkler maintenance operations since it limits the number of heads shut off.

*From "Sprinkler Piping Put in Deep Freeze for Repairs," by Gordon M. Betz.

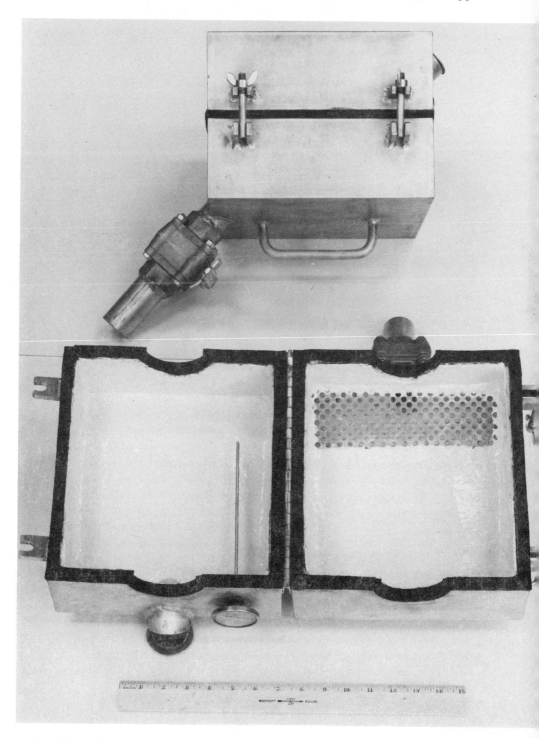

Fig. 3.12.   Pipe freeze container. (Gordon H. Betz, Western Electric Company, Inc.)

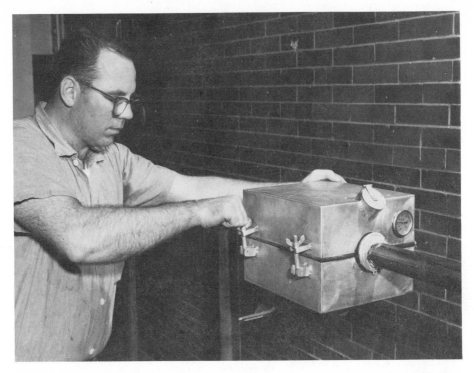

Fig. 3.13.   Pipe freeze container on a horizontal pipe. (Gordon H. Betz, Western Electric Company, Inc.)

SUMMARY

The development and improvement of the automatic sprinkler system reflects the development and progress of the theories, concepts, ideas, and procedures of fire protection. The reliability problem of the automatic sprinkler system during its period of development in the 18th century, is essentially the same today: the problem of water control valves being turned off before or during the fire. The economic advantages of automatic sprinkler systems have been established for many years, and are as valid now, if not more so, due to the need to conserve both raw and finished materials. The cost for installation of automatic sprinkler systems is approximately equal to what is spent on carpets and floor coverings in many office and institutional occupancies. The concept of the completely automatic sprinkler system, if incorporated in the design of a building, can be utilized to achieve savings in the expenditures required for passive fire protection. Consequently, an automatic sprinkler system will result in a totally improved fire safety environment for the building and, more importantly, for the occupants and the general public using the building.

SI Units

The following conversion factors are given as a convenience in converting to SI units the English units used in this chapter:

1 square foot	=	0.0929 m²
1 inch	=	25.400 mm
1 foot	=	0.305 m
1 pound (force)	=	4.448N
5/9(°F—32)	=	°C
1 gpm	=	3.785 litres/minute
1 gpm/sq ft	=	40.746 litres/minute m²

ACTIVITIES

1. Make a list of several of the disadvantages of the earlier sprinkler systems. Then, discuss how these disadvantages have been eliminated or refined in the automatic sprinkler systems in use today.
2. Write, in your own words, an operational definition for the basic concept of an automatic sprinkler system.
3. List three reasons why the perforated pipe sprinkler systems were replaced with the development of the Parmelee Sprinkler System.
4. Survey the opinions of the personnel in your fire department or organization in regard to their favoring or disfavoring fire suppression systems in apartment buildings, office buildings, high-rise buildings, and single-family dwellings. Compare the results to the opinions presented in Table 3.3. Then, discuss reasons why the results of your survey are similar or different than those presented in Table 3.3.
5. Determine the charges applied by the water company in your area to water service connections for private fire protection, including automatic sprinkler systems.
6. List some of the ways the installation of an automatic sprinkler system can help effect financial savings for a community.
7. Explain how programs such as a revolving building incentive fund or a fire service demand charge could help create substantial economic incentives for the installation of automatic sprinkler systems.
8. Obtain information relative to the cost of installation of sprinkler systems of various types in different occupancies in your area. Then examine the occupancy having the lowest installation cost, and write an explanation justifying the variance in costs.
9. Discuss four of the most common objections to the installation of automatic sprinkler systems. As each objection is discussed, present reasons why these objections may be refuted.
10. In outline form, present what you think is an appropriate procedure for the inspection of water control valves.

SUGGESTED READINGS

Burtner, Carrol E., "The Economics of a Fire Protection Program," *Fire Technology,* Vol. 2, No. 1, Feb. 1966, pp. 5–14.

Care and Maintenance of Sprinkler Systems, 1960, National Board of Fire Underwriters, Insurance Services Office, New York.

Dana, Gorham, *Automatic Sprinkler Protection*, John Wiley and Sons, Inc., New York, 1919.

Internal Cleaning of Sprinkler Piping, 1959, National Board of Fire Underwriters, Insurance Services Office, New York.

NFPA 13A, *Recommended Practice for the Care and Maintenance of Sprinkler Systems*, 1971, NFPA, Boston.

Powers, W. Robert, "Automatic Sprinkler Experience in High-Rise Buildings," *Fire Journal*, Vol. 66, No. 6, Nov. 1972, pp. 47–49.

Thompson, Norman J., *Fire Behavior and Sprinklers*, National Fire Protection Association, Boston, 1964.

BIBLIOGRAPHY

[1]Dana, Gorham, *Automatic Sprinkler Protection,* 2nd ed, Wiley, New York, 1919.

[2]Peterson, Marshall E., "Automatic Sprinkler Protection—Further Developments," *The Building Official and Code Administrator*, Vol. 5, No. 12, Dec. 1971, pp. 5–8, 31.

[3]Thompson, Norman J., *Fire Behavior and Sprinklers,* National Fire Protection Association, Boston, 1964.

[4]Rhodes, Jack, "Automatic Sprinklers, The Past, The Present, and a Glimpse Toward the Future," *Fire Journal,* Vol. 68, No. 6, Nov. 1974, pp. 42–47.

[5]Fowler, E. W., "The Development of Sprinkler Standards," *Fire Journal,* Vol. 62, No. 4, July 1968, pp. 75–77.

[6]Godfrey, Robert S., ed, *Building Construction Cost Data,* 1974, 32nd ed., Duxbury, Mass., Robert Snow Means Co., Inc., 1974, p. 185.

[7]Schirmer, Chester W., "Fire Safety for High-Rise Housing for the Elderly," *Fire Journal,* Vol. 67, No. 5, Sep. 1973, pp. 53–56.

[8]Casey, Raymond J., "Convincing Consumers to Install Automatic Sprinklers," *Fire Journal,* Vol. 65, No. 2, Mar. 1971, pp. 35–36, 41.

[9]Burtner, Carrol E., "The Economics of a Fire Protection Program," *Fire Technology,* Vol. 2, No. 1, Feb. 1966, pp. 5–14.

[10]Patton, Richard M., "Life Safety System vs. NFPA No. 13," *The Building Official and Code Administrator,* Vol. 6, No. 10, Oct. 1972, pp. 4–6.

[11]Reilly, Edward J., "Fire Protection Where There was None," *Fire Journal,* Vol. 59, No. 1, Jan. 1965, pp. 32–33.

[12]Duke, T. Seddon, "The Economics of Automatic Sprinkler Systems," *Fire Journal,* Vol. 59, No. 6, Nov. 1965, pp. 57–60.

[13]National Fire Protection Association, "Letters to the Association," *Fire Journal,* Vol. 61, No. 1, Jan. 1967, p. 87.

[14]*Private Fire Service Connections: Survey of Standby Charges and Meter Requirements,* Factory Mutual System, Norwood, Mass., 1960.

[15]Reilly, Edward J., *Private Fire Protection Serves a Public Purpose,* National Automatic Sprinkler and Fire Control Association, Inc., New York.

[16]Brouchard, John K., "A Study of Economic Constraining Factors Relative to the Cost of Required Meters, Meter Pits, and Standby Charges for Fire Protection Water Supply Connections," Undergraduate Paper, Fire Protection Curriculum, University of Maryland, College Park, 1971.

[17]Randall, J., "How the City of Fresno Achieved Better Fire Protection," *Fire Journal*, Vol. 69, No. 2, Mar. 1975, pp. 14–16, 27.

[18]Baldwin, R. and Thomas, P. H., "Passive and Active Fire Protection— The Optimum Combination," *Fire Technology*, Vol. 10, No. 2, May 1974, pp. 140–146.

[19]Ritz, Richard E., "The Georgia-Pacific Building, A High-Rise Fire-Resistive Office Building with Automatic Sprinklers Installed Throughout," *Fire Journal*, Vol. 63, No. 5, Sept. 1969, pp. 5–9.

[20]Cohn, Bert M. "The Validity of Trade-Offs for Automatic Sprinkler Protection," *Fire Protection News,* No. 74–3, Gage Babcock and Associates, Inc., Aug. 1974, pp. 1–4.

[21]Powers, W. Robert, "Automatic Sprinkler Experience in High-Rise Buildings," *Fire Journal*, Vol. 66, No. 6, Nov. 1972, pp. 47–49.

[22]Harrison, Gregory A. and Houser, James L., "A Survey for the Collection of Professional Opinion on Selected Fire Protection Engineering Topics," Technical Note 861, 1975, Center for Fire Research, National Bureau of Standards, Washington, D.C.

[23]Tryon, George H., and McKinnon, Gordon P., eds., *Fire Protection Handbook,* 13th ed., National Fire Protection Association, Boston, 1969, pp. 16–5.

[24]Stevens, Richard E., "Misconceptions on Sprinklers and Life Safety," *Fire Journal*, Vol. 60, No. 4, Jul. 1966, pp. 28–29.

[25]Betz, Gordon M., "Sprinkler Piping Put in Deep Freeze for Repairs," Plant Engineering, Apr. 1967, pp. 166–169.

Chapter 4

Water Specifications For Automatic Sprinkler Systems

DISTRIBUTION OF WATER

The fire suppression effectiveness of the automatic sprinkler system is dependent on the suppression and quenching action of water delivered to the fire area by the system of pipes and sprinkler heads that make up the system's principal components. As discussed in Chapter 2, "Fire Department Procedures for Automatic Spinkler Systems," the primary cause of unsatisfactory sprinkler performance involves sprinkler systems not being supplied with water at the time of fire occurrence. Thus, it is important to review the extinguishing capability of water and the suppression action that results from the distribution of water by standard sprinkler heads which project water in the form of finely divided water particles or fog. Layman[1] explained that the extinguishing action of water involves efficient conversion to steam in the following manner:*

One gallon of fresh water will absorb approximately 1,250 British thermal units (Btu) in the process of raising its temperature from 62°F to 212°F. This theoretical gallon of water has reached its boiling point (212°F) and is ready to start changing from liquid to a vapor. Absorption of additional heat will not increase the temperature of the water but will reduce the liquid volume by converting some of the water into steam which will escape into the surrounding atmosphere. In the process of vaporization (boiling), this gallon of water will absorb more than six times the volume of heat that is absorbed in the process of raising its temperature from 62°F to 212°F. When the last drop has been converted into steam, this gallon of water has absorbed a total of approximately 9,330 Btu; 1,250 Btu in the process of raising its temperature from 62°F to 212°F, and 8,080 Btu in the process of vaporization. Based upon these scientific facts, the following truism can be stated: The maximum cooling action of a given volume of water is obtained

*From Layman, Lloyd A., *Attacking and Extinguishing Interior Fires,* National Fire Protection Association, Boston, 1955, p. 24.

only when the entire volume has been converted into steam.

Vaporization of water results in the production of steam in ratio of 1 to over 1,600. Water in the volume of 1.05 cu in. measured at its boiling point will expand into 1,728 cu in. or 1 cu ft of steam. A gallon of water measured at 62°F will produce approximately 223 cu ft of steam.

In 1955 the National Board of Fire Underwriters (now the American Insurance Association) sponsored a scientific study conducted by the Underwriters Laboratories Inc. (UL), relative to the principles of fire extinguishment by water spray.[2] The study consisted of a total of 141 experiments with the water spray directed vertically downward onto the fire in 109 of the fire situations. Thus, the majority of the fire situations were similar to the standard upright or pendent sprinkler head operation. In the remaining 32 experiments, the water spray was directed horizontally onto the burning fuel area in a situation similar to the application of water from small hose streams or from standard sidewall sprinkler heads. Gasoline, kerosine, ethyl alcohol, and wood in the form of white pine strips were the fuels that were used in the study. In order to vary the ventilation effects, all of the experiments were conducted indoors with an assembly erected around the fuel area. Nineteen different nozzles were used in the study, the nozzle pressure was standardized at 100 psi, and the flow was varied from a low of 0.05 gpm to a high of 13.23 gpm. The results of the study indicated the variables of oxygen dilution by the generated steam, cooling by water droplets, and the ventilation rates in the fire area—each or all of which apparently affected the extinguishing efficiency. The principal conclusions from the NBFU study are summarized as follows:*

> The extinguishing action of sprays of finely divided water applied to commonly encountered fires appears to be due predominantly to dilution of the air (oxygen) supply in the zone of burning with vapor (steam) resulting from evaporation of water droplets in the heated area surrounding the fire. The cooling effects of the water may also be important factors in extinguishment in many cases. In order to obtain extinguishment, the water droplets comprising the spray must be relatively small, and the amount of water applied must be adequate in relation to the specific fire.
>
> The marked increase in water surface per gallon where water is dispersed as droplets in a spray is an advantage from the standpoint of extinguishment as shown in Figure 9, since the rate of evaporation of the water (mass transfer) and the cooling effects (heat transfer) are directly proportional to the surface area, other factors being constant. There is a lower limit to the size of droplets which are of practical use in fighting fire, however, since the droplets must be large enough to travel from the source of the spray to the heated area surrounding the fire, and to arrive there with sufficient energy to overcome the currents of hot and turbulent gases moving upwards and away from the fire. Droplets too large to evaporate completely on passing through the zone of combustion and its immediate surroundings may contribute to extinguishment in some cases by cooling the unburned fuel. Even though no extinguishment occurs, the application of water spray may control the rate of burning in fires to some extent.

*From "The Mechanism of Extinguishment of Fire by Finely Divided Water," Research Report No. 10, 1955, National Board of Fire Underwriters, American Insurance Association, New York, p. 70.

The size of the water droplets in the spray from automatic sprinkler heads is of importance in occupancies where fuel is arranged to produce high energy release rates, as in plastic and high-rack storage situations. The optimum sizes of the water droplets in the studies conducted by the National Board of Fire Underwriters varied from approximately 150 microns for the wood fires to approximately 300 microns in size for the gasoline fires. Figure 4.1 is a photomicrograph of typical water droplets in the spray from a nozzle, measured and collected in the center of the discharge 10 feet below the nozzle which was directed vertically downward.

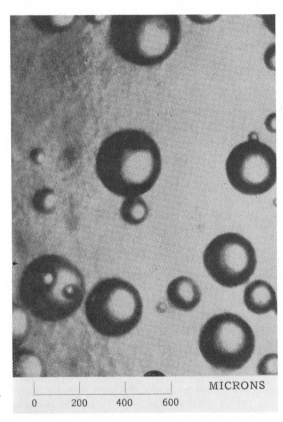

Fig. 4.1. Photomicrograph of typical droplets in water spray. (American Insurance Association)

MICRONS

0   200   400   600

CAPACITY AND DURATION SPECIFICATIONS

As previously discussed in Chapter 3, "Introduction to Automatic Sprinkler Systems," the need to specify minimum water flows relative to sprinkler system design procedures was of such concern that it helped bring about the formation of the National Fire Protection Association and the initial *NFPA Sprinkler System Standard.* Inherent in the establishment of water capacity and duration specifications for sprinkler systems was a procedure for categorizing or classifying the severity of expected fires in occupancies in order to enable a correlation of the water capacity with the severity of the expected fire situation. The history and exact origin of the classification of occupancies relative to fire hazard for the installation of automatic sprinkler systems has not been established. However, it is apparent that this procedure originated with the related sprinkler practices and standards being applied to the installation of sprinkler systems in occupancies differing from the initial textile mills in New England. Gomberg, in a paper titled "A Study of the Validity of Occupancy Hazard Classifications in Sprinklered Occupancies, and a Proposed Method for Occupancy Hazard Determination," stated the classification of the occupancies relative to fire hazard was included in the original sprinkler standards recommended at the second annual meeting of the NFPA in 1898.[3] Currently, occupancy classifications are reviewed by the NFPA Committee on Automatic Sprinklers. The classification of occupancies relative to the categories of Light Hazard, Ordinary Hazard (Groups 1,2, and 3), and Extra Hazard Occupancies as established in the 1975 edition of the *NFPA Sprinkler System Standard*,[4] is as follows:*

Light Hazard–Occupancies or portions of other occupancies where the quantity and/or combustibility is low and fires with relatively low rates of heat release are expected.

Light Hazard Occupancies include occupancies such as:

Churches	Museums
Clubs	Nursing or Convalescent Homes
Educational	Office, including Data Processing
Hospitals	Residential
Institutional	Restaurant seating areas
Libraries, except large stack rooms	Theaters and Auditoriums excluding stages and prosceniums

The requirements for the installation of sprinkler systems in light hazard occupancies may also apply to mercantile or similar occupancies that are incidental to light hazard properties provided such occupancies do not individually exceed 3,000 sq ft of floor area.

Ordinary Hazard (Group 1)–Occupancies or portions of other occupancies where combustibility is low, quantity of combustibles is moderate, stock piles of combustibles do not exceed 8 ft, and fires with moderate areas of heat release are expected.

Ordinary Hazard Occupancies (Group 1) include occupancies such as:

Automobile Parking Garages	Electronic Plants
Bakeries	Glass and Glass Products Manufacturing
Beverage Manufacturing	Laundries
Canneries	
Dairy Products Mfg. & Processing	

---

*From NFPA 13, *Standard for the Installation of Sprinkler Systems*, 1975, NFPA, Boston, pp. 7–9.

Ordinary Hazard (Group 2)–Occupancies or portions of other occupancies where quantity and combustibility of contents is moderate, stock piles do not exceed 12 ft, and fires with moderate rate of heat release are expected.

Ordinary Hazard Occupancies (Group 2) include occupancies such as:

Cereal Mills
Chemical Plants–Ordinary
Cold Storage Warehouses
Confectionary Products
Distilleries
Leather Goods Mfg.
Libraries–Large Stack
  Room Areas
Mercantiles
Machine Shops
Metal Working
Printing and Publishing
Textile Mfg.
Tobacco Products Mfg.
Wood Product Assembly

Ordinary Hazard (Group 3)–Occupancies or portions of other occupancies where quantity and/or combustibility of contents is high, and fires of high rate of heat release are expected.

Ordinary Hazard Occupancies (Group 3) include occupancies such as:

Exhibition Halls
Feed Mills
Paper and Pulp Mills
Paper and Process Plants
Piers and Wharves
Repair Garages
Tire Manufacturing
Warehouses (having moderate to higher combustibility of content, such as paper, household furniture, paint, general storage, whiskey, etc.)
Wood Machining

Extra Hazard–Occupancies or portions of other occupancies where quantity and combustibility of contents is very high, flammable liquids, dust, lint, or other materials are present introducing the probability of rapidly developing fires with high rates of heat release.

Extra Hazard Occupancies include occupancies such as:

Aircraft Hangars
Chemical Works
  (Extra Hazard)
Cotton Pickers and
  Opening Operations
Explosive and Pyrotechnics
Woodworking with Flammable
  Finishing

Under favorable conditions and subject to the approval of the authority having jurisdiction, a reduction of requirements to the next less restrictive occupancy classification may be applied to the following occupancies:

Cold Storage Warehouses
Cotton Picker & Opening
  Operations
Feed Mills
Leather Goods
  Manufacturing
Machine Shops
Mercantiles
Metal Working
Paper and Pulp Mills

In Gomberg's study (conducted in 1965), the records of sprinkler performance by occupancy are compared with the classification of the occupancy according to the procedure utilized for insurance purposes in the New England area. Gomberg found that adjustments in occupancy classification could be made on the basis of the sprinkler performance within the occupancy classes relative to the frequency of the occurrence of the situation of the fire overpowering the sprinkler system. Thus, the situations in which the sprinkler protection was not adequate for the occupancy were classified according to fire hazard. Gomberg also established that fire loading as a measure of the amount of combustibles per square foot of floor space was not suitable as a solitary criterion for the classification of occupancies relative to fire hazard for automatic sprinkler system installations.

The classification of the fire hazard of various occupancies relative to the installation of automatic sprinkler systems involves not only the water specifications, but other basic installation features such as sprinkler head spacing, pipe sizes, and schedules. Thus, the inclusion of the following NFPA specifications for water capacity and duration relative to automatic sprinkler systems will aid in determining the classifications.

NFPA Recommendations

The NFPA currently provides specifications for the minimum amount and duration of water flow relative to the design procedure followed for the automatic sprinkler system. The specifications are developed to be applied relative to the design of a system by the pipe schedules, or by the hydraulic design procedures contained in the *NFPA Sprinkler System Standard*. The water flow specifications established in most standards are usually minimum requirements and should always be applied with experienced professional judgment.

The NFPA recommended water capacity and duration specifications for pipe schedule sprinkler systems are presented in Table 4.1. These requirements were developed by committee members utilizing their professional judgment and experience, relative to the adequacy of the recommended water flow requirements and the length of time the flow should be maintained. Reference should be made to the Notes in the *NFPA Sprinkler System Standard* stating that the lower duration times may be applied only where the fire department is automatically notified of the operation of the sprinkler system. It should be kept in mind that the 15 psi minimum pressure requirement is the residual (flowing) pressure required at the highest and most remote sprinkler head on the system. Thus, the pressure required with the minimum recommended flow at the base of the riser will be 15 psi plus the head loss due to the system's height as installed in a particular building. It should also be kept in mind that the acceptable flow at the base of the riser is the flow required as indicated, including the flow specified for standpipe hose streams.

Schirmer stated the minimum specification of 250 gpm for an automatic sprinkler system in a high-rise building designed as housing for the elderly should be adequate in most situations with a combination standpipe and sprinkler system when the sprinklers are supplied from the wet pipe standpipes on each floor.[5] Schirmer also stated this minimum flow of 250 gpm was determined by anticipating a flow of 150 gpm for automatic sprinklers, and 100 gpm for standpipe hose streams. Thus, the expected flow estimates the utilization of one 1½-inch hose stream and the operation of approximately seven automatic sprinkler heads. It should be kept in mind that once the fire department arrives, the pressure and capacity of the system would be expected to be

Table 4.1. Guide to Water Supply Requirements for Pipe Schedule Sprinkler Systems*

OCCUPANCY CLASSIFICATION	RESIDUAL PRESSURE REQUIRED (See Note 1)	ACCEPTABLE FLOW AT BASE OF RISER (See Note 2)	DURATION IN MINUTES (See Note 4)
Light Hazard	15 psi	500–750 gpm (See Note 3)	30–60
Ordinary Hazard (Group 1)	15 psi or higher	700–1,000 gpm	60–90
Ordinary Hazard (Group 2)	15 psi or higher	850–1,500 gpm	60–90
Ordinary Hazard (Group 3)	Pressure and flow requirements for sprinklers and hose streams to be determined by authority having jurisdiction.		60–120
Warehouses	Pressure and flow requirements for sprinklers and hose streams to be determined by authority having jurisdiction. See Chapter 7 and NFPA 231 and NFPA 231 C.		
High-rise Buildings	Pressure and flow requirements for sprinklers and hose streams to be determined by authority having jurisdiction. See Chapter 8.		
Extra Hazard	Pressure and flow requirements for sprinklers and hose streams to be determined by authority having jurisdiction.		

NOTES:
1. The pressure required at the base of the sprinkler riser(s) is defined as the residual pressure required at the elevation of the highest sprinkler plus the pressure required to reach this elevation.
2. The lower figure is the minimum flow including hose streams ordinarily acceptable for pipe schedule sprinkler systems. The higher flow should normally suffice for all cases under each group.
3. The requirement may be reduced to 250 gpm if building area is limited by size or compartmentation, or if building (including roof) is noncombustible construction.
4. The lower duration figure is ordinarily acceptable where remote station water flow alarm service or equivalent is provided. The higher duration figure should normally suffice for all cases under each group.

*From NFPA 13, *Standard for the Installation of Sprinkler Systems.*

augmented and increased by the connection of pumper-supplied hose lines into the fire department connection on the system from the public water mains. Therefore, the application of the water supply guide as explained by Schirmer for the housing for the elderly occupancy can be better understood by realizing that residential occupancies are classified as Light Hazard Occupancies. Referring to Note 3 of Table 4.1 (see Light Hazard), the specification for water supply for a pipe schedule sprinkler system allows a minimum flow of 250 gpm where the building area is limited by size or compartmentation, or where the building is totally of noncombustible construction.

Bond, in an article on fire fighting problems in high-rise buildings, recommended a minimum water supply requirement of approximately 500 gpm, with the exception of buildings of Light Hazard and limited size being adequately protected with 250 gpm.[6] In the same article, Bond stated occupancies such as mercantile and restaurant occupancies in high-rise buildings could require a minimum flow specification of 1,000 gpm. Bond, like Schirmer, recommended the combination standpipe and automatic sprinkler system in high-rise buildings, and believed the installation of such systems is the only practical means of controlling high-rise building fires. Figure 4.2 illustrates the water supply arrangement recommended by Bond for a general design in high-rise buildings, with each system serving from 10 to 12 stories.

According to Bratlie, the NFPA specifications for the water flow required for pipe schedule sprinkler systems are conservative estimates.[7] Bratlie conducted an analysis of the NFPA automatic sprinkler performance records for a period of five years (from 1960 to 1964) relative to the average and the maximum number of sprinkler heads that operated at fires which were satisfactorily controlled or extinguished by automatic sprinkler systems. In his analysis, Bratlie recommended that reduced water supplies could be considered, based on the range of automatic sprinkler head operation for the five years of his study. His recommended flow requirements for the Light Hazard and Ordinary Hazard Group 1 and 2 Occupancies are presented in Table 4.2. Bratlie indicated that detailed analysis of the number of sprinklers operating to control fires in the various occupancies in all of the hazard classes could result in more realistic water flow specifications. Consideration of the variables of building construction relative to the number of sprinklers that operated, and identification of the type of sprinkler system involved in the occupancy would help make Bratlie's analysis procedure more valuable.

The *NFPA Sprinkler System Standard* recommendations for minimum water supplies for hydraulically designed sprinkler systems are graphically presented in Figure 4.3. The graph in Figure 4.3 indicates that the area of expected sprinkler operation becomes an important variable relative to the required sprinkler discharge density. In the *NFPA Sprinkler System Standard*, the largest single fire area or room in the building is defined as the area of operation. The openings in the room should be protected in an approved manner, and the enclosure must have a fire resistance rating equal to the water flow duration requirement. Thus, for an Ordinary Hazard Group 1 Occupancy, the room would be required to have a one-hour fire resistance rating. How-

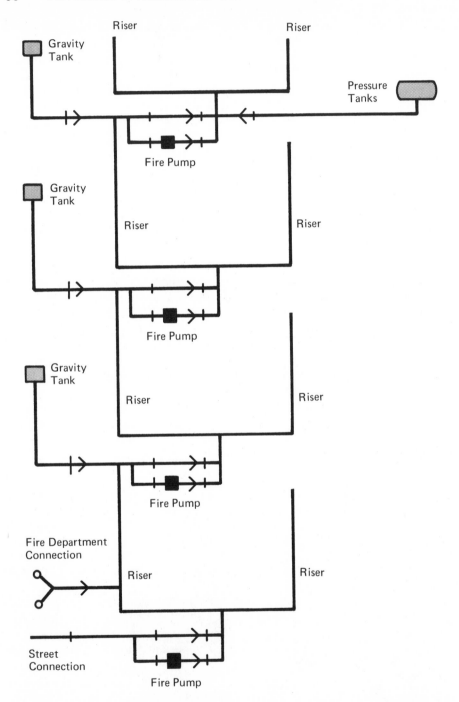

Fig. 4.2. Suggested water supply system design for combination standpipe and sprinkler system in high-rise building. (From "Water for Fire Fighting in High-Rise Buildings," by Horatio Bond)

Table 4.2.   Comparison of Recorded Water Flows and NFPA
Required Water Specifications for Sprinklers*

OCCUPANCY GROUP	RANGE OF SPRINKLERS EXPECTED TO OPEN[1]	MINIMUM ACCEPTABLE FLOW AT BASE OF RISER	
		5-YEAR STUDY[2]	NFPA 13
Light	10–15	220–330 gpm	500–750 gpm[3]
Ordinary (Group 1)	15–25	330–550 gpm	500–1,000 gpm
Ordinary (Group 2)	20–35	440–770 gpm	500–1,500 gpm

[1]Based on data in Tables 2–4 and including safety factors.
[2]Ideal flow (excluding friction loss) based upon 22 gpm per sprinkler, at uniform pressure of 15 psi, the top-line pressure required by NFPA No. 13.
[3]Reduction of this requirement to 250 gpm is allowed if building limited in area or if building (including roof) is of noncombustible construction.

ever, the *NFPA Sprinkler System Standard* also specifies that for areas of operation, less than 1,500 square feet, the density for 1,500 square feet must be used. The *NFPA Sprinkler System Standard* states: (1) when the design involves a dry pipe sprinkler system, the design area should be increased by 30 percent, and (2) should the building be of combustible construction, the minimum design area should be increased by 100 percent to a minimum of 3,000 square feet.

While the minimum water requirements relative to hose streams and sprinkler systems were combined in the pipe schedule system specifications, the guide for the hydraulically designed sprinkler systems has separated these requirements.

Factory Mutual System Recommendations

The Factory Mutual System (FM) recommendations for water flow and pressure for automatic sprinkler systems are generally predicated upon NFPA requirements.[8] However, in the application of the water flow specifications, some additional considerations have been developed due to FM's predominantly industrial, manufacturing, and storage type of operations. FM has developed tables based on the specific fire experience of various occupancies and processes within the occupancies relative to the area of demand to be considered, the water density rate in gpm per square foot, and the duration or the length of time the density should be available to control a fire in the various occupancy situations. Thus, FM's water flow specifications for automatic sprinkler systems are generally developed, using as a basis the density of water required to control a fire in the hazard area, and the minimum size of the hazard or water demand area. In addition, FM also considers the temperature rating of the automatic sprinkler heads, the type of automatic sprinkler system, and the water demand for hose streams (which is considered separate from the

*From "Automatic Sprinklers: How Much Water?," by Ernest H. Bratlie, Jr.

Minimum Water Supplies

Hazard Classification	Sprinklers GPM	Combined Inside & Outside Hose — GPM	Duration in Minutes
Light	150	100	30
Ord. — Gp. 1	400	250	*60–90
Ord. — Gp. 2	600	250	*60–90
Ord. — Gp. 3	750	500	*60–120

*NOTES: The lower duration figure is ordinarily acceptable where remote station water flow alarm service or equivalent is provided.

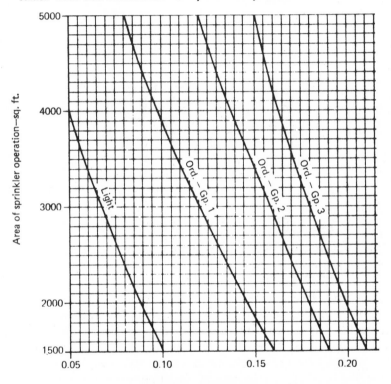

Density – GPM/sq. ft.

Fig. 4.3. Guide for determining density, area of sprinkler operation, and water supply requirements for hydraulically designed sprinkler systems. (From NFPA 13, Standard for the Installation of Sprinkler Systems)

water flow specified for the automatic sprinkler system). Table 4.3 is FM's Water Demand Table for Light Hazard Occupancies, developed by FM to help determine the specified water flow and duration for any general light hazard occupancy relative to both the automatic sprinkler system and hose streams. Eleven additional Factory Mutual System Water Demand Tables are included in Table 4.7 of this chapter.

Table 4.3.   FM Water Demand Table for Light Hazard
Occupancies*

TYPE SPRINKLER SYSTEM	SPRINKLER TEMPERATURE RATING	DENSITY, gpm/sq ft (mm/min)		AREA OF DEMAND, sq ft (m²)	
Any	Any	0.10	4	3,000	278.7

Where construction is noncombustible and areas are well subdivided,
the area of demand may be reduced to 1,500 sq ft (139.4 m²).

Hose stream demand: 250 gpm (0.95 m³/min)
Duration: 60 min

The FM Water Demand Tables exclude the water demand for deluge sprinkler systems, including open outside exposure sprinklers, and also exclude any
allowance for the storage of materials other than the limited processing type of
storage. The water demand specified in the FM Water Demand Table for Light
Hazard Occupancies (Table 4.3) may be increased up to 50 percent, based on
evaluations of conditions at specific industrial locations. Such evaluations are
usually made by inspection personnel. Some examples of typical conditions
that could require an increase in the water flow specifications for an automatic
sprinkler system are: poor housekeeping, shielded sprinkler heads, inadequate
cut-offs, lack of drainage in flammable liquid areas, and dry pipe sprinkler systems in hazardous areas. The following industrial conditions could also result
in an increase of up to 50 percent of the water demand specified for hose
streams and the duration of the water flow (both of which are based on an
analysis of previous fire loss experience): inaccessible areas for manual fire
fighting; lack of ventilation to reduce untenable conditions such as heat and
smoke; combustible concealed spaces in walls, floors, or ceiling areas; shielded
areas from the automatic sprinkler system discharge (including shelves or work
benches); and locations where the fire department will need to use an excessive
number of hose streams. Inspection personnel may also require an additional
250 gpm for buildings predominantly of wood frame construction. Specific
water density specifications as developed by FM for various types of occupancies will be examined later in this chapter in the section concerned with
discharge density.

BOCA 100 Recommendations

In 1974—after approximately four years of research effort conducted by the
Copper Development Association and Patton Fire Protection and Research,
Inc. concerning the development of a fire suppression system for life safety—
BOCA 100, *Standard for the Design and Installation of the Fire Suppression*

---

*From "Water Demand for Private Fire Protection," Factory Mutual System.

System for Life Safety, was adopted as a standard by the Building Officials and Code Administrators International, Inc. (BOCA).[9] The life safety system was primarily designed for use in occupancies where the critical fire protection problem involved the life safety of the occupants. Thus, the life safety system is primarily found in health care, residential, educational, and office occupancies. It should be noted, the *NFPA Sprinkler System Standard* classifies these facilities as light hazard occupancies. The life safety fire suppression system is derived from the following basic concepts and principles, as contained in the *BOCA Life Safety Fire Suppression System Standard:**

- The use of wide angle deflectors to cover large areas, and the fitting of the spray pattern to the plat (nozzle discharge area) protected. Room partitioning as a definite limiting factor on the number of sprinklers that will open.
- The effectiveness of using small water supplies and light densities.
- The effectiveness of using low cost soldered copper tubing.
- The effectiveness of varying orifice sizes to fit water density demands, instead of changing head spacing.
- The use of a total systems design approach, including the selection of spray pattern, orifice and piping sizes, water supply and supervision or monitoring.

In an article titled "Seventy Years to Nowhere, or Water's Cheap," R. M. Patton stated he believed there is an ideal water discharge density for application from the sprinkler system to every occupancy.[10] He indicated ideal density is the water discharge density where extraneous sprinkler head operation is controlled. Thus, sprinkler head operation beyond the immediate fire zone is eliminated. Patton believed that the designer of the sprinkler system should be allowed the flexibility to adjust and fit the system to the occupancy involved, with consideration of the total design problem.

Patton has identified the following design principles of the fire suppression system for life safety (from "Seventy Years to Nowhere or Water's Cheap") which he has incorporated into the recommended standards relative to the water flow specifications:**

- The second principle of sprinkler design is that fire control improves with an increase in density of water discharge (gpm/sq ft) of floor area.
- The third principle of sprinkler design is that an increase in density reduces extraneous head operation (operation of heads outside the fire zone).
- The fourth principle of sprinkler design is that, with pressure given, the density varies as the square of the orifice diameter.
- The fifth principle of sprinkler design is that water flow in the pipe, other factors being held constant, varies as the square of the orifice diameter.

---

*From *Standard for the Design and Installation of the Fire Suppression System for Life Safety,* BOCA 100, 1st ed., 1975, Building Officials and Code Administrators International, Chicago, p. viii.
**From Patton, R. M., "Seventy Years to Nowhere or Water's Cheap," *The Building Official,* Vol. 3, No. 8, Aug. 1969, p. 8.

Patton developed a total of seven principles of sprinkler design; in the preceding excerpt only the four principles that are directly related to the water specifications for the installation of the fire suppression system for life safety are given. The water specifications for the fire suppression system for life safety are developed on the concept of a hydraulically designed sprinkler system. The initial water demand determination procedure is to examine and divide the building or structure to be protected into plats, with a nozzle located to entirely cover each plat with the recommended water discharge density. In the *BOCA Life Safety Fire Suppression System Standard,* the recommended water discharge density is specified according to the fire loading of the building. (The fire loading is the pounds of combustibles relative to the square feet of floor space in the building.) Generally, the system's water discharge density is prescribed on an equivalent basis for the fire loading, as illustrated in Table 4.4.

Table 4.4.   Fire or Fuel Loading Versus System Water Discharge Density*

POUNDS OF COMBUSTIBLES PER SQUARE FOOT (Wood Equivalent)	BTU'S PER SQUARE FOOT	DENSITY GPM/SQ FT
10 or less	80,000 or less	0.10
15	120,000	0.15
20	160,000	0.20
25	200,000	0.25
30	240,000	0.30

After the plats have been identified and the specific sprinkler system nozzles selected, the sectional water requirements are determined by an estimation of the maximum number of nozzles expected to operate in a portion of the building or structure. Nozzles are usually not expected or estimated to be opened outside the section or room of fire origin. The section areas or sizes should be enlarged for unusual conditions, such as fires involving extreme flame spread behavior or high rates of heat release. Aside from doors and windows, when the area is fully enclosed as a room or compartment, the section size would not be expected to exceed the size of the room unless movable walls or partitions were involved. Open areas should include a minimum of four plats; the minimum design area of the section, specified relative to the ceiling height of the open area, is illustrated in Table 4.5. (In the *BOCA Life Safety Fire Suppression System Standard,* this table is referred to as Table FS–702.1.)

The sprinkler system piping is specified to supply the flow of water required by the section. The design of the total fire suppression system should provide

*From BOCA 100, *Standard for the Design and Installation of the Fire Suppression System for Life Safety,* Building Officials and Code Administrators International.

Table 4.5. Ceiling Height as a Determinant of Section Area*

CEILING HEIGHT VS. SECTION AREA	
CEILING HEIGHT (Not Exceeding, in Feet)	MINIMUM DESIGN[1] SECTION AREA (Square Feet)
10'0"	800
15'0"	1,000
20'0"	1,250
25'0"	1,500
50'0"	2,000
75'0"	3,000
100'0"	4,000

[1]Expected area of operation of nozzles.

an appropriate amount of water to the most remote key flow and pressure sections. The key sections are those sections requiring the greatest flow and pressure; thus, with the water flow and pressure adequate to suppress a fire in the most remote key sections, the system will be able to supply the remaining sections with adequate water flow and pressure. To assure adequate water capacity and pressure in the system, it is recommended the design of the piping arrangement include looped mains or gridded pipe layouts. Dead end mains should be avoided whenever possible.

The two basic types of systems designated for the life safety fire suppression system are the "Supervised" and the "Monitored" systems. The *BOCA Life Safety Fire Suppression System Standard* defines these types of systems as follows:**

> A "Supervised" system shall be supervised as required by sections FS–901.1 and FS–902.0 and shall be connected to the servicing fire department by direct electrical or radio communication, or through central station.
> A "Monitored" system shall be electrically monitored as required by sections FS–901.2 and FS–902.0, but without connection to the servicing fire department or central station.

It should be remembered the sprinkler system, as specified by the *BOCA Life Safety Fire Suppression System Standard,* is designed for the suppression of fire in life safety occupancies. Consequently, we are concerned with a wet pipe type of system in occupancies that would generally be considered to be light hazard occupancies, according to the *NFPA Sprinkler System Standard* classification of occupancies for the installation of automatic sprinkler systems.

---

*From BOCA 100, *Standard for the Design and Installation of the Fire Suppression System for Life Safety,* Building Officials and Code Administrators International.
**Ibid, p. 2.

Due to its design objective for life safety occupancies, the *BOCA Life Safety Fire Suppression System Standard* requires the life safety fire suppression system to have an extensive degree of electrical monitoring of conditions within the system.

The system water pressure must be adequate to provide the design orifice pressure on the nozzles in the key pressure section; thus, pressure is sufficient in all sections with two hose streams in operation, if hose lines are provided. With a supervised system, the duration of the water flow must be adequate for a minimum period of 30 minutes. With a monitored system, water flow duration should be sufficient for a minimum of 60 minutes. The flow capacity of the system must be adequate to supply all the nozzles in the key flow section. Thus, the water flow specifications in the *BOCA Life Safety Fire Suppression System Standard* are determined by the pressure and flow requirements in the key sections. These requirements, in turn, are derived from the requirements in the plats, which are determined by the requirements of the single nozzle which defines the parameters of each plat.

Comparison of the Various Sprinkler System Water Specifications

The water capacity and duration requirements have been reviewed for the NFPA standards for both the pipe schedule sprinkler systems and the hydraulically calculated systems. The schematic of a pipe schedule sprinkler system, with a single riser feeding to a cross or feed main with 6 to 8 automatic sprinkler heads on each branch line, is illustrated in Figure 4.4.

Obviously, criteria developed for the determination of the water specifications for automatic sprinkler systems have many similarities, although each group or organization has included some unique features relative to their specific water specification needs. In an article titled "What are the Real Water

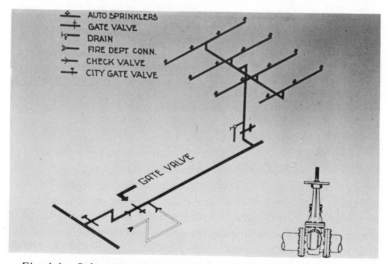

Fig. 4.4. Schematic of pipe schedule sprinkler system. (International Fire Training Association, Oklahoma State University)

Flow Requirements?," Paul R. Smith explained completely and adequately the differences in the water flow specifications often referred to as fire flow requirements.[11] Smith has identified the water flows, specified by the insurance companies and the private protection sections of the Insurance Services Office (ISO), as different from the fire flow requirements developed by the public protection section of the ISO. The apparent confusion resulting from this differentiation is likely to increase as more cities adopt public protection fire flow requirements into ordinances. These ordinances require that water flows, if not available, be provided by the property owner before the issuance of a building permit. According to Smith, the essential difference between the various water flows is a disparity between the purpose for which the flows are developed, and the need for the water relative to a fire occurrence in the occupancy. Smith explained and summarized this situation as follows:*

> In summary, the fire flow figures required by the insurance groups present the minimum necessary for good automatic sprinkler protection, with the addition of one or two hose streams for final control or immediate exposures. They are principally based on the individual property, building, or complex being considered by the insurance group. On the other hand, the public protection requirements are based on the assumption the sprinklers will fail, or will only partially control a fire, and that a fire will develop that endangers the exposures, a block, or the community, so sufficient water must be provided to prevent this situation under ordinary conditions. The quantity required is greater for the simple reason that application of water by hose streams is not as efficient or effective as application of water by automatic sprinkler systems.

Smith illustrated these differences by an example of a typical mercantile building. The building is an ordinary construction one-story building with 40,000 square feet of floor area, protected with a wet pipe automatic sprinkler system with ordinary temperature rated heads. The two exposures are of similar construction, sprinklered and separated from the mercantile occupancy by four-hour fire walls with parapets. The required fire flow water specifications for the mercantile building as calculated by Smith's utilization of the NFPA, FM, and ISO's Public Protection Section's specifications are shown in Figure 4.5. The variables identified by Smith in Figure 4.5 as being critical for the determination of the fire flow requirements are identified in Figure 4.6.

AUTOMATIC SPRINKLER DISCHARGE DENSITY RECOMMENDATIONS

According to the *BOCA Life Safety Fire Suppression System Standard*, the recommended sprinkler discharge density is directly dependent on the fire loading present in the occupancy (as indicated in Table 4.4 earlier in this Chap-

---

*From Smith, Paul D., "What are the Real Fire Flow Requirements?," *Fire Journal,* Vol. 69, No. 2, March 1975, p. 96.

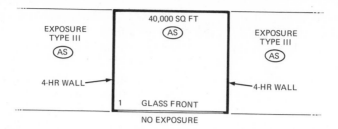

OCCUPANCY: RETAIL STORE.

CONSTRUCTION: ORDINARY (TYPE III-1), 30-IN. PARAPETS, ONE STORY ONE HOUR ROOF.

SPRINKLERS: WET PIPE, ORDINARY TEMPERATURE HEADS.

METHOD	REQUIRED FLOW (GPM)	MINIMUM RES. PRESS.	FLOW DURATION
NFPA #13	850-1500	15 PSI	1 - 1-1/2 HOURS
NFPA #13 HYDRAULIC CURVES	850-1200	DEPENDS ON DESIGN	1 - 1-1/2 HOURS
FACTORY MUTUAL	625	20 TO 30 PSI	1 HOUR
ISO "PUBLIC PROTECTION"	3000	10 OR 20 PSI	3 HOURS

Fig. 4.5. Determination of fire flow water requirements for mercantile occupancy by various procedures. (From "What are the Real Fire Flow Requirements?" by Paul D. Smith)

DETERMINATION OF "REQUIRED" FIRE FLOW

METHOD ⟶	NFPA #13	F.M.	ISO "PUBLIC PROTECTION"
VARIABLES CONSIDERED ⟶			
TYPE OF CONSTRUCTION	X	X	X
HEIGHT AND FIRE AREA	X	X	X
OCCUPANCY	X	X	X
EXPOSURES			X
SPRINKLERS OR NOT			X
TEMPERATURE RATING OF SPRINKLERS		X	
WET OR DRY PIPE SYSTEM	X	X	
PIPING CONFIGURATION & AREA		X	
JUDGMENT	X	X*	X

*INCREASE DEMAND UP TO 50% FOR INACCESSIBLE AREAS, LACK OF DRAINAGE FOR FLAMMABLE LIQUIDS, CONCEALED SPACES, CONDITIONS WHICH WOULD CAUSE RAPID FIRE SPREAD OVER LARGE AREA (I.E.: DUST), OR IF THE FIRE DEPARTMENT NORMALLY USES LOTS OF WATER.

Fig. 4.6. Variables considered in the determination of the fire flow water requirements by various procedures. (From "What are the Real Fire Flow Requirements?" by Paul D. Smith)

ter). The relationship of the type of fuel, fuel geometry, heat release thermal characteristics of the fuel, and the distance of the fuel from the sprinkler head are important variables critical to the water discharge density of the sprinkler system.

In "An Approach to Evaluating and Maintaining Sprinkler Performance," Harry E. Hickey indicated the storage of cordage fiber in 2,000 square foot piles (50 × 40 feet) with 16 feet heights required a discharge density of 0.4 gpm/sq ft.[12] Eighteen sprinklers were assumed to be the total number of heads to operate over the 2,000 square foot area, which required a total flow of 1,100 gpm with 96 psi at the sprinkler system riser valve. The water demand curve for the cordage pile in a warehouse situation with both a 0.4 and a 0.3 gpm/sq ft sprinkler discharge density is illustrated in Figure 4.7 for areas up to approximately 6,000 square feet.

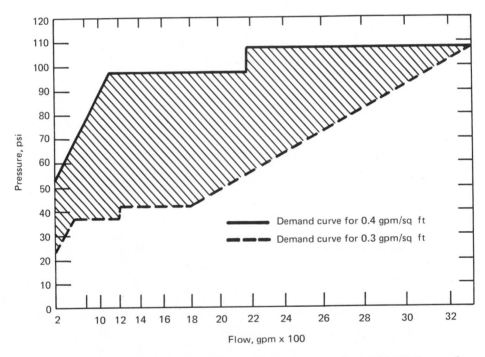

Fig. 4.7. Sprinkler system pressure and flow requirements with discharge densities of 0.4 and 0.3 gpm/sq ft for areas up to 6,000 square feet. (From "An Approach to Evaluating and Maintaining Sprinkler Performance," by Harry E. Hickey)

Approximately ten years ago, fire tests relative to the effectiveness of sprinklers for high-rack storage situations were initially conducted by manufacturers, sprinkler representatives, and the insurance industry. In 1968 the NFPA Committee on Rack Storage of Materials was organized. In 1971 NFPA 231C,

Standard for Rack Storage of Materials, was adopted.[13] In the standard the materials found in rack storage are divided into four classes according to hazard, with specialized high hazard materials being considered by the committees responsible for the basic materials. The standard gives the following definitions of the commodities in the various classifications:

- Class I commodity is defined as essentially noncombustible products on wood pallets, or in ordinary corrugated cartons with or without single thickness dividers, or in ordinary paper wrappings; on wood pallets. Such products may have a negligible amount of plastic trim, such as knobs or handles.
- Class II commodity is defined as Class I products in slatted wooden crates, solid wooden boxes, multiple thickness paperboard cartons, or equivalent combustible packing materials on wood pallets.
- Class III commodity is defined as wood, paper, natural fiber cloth, or products thereof, on wood pallets. Products may contain a limited amount of plastics. Metal bicycles with plastic handles, pedals, seats, and tires are an example of a commodity with a limited amount of plastic.
- Class IV commodity is defined as Class I, II, and/or III products containing an appreciable amount of plastics in a paperboard carton, or Class I, II, and/or III products with plastic packing in paperboard cartons on wood pallets.

The sprinkler discharge density requirements are based on the hazard of the commodity stored, the height of the storage arrangement, the width of the aisles, the arrangement of the storage racks, and the temperature rating of the sprinkler heads. The sprinkler discharge densities vary from a low of 0.14 gpm/sq ft over a 6,000 square foot area for Class I commodities in 20-foot storage with 8-foot aisles and 265°F temperature-rated sprinklers at the ceiling and 165°F temperature-rated sprinklers in the racks, as illustrated in Figure 4.8.

The highest ceiling sprinkler discharge densities are indicated in the *NFPA Standard for Rack Storage of Materials* for a 3,000 square foot area with a discharge density of 0.6 gpm/sq ft for 165°F temperature-rated sprinklers at the ceiling with 4-foot aisles. The storage materials are Class IV nonencapsulated commodities on double row racks, 20-foot high-racked storage with conventional pallets. These ceiling sprinkler discharge density design curves are illustrated in Figure 4.9.

Navy Department Recommended Sprinkler Discharge Densities

The U.S. Department of the Navy recommended for warehouse areas exceeding 4,000 square feet in area, and areas considered to be of extra hazardous occupancy as classified by the *NFPA Sprinkler System Standard,* are to be designed on a hydraulic basis.[14] The hydraulic design should consider the area of application to be equal to 25 percent of the sprinkler heads in the fire area. The recommended sprinkler discharge densities, as developed by the Navy, are presented in Table 4.6.

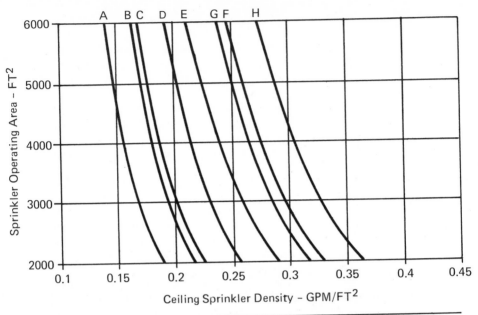

Curve	Legend	Curve	Legend
A — 8 ft aisles with 286°F ceiling sprinklers and 165°F in-rack sprinklers		E — 8 ft aisles with 286°F ceiling sprinklers	
B — 8 ft aisles with 165°F ceiling sprinklers and 165°F in-rack sprinklers		F — 8 ft aisles with 165°F ceiling sprinklers	
C — 4 ft aisles with 286°F ceiling sprinklers and 165°F in-rack sprinklers		G — 4 ft aisles with 286°F ceiling sprinklers	
D — 4 ft aisles with 165°F ceiling sprinklers and 165°F in-rack sprinklers		H — 4 ft aisles with 165°F ceiling sprinklers	

Fig. 4.8.   Ceiling Sprinkler discharge density, double row racks, 20-foot high-rack storage, Class I nonencapsulated commodities, conventional pallets. (From NFPA 231C, Standard for Rack Storage of Materials)

In *A Study of the Performance of Automatic Sprinkler Systems*, O'Dogherty, Nash, and Young have collected data on the minimum sprinkler discharge densities required for the suppression of fire situations involving various fuels and fuel arrangements.[15] This sprinkler discharge density information is presented in Figure 4.10. The highest discharge density reported is for 0.45 to 0.55 gpm/sq ft for high-piled stock, as recommended by the Factory Insurance Association (FIA).[16] In experiments conducted by the British Joint Fire Research Organization, the lowest application rate was 0.1 gpm/sq ft for the extinguishment of wooden cribs.

In his book titled *Fire Behavior and Sprinklers*, Norman J. Thompson stated

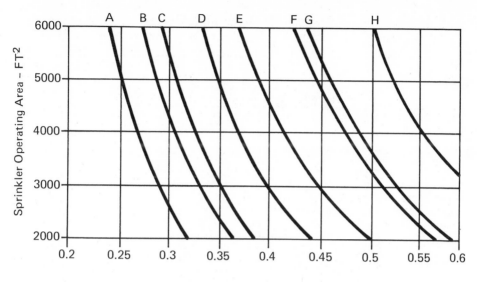

Curve	Legend	Curve	Legend
A — 8 ft aisles with 286°F ceiling sprinklers and 165°F in-rack sprinklers		E — 8 ft aisles with 286°F ceiling sprinklers	
B — 8 ft aisles with 165°F ceiling sprinklers and 165°F in-rack sprinklers		F — 8 ft aisles with 165°F ceiling sprinklers	
C — 4 ft aisles with 286°F ceiling sprinklers and 165°F in-rack sprinklers		G — 4 ft aisles with 286°F ceiling sprinklers	
D — 4 ft aisles with 165°F ceiling sprinklers and 165°F in-rack sprinklers		H — 4 ft aisles with 165°F ceiling sprinklers	

Fig. 4.9. Ceiling sprinkler discharge density, double row racks, 20-foot high-rack storage, Class IV nonencapsulated commodities, conventional pallets. (From NFPA 231C, *Standard for Rack Storage of Materials*)

that water application rates for sprinkler discharge densities from 0.12 to 0.15 gpm/sq ft would usually be sufficient for ordinary hazard occupancies, while extra hazard occupancies would require from 0.3 to 0.6 gpm/sq ft of sprinkler discharge density for effective control and extinguishment.[17]

As previously shown in Table 4.4, the *BOCA Life Safety Fire Suppression System Standard* relates the sprinkler discharge density to the fire load of the occupancy with a range from 0.1 to 0.3 gpm/sq ft. An increase in sprinkler discharge density of 0.1 gpm/sq ft for nozzle heights of 50 feet or higher above the floor, and an increase of 0.2 gpm/sq ft for nozzle heights 100 feet or higher above the floor is specified in the standard.

Table 4.6.   Naval Department Recommended Sprinkler Discharge Densities for Large Warehouses and Extra Hazard Occupancies*

OCCUPANCY	DISCHARGE DENSITY (gal per sq ft per min)
Crude rubber storage	0.4
Cordage fiber storage	0.25
General supplies storage	0.18
Overhaul and repair hangars	0.2
Paint hangar (all types)	0.22
Storage hangars	0.18
Paint and paint-spray shop	0.2
Woodworking shops (power)	0.18
Paint and oil storage (large)	0.19
Windowless buildings (except in above categories)	0.12
Ordnance (explosives, propellants, pyrotechnics, etc.)	(see Paragraph 1.5, Chapter 9)

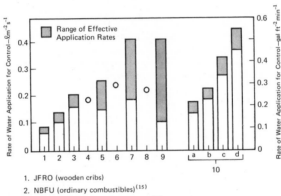

1. JFRO (wooden cribs)
2. NBFU (ordinary combustibles)[15]
3. Factory Mutual (cardboard cartons)[7]
4. Mather and Platt Ltd. (hanging and boxed textiles)[12]
5. NBFU (racked whisky storage - 6 barrels high)[15]
6. Factory Mutual (palletized whisky storage - 4•1-m(13-1/2 ft) high)[14]
7. NBFU (rubber storage: 1•8-4•3-m(6-14-ft) stacks)[16]
8. Bray and Hoyle (foamed rubber and plastics)[13]
9. Underwriters Laboratories (foamed polystyrene packages:
      2•4-6•4-m(8-21-ft) stacks)[17]
10. FIA: recommendations for high piled stock 4•9-6•4-m(16-21 ft)
      — plank on timber ceiling or roof construction)[11]
   a. low hazard
   b. average hazard
   c. high hazard
   d. very high hazard

Fig. 4.10.   Effective application rates of water relative to sprinkler discharge densities for various fuels and arrangements. (From *A Study of the Performance of Automatic Sprinkler Systems,* by O' Dogherty, Nash, and Young)

*From *Fire Protection Engineering,* Navdocks Design Manual 8, U.S. Department of the Navy.

Factory Mutual System Recommended Sprinkler Discharge Densities

As indicated earlier in this chapter, FM has developed discharge densities for sprinklers relative to the hazard of the material, the type of sprinkler system, and the temperature rating of the sprinkler heads (see Table 4.3). FM's Engineering Division recommends that discharge densities vary from a low of 0.1 gpm/sq ft for light hazard occupancies[8] to as high as 0.6 gpm/sq ft for the storage of plastic materials in warehouses.[17] The FM recommended sprinkler discharge densities with the area of demand and the recommended duration of the supply are presented as Table 4.7.[18]

Table 4.7. Factory Mutual System Recommended Sprinkler Discharge Densities with Areas of Application and Duration of Demand*

FM Table 3. Flammable Liquid Spraying

TYPE SPRINKLER SYSTEM	SPRINKLER TEMPERATURE RATING	DENSITY,[2] gpm/sq ft	(mm/min)	AREA OF DEMAND	HOSE STREAM DEMAND, gpm	(m³/min)	DURA-TION, min
Wet	Int. & High 212°–286°F (100°–141°C)	0.30	(12)	Spraying area,[1] or air drying area & 3000 sq ft (278.7m²)	500	(1.9)	60
Wet	Ord. 160°F (71.1°C)	0.30	(12)	Spraying area,[1] or air drying area & 4000 sq ft (371.6m²)	500	(1.9)	60

[1]The spraying area is defined as the area where spraying residues are present, and adjacent mixing, dispensing, or solvent cleaning areas.

[2]Spraying operations vary and judgment must be used when determining the area of demand for this density. For example:

1. Where the spraying area is enclosed by spray booth and duct construction and where flammable liquids are piped to the booth or located within the booth, there is no need to apply this density to the ceiling sprinkler system, the density for this condition being determined by the surrounding occupancy. This assumes spray booth and duct work completely sprinklered.

2. Where the spraying area extends beyond the confines of spray booth and duct construction, then this density should be applied to that portion of the ceiling sprinkler system within the spraying area.

3. Where there is air drying material beyond the spraying area, this density should be applied to that portion of the ceiling sprinkler system where there is likely to be a continuity of combustion due to the flammable solvents. Beyond this portion, the density is determined by the surrounding occupancy.

For example, the spraying area is enclosed by spray booth and duct construction, but wood furniture is air dried over a total area of 5000 sq ft (464.5 m²). Of this 5000 sq ft, the furniture is judged to have a continuity of combustion due to the flammable solvents for 1000 sq ft (92.9 m²). This density would be applied to that portion of the ceiling sprinkler system over the 1000 sq ft, the density beyond this area being determined from FM Table 9.

---

*From "Water Demand for Private Fire Protection," Factory Mutual System.

Table 4.7 (Continued)

FM Table 4. Flammable Liquids in Open Tanks

FLASH POINT OF LIQUID	TYPE SPRINKLER SYSTEM	SPRINKLER TEMPERATURE RATING	DENSITY,[1] gpm/ sq ft	(mm/ min)	AREA OF DEMAND,[2] sq ft	(m²)	HOSE STREAM DEMAND, gpm	(m³/ min)	DURA- TION, min
Below 20°F (−6.7°C) and nitrocellulose lacquer	Wet	Int. & High 212°–286°F (100°–141°C)	0.30	(12)	6,000	(557.4)	1000	(3.8)	120
		Ord. 160°F (71.1°C)	0.30	(12)	8,000	(743.2)			
20°F–80°F (−6.7°–26.7°C)	Wet	Int. & High 212°–286°F (100°–141°C)	0.30	(12)	4,000	(371.6)	500	(1.9)	60
		Ord. 160°F (71.1°C)	0.30	(12)	6,000	(557.4)			
80°F–200°F (26.7°C–93.3°C)	Wet	Int. & High 212°–286°F (100°–141°C)	0.25	(10)	4,000	(371.6)	500	(1.9)	60
		Ord. 160°F (71.1°C)	0.25	(10)	6,000	(557.4)			
Over 200°F (93.3°C) (unheated)	Wet	Int. & High 212°–286°F (100°–141°C)	0.20	(8)	3,000	(278.7)	500	(1.9)	60
		Ord. 160°F (71.1°C)	0.20	(8)	4,000	(371.6)			
200°F–400°F (93.3°–204.4°C) (heated)	Wet	Int. & High 212°–286°F (100°–141°C)	0.25	(10)	4,000	(371.6)	500	(1.9)	60
		Ord. 160°F (71.1°C)	0.25	(10)	6,000	(557.4)			
Over 400°F (204.4°C) (heated)	Wet	Int. & High 212°–286°F (100°–141°C)	0.25	(10)	3,000	(278.7)	500	(1.9)	60
		Ord. 160°F (71.1°C)	0.25	(10)	4,000	(371.6)			

[1]The density specified is the density needed in the flammable liquid area. If the flammable liquid area is enclosed by a wall or curb, it is the area within the enclosure. If there is no wall or curb, the flammable liquid area should extend 20 ft (6.1 m) beyond the tank or other equipment. In the area beyond, the density needed is determined by the occupancy in that area.

If there are automatic or open sprinklers, automatically operated, for local protection of the equipment and there is no chance of a large spill fire, the ceiling density may be reduced to 0.20 gpm/sq ft (8 mm/min).

[2]If the flammable liquid occupancy is in a cut-off room, the area of demand will be the area specified or the area of the room, whichever is smaller. Where there are accumulations near ceiling level of asphalt, lint, or oil sufficient to produce a rapidly spreading fire as in asphalt saturating, the area of demand should cover the full extent of such accumulations.

Table 4.7 (Continued)

FM Table 5. Flammable Liquids in Closed Containers

OCCUPANCY	TYPE SPRINKLER SYSTEM	SPRINKLER TEMPERATURE RATING	DENSITY,* gpm/ sq ft	(mm/ min)	AREA OF DEMAND, sq ft	(m²)	HOSE STREAM DEMAND gpm	(m³/ min)	DURA-TION, min
Room explosion hazard. See flamm. liquid processing standards.	Wet	Int. & High 212°–286°F (100°–141°C)	0.25	(10)	6,000	(557.4)	1000	(3.8)	120
		Ord. 160°F (71.1°C)	0.25	(10)	8,000	(743.2)			
Enclosed quench tank with water cooling & steam eruption potential.	Wet	Int. & High 212°–286°F (100–141°C)	0.15	(6)	7,500	(696.8)	1000	(3.8)	90
	Wet	Ord. 160°F (71.1°C)	0.15	(6)	10,000	(929)			
	Dry	Int. & High 212°–286°F (100–141°C)							
	Dry	Ord. 160°F (71.1°C)	0.15	(6)	13,000	(1207.7)			
Other flammable liquids, flash point below 200°F (93.3°C) or heated above the flash point; no room explosion hazard.	Wet	Int. & High 212°–286°F (100°–141°C)	0.25	(10)	3,000	(278.7)	500 (1.9)		60
	Wet	Ord. 160°F (71.1°C)	0.25	(10)	4,000	(371.6)			
	Dry	Int. & High 212°–286°F (100°–141°C)	0.25	(10)	5,000	(464.5)			
	Dry	Ord. 160°F (71.1°C)	0.25	(10)	6,000	(557.4)			
Flash points above 200°F (93.3°C) not heated above flash point.	Wet	Int. & High 212°–286°F (100°–141°C)	0.20	(8)	3,000	(278.7)	500	(1.9)	60
	Wet	Ord. 160°F (71.1°C)	0.20	(8)	4,000	(371.6)			
	Dry	Int. & High 212°–286°F (100°–141°C)	0.20	(8)	5,000	(464.5)			
	Dry	Ord. 160°F (71.1°C)	0.20	(8)	6,000	(557.4)			

*The density specified is the density needed in the flammable liquid area. If the flammable liquid area is enclosed within a wall or curb, it is the area within the enclosure. If there is no wall or curb, the flammable liquid area would extend 20 ft (6.1 m) beyond the area where a spill fire hazard exists. In the area beyond, the density needed should be determined by the occupancy in that area.

**If the flammable liquid occupancy is in a cutoff room, the area of demand will be the area specified

Table 4.7 (*Continued*)

or the area of the room, whichever is smaller. Where more than 2,000 gallons (7.6 m³) of flammable liquid is present or a flow in excess of 50 gpm (0.19 m³/min) is possible, the area specified should be increased 50 percent and the duration specified should be increased 100 percent.

FM Table 6.   Flammable Hydraulic Fluids

TYPE SPRINKLER SYSTEM	SPRINKLER TEMPERATURE RATING	DENSITY,		AREA OF DEMAND,	
		gpm/sq ft	(mm/min)	sq ft	(m²)
Wet	Int. & High 212°–286°F (100°–141°C)	0.15	(6)	7,500	(696.8)
Wet	Ord. 160°F (71.1°C)	0.15		10,000	
Dry	Int. & High 212°–286°F (100°–141°C)		(6)		(929)
Dry	Ord. 160°F (71.1°C)	0.15	(6)	13,000	(1207.7)

Hose stream demand: 250 gpm (0.95 m³/min) Duration: 60 min

FM Table 7.   Various Occupancies

TYPE SPRINKLER SYSTEM	SPRINKLER TEMPERATURE RATING	DENSITY,		AREA OF DEMAND,	
		gpm/sq ft	(mm/min)	sq ft	(m²)
Wet	Int. & High 212°–286°F (100°–141°C)	0.25	(10)	3,000	(278.7)
Wet	Ord. 160°F (71.1°C)	0.25	(10)	4,000	(371.6)
Dry	Int. & High 212°–286°F (100°–141°C)	0.25	(10)	5,000	(464.5)
Dry	Ord. 160°F (71.1°C)	0.25	(10)	6,000	(557.4)

Hose stream demand: 250 gpm (0.95 m³/min) Duration: 60 min

FM Table 8.   Various Occupancies

TYPE SPRINKLER SYSTEM	SPRINKLER TEMPERATURE RATING	DENSITY,		AREA OF DEMAND,	
		gpm/sq ft	(mm/min)	sq ft	(m²)
Wet	Any	0.15	(6)	2,500	(232.2)
Dry	Any	0.15	(6)	3,500	(325.2)

Where there are accumulations of combustible oil, resin, lint, or dust deposits sufficient to produce a rapidly spreading fire near ceiling level, the area of demand should include the full extent of such deposits. (See FM Table 1.)

In glass plants where there is oil on the roof over furnaces or forming machines, the area of demand is 5,000 sq ft (464.5 m²) or the area of oil deposits, whichever is less. Where sprinklers are on open-head deluge systems, all systems within the area of oil deposits may be expected to operate.

Hose stream demand: 250 gpm (0.95 m³/min) Duration: 60 min

Table 4.7 (Continued)

FM Table 9. Various Occupancies

TYPE SPRINKLER SYSTEM	SPRINKLER TEMPERATURE RATING	DENSITY,		AREA OF DEMAND,	
		gpm/sq ft	(mm/min)	sq ft	(m²)
Wet	Int. & High 212°–286°F (100°–141°C)	0.20	(8)	3,000	(278.7)
Wet	Ord. 160°F (71.1°C)	0.20	(8)	4,000	(371.6)
Dry	Int. & High 212°–286°F (100°–141°C)				
Dry	Ord. 160°F (71.1°C)	0.20	(8)	5,000	(464.5)

Where there are accumulations of combustible oil, resin, lint, or dust deposits sufficient to produce a rapidly spreading fire near ceiling level, the area of demand should include the full extent of such deposits. (See FM Table 1.)

Hose stream demand: 500 gpm (1.9 m³/min)
Duration, woodworking areas: 120 min
Duration, other areas: 60 min

FM Table 10.   Mobile Homes, Certain Exhibit Halls

TYPE SPRINKLER SYSTEM	SPRINKLER TEMPERATURE RATING	DENSITY,		AREA OF DEMAND,	
		gpm/sq ft	(mm/min)	sq ft	(m²)
Wet	High 286°F (141°C)	0.30	(12)	5,000	(464.5)
Dry	High 286°F (141°C)	0.30	(12)	8,000	(743.2)

Hose stream demand: 500 gpm (1.9 m³/min)
Duration: 120 min

FM Table 11.   Woodworking—Flat Panels

TYPE SPRINKLER SYSTEM	SPRINKLER TEMPERATURE RATING	DENSITY,		AREA OF DEMAND,	
		gpm/sq ft	(mm/min)	sq ft	(m²)
Wet	Int. & High 212°–286°F (100°–141°C)	0.20	(8)	4,000	(371.6)
Wet	Ord. 160°F (71.1°C)	0.20		6,000	
Dry	Int. & High 212°–286°F (100°–141°C		(8)		(557.4)
Dry	Ord. 160°F (71.1°C)	0.20	(8)	8,000	(743.2)

Hose stream demand: 500 gpm (1.9 m³/min)
Duration: 120 min

Table 4.7 (Continued)

FM Table 12.   Plastics

TYPE SPRINKLER SYSTEM	SPRINKLER TEMPERATURE RATING	DENSITY, gpm/sq ft	(mm/min)	AREA OF DEMAND, sq ft	(m²)
Wet	High 250°–286°F (100°–141°C)	0.30	(12)	2,500	(232.2)
Dry	High 250°–286°F (100°–141°C)	0.30	(12)	3,500	(325.2)

Where there are accumulations of combustible oil, resin, lint, or dust deposits sufficient to produce a rapidly spreading fire near ceiling level, the area of demand should include the full extent of such deposits. (See FM Table 1.)

Hose stream demand: 500 gpm (1.9 m³/min)
Duration: 60 min

FM Table 13.   Some Textile Occupancies

TYPE SPRINKLER SYSTEM	SPRINKLER TEMPERATURE RATING	DENSITY, gpm/sq ft	(mm/min)	AREA OF DEMAND, sq ft	(m²)
Wet	Any	0.15	(6)	3,500	(325.2)
Dry	Any	0.15	(6)	5,000	(464.5)

Where there are accumulations of combustible oil, resin, lint, or dust deposits sufficient to produce a rapidly spreading fire near ceiling level, the area of demand should include the full extent of such deposits. Include also the full extent of dust collectors or dust rooms connected to the operation. (See FM Table 1.)

Hose stream demand: 500 gpm (1.9 m³/min)
Duration: 120 min

WATER FLOW DETERMINATION FOR SPRINKLER SYSTEMS

There are two basic procedures for determining the availability of an adequate flow of water from a public or private water main to supply an automatic sprinkler system. These two procedures consist of: (1) the relatively easy and less precise method of main drain testing, and (2) the more elaborate and relatively accurate procedure of hydrant flow testing.

Sprinkler System Main Drain Flow Testing

In a paper titled "A Study of Flow from Two-Inch Sprinkler Drains," John B. Dietz stated the value of the main drain flow test is its indication of any change in the water main's supplying of sprinkler system by closed or partially closed valves, or other obstructions.[19] The main drain test procedure involves an initial and recorded observation, usually determined with a Pitot tube, of the pressure drop at the sprinkler riser from a measured flow of water at a hydrant adjacent to the building protected by the sprinkler system. The pressure drop noted at the sprinkler riser is observed by the drop in water pressure from the static pressure with the hydrant closed, and the residual pressure with the hydrant flowing. Thus, an approximate flow curve may be constructed. In effect, the sprinkler riser pressure gages are acting as the pressure hydrant (to be discussed later in this Chapter under "Hydrant Flow Testing").

Once the pressure drop and flow curve have been plotted for the sprinkler riser, subsequent readings can be compared with the original reading to indicate any change in the flow of water to the system. These readings are usually taken with the flow from the 2-inch main drain on the sprinkler riser. Thus, this flow test procedure is usually identified as a main drain test.

Dietz claimed the original main drain test procedure appeared to have been published by Benjamin Richards of the New England Underwriting Bureau in 1918.[20] In the 1950s the Factory Insurance Association made flow curves for both long and short sprinkler drain lines available to their inspection and engineering personnel. William D. Milne, in his book published in 1959, indicated a flow curve for the sprinkler system main drain line with the pressure gage located on the riser in close proximity to the outlet of the drain from the riser.[21]

Also in the 1950s the Federation of Mutual Fire Insurance Companies (now the American Mutual Insurance Alliance) published a flow curve indicating the pressure and volume relationship for the discharge from the main drain of the sprinkler system. It should be kept in mind that all the main drain flow curves developed by various sources have been utilized on 2-inch drain lines as approximate indicators of the flow to be expected, given the observed residual pressure on the sprinkler system riser. Dietz, in his 1963 study, plotted the various main drain flow curves for comparison; these curves are shown in Figure 4.11.[19]

Dietz conducted detailed experiments which indicated the specific type of sprinkler valve is an important factor in the observed flow through the 2-inch main drain line. Thus, he developed a flow constant value for each particular valve he tested; these flow constant values attempted to equate the differences found on the various sprinkler valves relative to the location of the main drain outlet on the valve housing and the location of the pressure gage on the valve. At the present time the principal use of the main drain test is to determine obstructions in the supply mains to the sprinkler system, with a basis of comparison to the original flow tests. Therefore, it is critically important that the original indication of the pressure-volume flow relationship on the sprinkler system is an accurate observation of the system with no obstructions to the flow of water in the system. Figure 4.12 illustrates a main drain test being conducted on a sprinkler system riser. The procedure is relatively simple, with the static pressure initially recorded. Then, the main drain valve is fully opened, and the residual pressure is noted once the pressure has stabilized on the gage. Note that the conduct of a main drain test as well as other flow tests may activate water flow devices on the sprinkler system. In order to prevent unnecessary fire department response, the fire department and private central station or alarm companies should always be notified prior to conducting such tests.

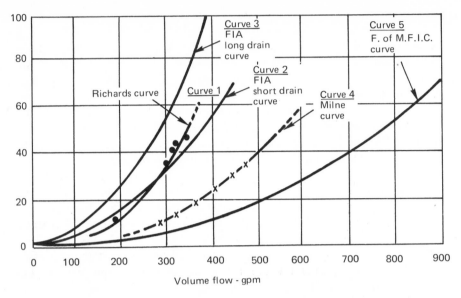

Fig. 4.11. *Various pressure-volume relationships for two-inch sprinkler system main drain lines.* (From *A Study of Flow From Two-Inch Sprinkler Drains,* by John B. Dietz)

Fig. 4.12. *Obtaining the residual pressure reading during a main drain test.* (Stephen C. Leahy, Environmental Safety Services Division, University of Maryland)

Hydrant Flow Testing

Presently, hydrant flow testing is the most accurate method that is practical for use by fire department officers, insurance engineers, and industrial personnel to determine the flow of water that is available for a sprinkler system at a given location or building. It is important for fire department personnel to know the flow of water available from public or private water systems in order for them to determine the proper placement of apparatus. As indicated in the study of prefire survey procedures discussed in Chapter 2, "Fire Department Procedures for Automatic Sprinkler Systems," this information will help eliminate taking water from the mains that will be supplying the automatic sprinkler systems.

The hydrant flow testing procedure involves the selection of a minimum of two hydrants on the supply main: (1) the pressure hydrant, and (2) the flow hydrant. In situations where the supply is reasonably good, as in most urban areas, utilization of more than one hydrant as the flow hydrant may be necessary in order to obtain a suitable pressure drop of approximately 75 to 50 percent of the static pressure. Figure 4.13 illustrates a number of flow hydrants being utilized to obtain a suitable pressure drop.

The pressure hydrant is usually the hydrant that is closest to the building containing the automatic sprinkler system; often it is the hydrant which the fire department utilizes initially, or the hydrant which is essential to their initial attack. The pressure hydrant is the hydrant to which the pressure gage is attached in order to obtain a determination of the static pressure and the resid-

Fig. 4.13. Multiple flow hydrants being utilized in water flow determination. (L. Murray Young)

ual pressure, and, therefore, the pressure drop with a measured flow of water from the flow hydrant or hydrants. The pressure hydrant has a cap with a gage attached for measuring static and residual pressures. Figure 4.14 illustrates the pressure hydrant with the gage attached.

Fig. 4.14.   Pressure hydrant with gage attached. (Stephen C. Leahy, Environmental Safety Services Division, University of Maryland)

The flow hydrant(s) is adjacent to the pressure hydrant, and in the direction of the main's normal flow of water which is from the larger main to the smaller main. Compare Figures 4.13 and 4.15. Figure 4.13 illustrates a number of flow hydrants being operated, and Figure 4.15 illustrates a single flow hydrant with the Pitot tube being utilized to determine the residual pressure from the hydrant outlet in order to determine the flow of water from the flow hydrant.

The residual pressure obtained with the Pitot tube at the flow hydrant measures the flow of water from the flow hydrant by converting the velocity measurement of the residual pressure to a corresponding flow in gpm from the NFPA's hydraulic tables for the type and size of discharge orifice.[22] As shown in Figure 4.15, the flow orifice is the hydrant outlet; however, with some hydrants with low residual pressure, a suitable nozzle may be utilized to obtain a reduced orifice and thus a more accurate residual pressure measurement with the Pitot tube, as shown in Figure 4.16.

Fig. 4.15. Flow hydrant with residual pressure being determined by Pitot tube measurement. (Stephen C. Leahy, Environmental Safety Services Division, University of Maryland)

When the hydrant orifice is utilized as illustrated in Figure 4.15, the coefficient of discharge for the orifice must be determined by checking the outlet connection from the hydrant's barrel by finger. The coefficients for the various hydrant outlets are illustrated in Figure 4.18. It should be noted the hydrant coefficients vary from 0.7 to 0.9, relative to the design of the hydrant barrel and outlet connection.

When the residual pressure has been obtained at the flow hydrant with the Pitot tube, the residual pressure should be read simultaneously at the pressure hydrant from the attached pressure gauge. Hand signals or walkie-talkie radios should be utilized to coordinate the observation of the residual pressure readings at the flow and pressure hydrants. Given the measurement of the residual pressure with the Pitot tube at the flow hydrant, the pressure should be converted to a flow for the type and size of discharge orifice by referring to the NFPA's hydraulic tables, or to other sources (see Table 4.8).[22,23]

A pressure drop has been obtained at the pressure hydrant from the static pressure to the residual pressure when the flow in gpm has been obtained from the flow hydrant. In obtaining the gpm flows from Table 4.8, if the fire department nozzle is utilized as shown in Figure 4.16, the nozzle will have a

Fig. 4.16. Flow hydrant with nozzle being utilized to reduce the discharge orifice and increase the residual pressure. (Stephen C. Leahy, Environmental Safety Services Division, University of Maryland)

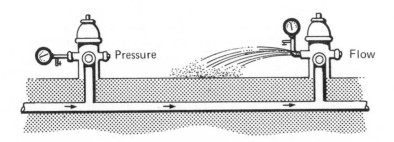

Fig. 4.17. Relationship of the positions of the pressure and flow hydrants relative to the water main flow. (From *Simplified Water Supply Testing for Fire Departments and Insurance Engineers*, American Mutual Insurance Alliance)

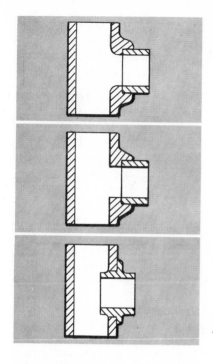

Fig. 4.18. Hydrant flow coefficient conditions. (From *Simplified Water Supply Testing for Fire Departments and Insurance Engineers*, American Mutual Insurance Alliance)

flow coefficient of 0.97 which should be applied to the values in the table for the flow from nozzle orifices.

Obtaining the residual pressure from the flow hydrant requires several precautions. Care should always be taken to avoid hydrant discharge damage to pedestrians, vehicles, shrubs, grass, or other property. The Pitot tube should not be inserted into the flow from the hydrant until flow is steady and the water is clear. (Sediment in the water can damage the gage.) The preferred range of pressures for the most accurate measurements will vary from 10 to 30 psi. Pressures may be obtained in these ranges by utilizing a fire department nozzle to reduce the orifice and thus increase the pressure, and by utilizing large hydrant outlets or multiple hydrants to decrease the pressure. The proper application of the Pitot tube to a hydrant outlet is of utmost importance (see Figure 4.19).

The flow available at the pressure hydrant can now be determined for the desired residual pressure by calculation or by plotting the pressure volume relationship on semiexponential paper often referred to as $N^{1.85}$ paper or hydraulic paper. The basis for the determination of the water flow available at the pressure hydrant for any assumed residual pressure is based on the established observed relationship in the measured flow from the flow hydrant for the observed pressure drop at the pressure hydrant. (The pressure drop is the difference in pressure from the static pressure with no flow to the observed residual pressure with the measured flow at the flow hydrant.) Thus, the amount of water or flow obtainable at a given drop in pressure is proportional to the amount of water or flow obtained at other drops in pressure.

Table 4.8.   GPM Flow Tables for Nozzle and Hydrant Orifices*

(Coefficients from Fig. 4.18 must be applied to flows listed for hydrant butt orifices.

GPM FLOW TABLES

PITOT PRESSURES (psi)	NOZZLE ORIFICE DIAMETERS					PITOT PRESSURES (psi)	HYDRANT BUTT ORIFICE DIAMETERS				
	$1''$	$1\frac{1}{8}''$	$1\frac{1}{4}''$	$1\frac{1}{2}''$	$1\frac{3}{4}''$		$2\frac{3}{8}''$	$2\frac{1}{2}''$	$2\frac{5}{8}''$	$4''$	$4\frac{1}{2}''$
6	73	93	114	164	224	6	412	457	500	1170	1480
8	84	107	132	190	259	8	475	528	575	1351	1710
10	94	119	148	212	289	10	538	590	650	1510	1910
12	103	131	162	233	317	12	588	646	712	1655	2100
14	112	141	175	251	342	14	625	698	775	1787	2260
16	120	151	187	269	366	16	662	746	825	1910	2420
18	127	160	198	285	388	18	712	791	875	2026	2570
20	134	169	209	300	409	20	750	834	925	2136	2710
22	140	177	219	315	429	22	788	875	962	2240	2840
24	146	185	229	329	448	24	825	914	1012	2340	2970
26	152	193	238	343	466	26	850	951	1050	2435	3090
28	158	200	247	356	484	28	888	987	1088	2527	3210
30	164	207	256	368	501	30	925	1022	1125	2616	3320
35	177	223	276	398	541	32	950	1055	1062	2702	3430
40	189	239	295	425	578	34	975	1088	1200	2785	3540
45	201	254	314	450	613	36	1012	1119	1238	2866	3640
50	211	267	330	475	646	38	1038	1150	1262	2944	3740
60	231	293	362	520	708	40	1062	1180	1300	3021	3840
70	250	316	391	562	765	42	1088	1209	1338	3095	3935
80	267	338	418	601	818	44	1112	1237	1362	3168	4030
90	283	358	443	637	867	46	1137	1265	1400	3239	4120
100	299	378	467	672	914	48	1162	1293	1425	3309	4205
						50	1188	1319	1450	3377	4290

Calculation of Water Flow Results

Assume that flow tests have been conducted with a pressure hydrant and a flow hydrant, and a static pressure of 86 psi has been obtained at the gage attached to the pressure hydrant. Also assume that the flow hydrant discharged through a 4-in. hydrant orifice, and that a residual pressure of 14 psi has been obtained at the Pitot tube. Therefore, referring to Table 4.8, a flow of 1,787 gpm was obtained, and corrected with a hydrant coefficient of 0.9, thus providing a flow of 1,608 gpm. This 1,608 gpm flow was obtained at the flow hydrant with a residual pressure of 44 psi on the gage attached to the pressure hydrant, or a pressure drop of 42 psi (static pressure 86 − 44 residual pressure = 42 pressure drop).

The water flow available to supply both the sprinkler system and the manual fire fighting efforts with hose streams when a fire department pumper is attached to the pressure hydrant (and thus utilizes the water flow until a 20 psi residual pressure remains at the hydrant) can be determined by using the following American Mutual Insurance Alliance calculation procedure.[23] The

---

*From *Simplified Water Supply Testing for Fire Departments and Insurance Engineers*, 4th ed., American Mutual Insurance Alliance.

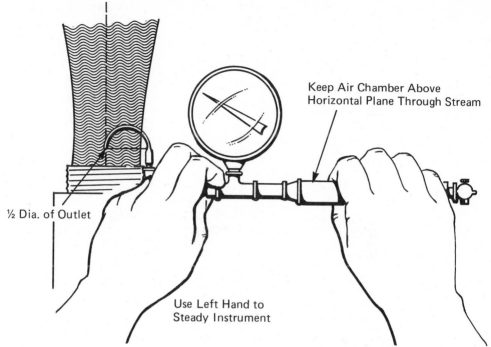

½ Dia. of Outlet

Keep Air Chamber Above
Horizontal Plane Through Stream

Use Left Hand to
Steady Instrument

Fig. 4.19.   Proper application of Pitot tube to hydrant outlet. (Insurance Services
Office)

procedure is based on the pressure volume relationships of flow through
orifices:*

$$Q_2 = \frac{Q_1 \times PD_2}{PD_1} = \frac{Q_1}{K_1} \times K_2$$

Q is the flow, or quantity, of water; thus, Q_1 is the measured flow of 1,608
gpm, and Q_2 is the unknown flow at 20 psi residual pressure. PD_1 is the observed pressure drop of 42, corresponding to the measured flow of 1,608 gpm.
PD_2 is the pressure drop that will be obtained when the residual pressure at
the pressure hydrant is reduced to 20 psi by the fire department pumper. Thus,
PD_2 assumes the value of 66 (86 psi static pressure − 20 psi residual pressure = 66 pressure drop). The pressure drops of 66 and 42 can be converted to
constant factors by using the constant factors K for the corresponding pressure
drops (PD), as shown in Table 4.9. The pressure volume equation can be
solved to obtain the flow at the pressure hydrant with an assumed residual
pressure of 20 psi:

$$Q_2 = \frac{1,608}{7.53} \times 9.61 = 2,052 \text{ gpm}$$

The determination of a flow curve by plotting on semiexponential paper is
illustrated in Figure 4.20, with the flow curve determined at five points. Note

---

*From *Simplified Water Supply Testing for Fire Departments and Insurance Engineers,* 4th ed., American
Mutual Insurance Alliance, Chicago, 1970, p. 21.

Table 4.9.   Pressure Drop Ratio Tables*

PD	K	PD	K	PD	K	PD	K	PD	K
1	1.00	21	5.18	41	7.43	61	9.21	81	10.73
2	1.45	22	5.31	42	7.53	62	9.29	82	10.80
3	1.81	23	5.44	43	7.62	63	9.37	83	10.87
4	2.11	24	5.56	44	7.72	64	9.45	84	10.94
5	2.39	25	5.69	45	7.81	65	9.53	85	11.01
6	2.63	26	5.81	46	7.91	66	9.61	86	11.08
7	2.86	27	5.93	47	8.00	67	9.69	87	11.15
8	3.07	28	6.05	48	8.09	68	9.76	88	11.22
9	3.28	29	6.16	49	8.18	69	9.84	89	11.29
10	3.47	30	6.28	50	8.27	70	9.92	90	11.36
11	3.65	31	6.39	51	8.36	71	9.99	91	11.43
12	3.83	32	6.50	52	8.44	72	10.07	92	11.49
13	4.00	33	6.61	53	8.53	73	10.14	93	11.56
14	4.16	34	6.71	54	8.64	74	10.22	94	11.63
15	4.32	35	6.82	55	8.71	75	10.29	95	11.69
16	4.48	36	6.93	56	8.79	76	10.37	96	11.76
17	4.62	37	7.03	57	8.88	77	10.44	97	11.83
18	4.76	38	7.13	58	8.96	78	10.51	98	11.89
19	4.90	39	7.23	59	9.04	79	10.59	99	11.96
20	5.04	40	7.33	60	9.12	80	10.66	100	12.02

NOTE: $K = PD^{.54}$

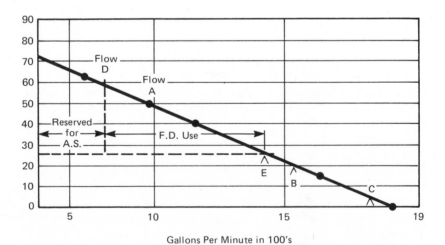

Gallons Per Minute in 100's

Fig. 4.20.   A flow curve plotted on semiexponential paper. (From *Simplified Water Supply Testing for Fire Departments and Insurance Engineers,* American Mutual Insurance Alliance)

---

*From *Simplified Water Supply Testing for Fire Departments and Insurance Engineers,* American Mutual Insurance Alliance.

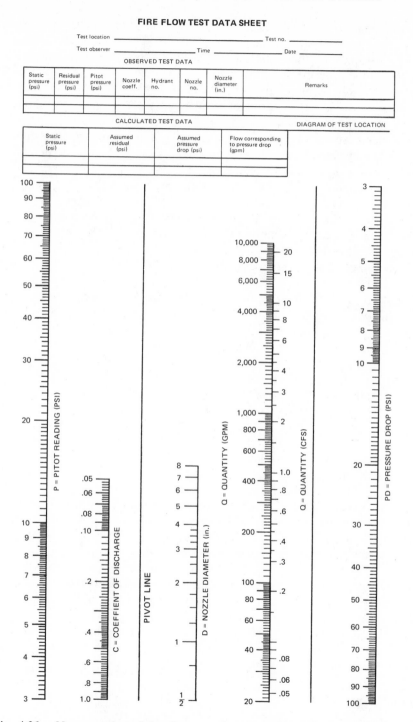

FIRE FLOW TEST DATA SHEET

Test location _____ Test no. _____

Test observer _____ Time _____ Date _____

OBSERVED TEST DATA

Static pressure (psi)	Residual pressure (psi)	Pitot pressure (psi)	Nozzle coeff.	Hydrant no.	Nozzle no.	Nozzle diameter (in.)	Remarks

CALCULATED TEST DATA

DIAGRAM OF TEST LOCATION

Static pressure (psi)	Assumed residual (psi)	Assumed pressure drop (psi)	Flow corresponding to pressure drop (gpm)

Fig. 4.21. Nomograph and fire flow test data sheet. (From "Nomograph for the Solution of Fire Flow Equations," by Carl M. Seifried and Robert J. Alban)

that the amount of water required for the use of the automatic sprinkler system and the fire department is indicated on the curve. These required flows supply essential information that should be provided to the fire department for consideration in the prefire planning activities for any sprinklered building. The amount of water required for the sprinkler system is determined by an estimation of the number of sprinkler heads that will operate, considering the spacing of the heads, the size of fire areas, the type of occupancy, and the type of sprinkler system. The determination of the amount of water required for the sprinkler system can also be estimated by utilizing the sprinkler density discharge specifications for various occupancies, as previously presented in Table 4.7 of this Chapter. The sprinkler discharge densities can then be applied to the recommended areas of application.

To assist in the solutions of the preceding fire flow equations, Carl M. Seifried and Robert J. Alban developed a nomograph relative to the determination of estimated flows at assumed pressures from observed flows at measured pressures.[24] Their nomograph and fire flow data sheet are shown in Figure 4.21.

The determination of the water available from the public or private water mains includes the total water available for the sprinkler system, for other water utilizing suppression systems, and for hose streams. Because of these systems, the amount of water available from the water mains and hydrants is critical information for the fire department in the prefire survey and planning activities, and is essential at the time of the fire occurrence.

Color coding or other means of marking fire hydrants can be of immense value to fire departments. The NFPA recommends the following color coding system: green to represent hydrants having flows of 1,000 gpm and above, orange to represent hydrants having flows between 500 and 999 gpm, and red for hydrants having flows of less than 500 gpm.[22] All of these flows are at the recommended fire flow residual pressure of 20 psi, and the color code is applied to the caps and the bonnets of the hydrants. A high visibility color, such as NFPA's recommendation of chrome yellow, should be utilized for the remainder of the hydrant. Some cities have identified the relative strength of hydrants by painting on the hydrant barrels the number for the size of the water main, or the numbers for the amount of the flow. Other areas have used symbols to indicate the relative flow of hydrants. All of these procedures are designed to help fire departments match their pumping capacities with the capacities of water mains and hydrants in order to best utilize the available water flow and pressure with the installed fire suppression systems, including automatic sprinklers and standpipes. Thus, it is in the best interests of fire departments to know the flow capabilities of the hydrants in their area and to identify the hydrants with suitable markings.

SUMMARY

Automatic sprinkler systems have attained an effective record of fire suppression and control through the application of sufficient quantities of water, in finely divided form, into areas of intense fire involvement. Thus, the provision of specifications and procedures to assure the automatic sprinkler system will deliver enough water to overcome the heat release rates of the fuels involved in the fire.

It should be remembered from Chapter 3, "Introduction to Automatic Sprinkler Systems," the NFPA was formed in 1896 to provide a permanent organization for the development and propagation of automatic sprinkler system standards. The water specifications contained in the *NFPA Sprinkler System Standard* apply to both pipe schedule systems and hydraulically designed systems. The *BOCA Life Safety Fire Suppression System Standard,* adopted in 1974, applies only to hydraulically designed sprinkler systems. The *BOCA Life Safety Fire Suppression System Standard* was primarily designed as a suppression system for life safety; thus, it is intended for use in occupancies where the critical fire protection problem involves the life safety of the occupants.

The water specifications prepared by the Factory Insurance and Factory Mutual Insurance organizations involve the water required in a variety of ordinary and special hazard situations primarily found in manufacturing, processing, and storage industries. The water specifications are primarily related to the hydraulically designed systems, and involve sprinkler discharge density specifications varying from 0.1 gpm/sq ft to 0.6 gpm/sq ft.

The water specifications presented in this chapter must be considered and applied only after extensive study and examination of the specific building and occupancy to be protected by the sprinkler system.

SI Units

The following conversion factors are given as a convenience in converting to SI Units the English units used in this chapter:

1 square foot	=	0.0929 m²
1 Btu	=	1.055 kJ
1 foot	=	0.305 m
1 inch	=	25.400 mm
1 psi	=	6.895kPa
1 gallon	=	3.785 litres
1 cubic foot	=	0.0283 m³
1 gpm	=	3.785 litres/min
1 gpm/sq ft	=	40.746 litres/min. m²

ACTIVITIES

1. Explain the identification or marking procedure utilized in your area for the identification of the flow of water available from various fire hydrants.
2. Describe the two basic procedures for determining the availability of an adequate flow of water from a public or private water main to supply an automatic sprinkler system.
3. Review one of the *NFPA Sprinkler System Standard* occupational categories (Light Hazard, Ordinary Hazard, or Extra Hazard), and present

some examples of occupancies for each category. Discuss appropriate designs for some of the occupancies.

4. NFPA 231C, *Standard for Rack Storage of Materials*, divides those materials found in rack storage into four classes according to hazard. Write a description for each classification. Then, write a brief statement describing how sprinkler discharge density requirements are determined.
5. List at least five variables that are considered in the establishment of water specifications for sprinkler systems by the Factory Mutual Engineering Division.
6. Explain the test procedures at both the pressure and the flow hydrant in hydrant flow testing to determine the water available at the sprinkler system.
7. In order to best utilize available water flow and pressure with installed fire suppression systems (including automatic sprinklers and standpipes), procedures have been developed to help fire departments match their pumping capacities with the capacities of water mains and hydrants. What are some of these procedures? What specific procedures are used in your area?
8. Water specifications for automatic sprinkler systems vary between agencies or organizations. How do you explain the differences in the water specifications between the *BOCA Life Safety Fire Suppression System* and the *NFPA Sprinkler System Standard?*

SUGGESTED READINGS

Baldwin, R., and North, M. A., "The Number of Sprinklers Opening and Fire Growth," *Fire Technology*, Vol. 9, No. 4, Nov. 1973, pp. 245–253.

Layman, Lloyd A., *Attacking and Extinguishing Interior Fires*, National Fire Protection Association, Boston, 1955, p. 24.

Simplified Water Supply Testing for Fire Departments and Insurance Engineers, 4th ed., American Mutual Insurance Alliance, Chicago, 1970.

The Mechanism of Extinguishment of Fire by Finely Divided Water, National Board of Fire Underwriters, New York: American Insurance Association, 1955.

BIBLIOGRAPHY

[1]Layman, Lloyd A., *Attacking and Extinguishing Interior Fires*, National Fire Protection Association, Boston, 1955, p. 24.

[2]"The Mechanism of Extinguishment of Fire by Finely Divided Water," Research Report No. 10, 1955, National Board of Fire Underwriters, American Insurance Association, New York, p. 70.

[3]Gomberg, Alan I., "A Study of the Validity of Occupancy Hazard Classifications in Sprinklered Occupancies, and a Proposed Method for Occupancy Hazard Determination," Undergraduate Paper, 1965, Fire Protection Curriculum, University of Maryland, College Park.

[4]NFPA 13, *Standard for the Installation of Sprinkler Systems*, 1975, NFPA, Boston, pp. 8–10.

[5]Schirmer, Chester W., "Fire Safety for High-Rise Housing for the Elderly," *Fire Journal*, Vol. 67, No. 5, Sept. 1973, pp. 53–56.

[6]Bond, Horatio, "Water for Fire Fighting in High-Rise Buildings," *Fire Technology*, Vol. 2, No. 2, May 1969, pp. 159–163.

[7]Bratlie, Ernest H., Jr., "Automatic Sprinklers: How Much Water?," *Fire Journal*, Vol. 62, No. 1, Jan. 1968, pp. 27–31.

[8]"A Guide for Determining the Adequacy of Fire Protection Water Supplies for F.I.A. Insured Risks," Factory Insurance Association, Hartford, Conn., 1961.

[9]*Standard for the Design and Installation of the Fire Suppression System for Life Safety*, BOCA 100, 1st ed., 1975, Building Officials and Code Administrators International, Chicago.

[10]Patton, R. M., "Seventy Years to Nowhere or Water's Cheap," *The Building Official*, Vol. 3, No. 8, Aug. 1969, pp. 6–11.

[11]Smith, Paul D., "What are the Real Fire Flow Requirements?," *Fire Journal*, Vol. 69, No. 2, March 1975, pp. 93–96.

[12]Hickey, Harry E., "An Approach to Evaluating and Maintaining Sprinkler Performance," *Fire Technology*, Vol. 4, No. 4, Nov. 1968, pp. 292–303.

[13]NFPA 231C, *Standard for Rack Storage of Materials*, 1975, NFPA, Boston.

[14]"Fire Protection Engineering," Navdocks Design Manual 8, June 1961, Department of the Navy, U.S. Department of Defense, Washington, D.C.

[15]O'Dogherty, M. J., Nash, P., Young, R. A., "A Study of the Performance of Automatic Sprinkler Systems," Fire Research Technical Paper No. 17, 1967, Department of Environment, Building Research Establishment, Fire Research Station, Borehamwood, Herts., England.

[16]"Guide for the Protection of Racked Storage," *The Sentinel*, Vol. 23, No. 4, Factory Insurance Association, July–Aug. 1967.

[17]Thompson, Norman J., *Fire Behavior and Sprinklers*, National Fire Protection Association, Boston, 1964.

[18]"Stored Plastics I. The New High Challenge Risk," *Factory Mutual Record*, Vol. 51, No. 6, Factory Mutual Systems, Nov.–Dec. 1974, pp. 8–10.

[19]"Water Demand for Private Fire Protection," Loss Prevention Data Sheet 3-26, July 1974, Factory Mutual System, Norwood, Mass.

[20]Deitz, John B., "A Study of Flow from Two Inch Sprinkler Drains," Undergraduate Paper, 1963, Fire Protection Curriculum, University of Maryland, College Park.

[21]Richards, Benjamin, "Experiments on Drip Valve Tests," NFPA *Quarterly*, Vol. 12, No. 2, Oct. 1918, pp. 163–167.

[22]Milne, William D., *Factors in Special Fire Risk Analysis*, Chilton Company, Phila., Pa., 1959.

[23]Tryon, George H., and McKinnon, Gordon P., eds., *Fire Protection Handbook*, 14th ed., National Fire Protection Association, Boston, 1976.

[24]*Simplified Water Supply Testing for Fire Departments and Insurance Engineers*, 4th ed., American Mutual Insurance Alliance, Chicago, 1970.

[25]Seifried, Carl M., and Alban, Robert J., "Monograph for the Solution of Fire Flow Equations," *Water and Wastes Engineering*, Nov. 1968, pp. 59–62.

Chapter 5

Basic Design of
Automatic Sprinkler Systems

WATER SUPPLY FACILITIES

The automatic sprinkler system primarily consists of a system of pipes with automatic sprinkler heads spaced along the pipes. When activated by the thermal environment of a fire occurrence, the sprinkler heads distribute water throughout the area covered by the system.

The sprinkler system must be provided with a source of water of sufficient capacity to supply the number of sprinkler heads that will be opened by the fire, and the water must have sufficient pressure in order to be adequately distributed from the highest and farthest sprinkler head on the system. Chapter 4, "Water Specifications for Automatic Sprinkler Systems," reviewed the various water requirements relative to capacity and pressure for automatic sprinkler systems; this chapter is concerned with water supply facilities, piping materials, piping arrangement and schedules, hydraulic sprinkler system design, and building modifications for sprinkler systems.

Types of Water Supply

As previously indicated in Chapter 1, "Standpipe and Hose Systems," types of water supply that are acceptable for standpipe systems will also be acceptable for automatic sprinkler systems. Depending on the reliability and capability of the water supply, some situations will require provision for both a primary and a secondary water supply. However, the type of water supply provided as a primary or secondary supply will be selected from the various types of water supply arrangements as reviewed for automatic sprinkler systems.

Public or Private Water Systems

Connecting the sprinkler system directly to a public or a private water main is one of the most reliable and most consistently utilized means of providing an automatic sprinkler system with water. When possible, water meters should not be installed on the sprinkler system water main connections. However, as previously indicated in Chapter 3, "Introduction to Automatic Sprinkler Systems," many governmental jurisdictions require water meters to be installed on fire protection water system connections. When the meter is required, it should be an approved detector check type of fire flow meter. Figure 5.1 illustrates a recommended and acceptable procedure for the connection of the water main to the sprinkler riser.

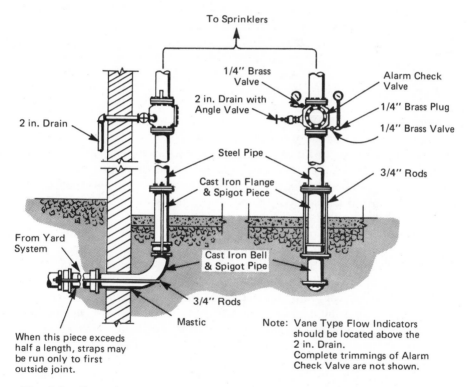

Fig. 5.1. Typical connection arrangement of water main to automatic sprinkler system riser. (From "Installing Sprinkler Equipment," Factory Mutual System)

When the connection to the public or private water system is unable to provide adequate pressure for the requirements of the automatic sprinkler system, a fire pump is usually provided. The details and provisions for fire pumps were covered in Chapter 1, "Standpipe and Hose Systems," and should be reviewed. A fire pump, when installed on the water main supplying a sprinkler system, must be arranged with an automatic start controller in order to be considered suitable for the primary water supply arrangement. Where a fire pump is installed to provide the water pressure for the sprinkler system as a portion of the primary water system, the pump should be provided with supervisory service to indicate nonoperable conditions.

Pressure, Gravity, and Suction Tanks

Pressure tanks on automatic sprinkler systems should have reliable and approved means for maintaining the air pressure on the tank. Pressure tanks may be used as the primary water supply to Light Hazard and some Ordinary Hazard Occupancies located in areas not having public water main systems. The minimum available water capacity of the pressure tanks recommended by the *NFPA Sprinkler System Standard*[2] for a Light Hazard Occupancy is 2,000 gallons, and 3,000 gallons for Groups 1 and 2 of the Ordinary Hazard classification. When the pressure tank is located above the topmost sprinklers served by the tank, a minimum air pressure of 75 psi is required. When the bottom of the tank is below the highest sprinklers served, the head pressure in the system must be considered. The minimum pressure should be 75 psi, plus three times the head pressure created by the height of the sprinkler heads above the bottom of the pressure tank (2.304 foot of height, or head, equals 1 psi of water pressure or, conversely, 1 foot of height equals .434 psi). A pressure tank should normally be maintained with two thirds of the tank filled with water, with the remaining one third of the tank being utilized to contain air pressure.

Because of the reliability of the head pressure created by the height of the water stored in the tank, gravity tanks are an approved and accepted water supply for automatic sprinkler systems.

Suction tanks may be provided in combination with fire pumps as a water supply to the sprinkler system and other fire protection facilities, such as standpipes and yard hydrants. Suction tanks are located at or above ground level, in some cases with reservoirs or pump sumps at below ground level. Figures 5.2 and 5.3 indicate typical fire pump and suction tank arrangements.

When a fire pump supplied by a suction tank is provided with an automatic controller and the flow of water from the suction tank is under head pressure, the water supply arrangement may be considered to be a primary water supply. However, should the pump be under a manual-start procedure, or should the suction tank and fire pump arrangement require that the pump be primed or utilize a priming device, the water supply source would often be considered to be a secondary water supply. It should be kept in mind, however, that the vertical type of fire pump may be utilized with a submersible pump impellor which is, in effect, operating under nonsuction requirements. The use of this type of fire pump is recommended for both underground and ground-level reservoirs, as well as for natural water sources such as ponds, lakes, and rivers. The vertical type of fire pump has made possible the dual use of water sources for operational and functional purposes, such as air conditioning and fire protection. Figure 5.4 illustrates an acceptable arrangement with a vertical

Fig. 5.2.   Suction tank and pump house. (Duane McSmith)

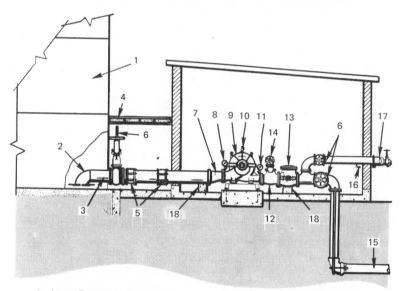

1. Above-Ground Suction Tank.
2. Entrance elbow and 4' x 4' square vortex plate. Distance above bottom of tank—one-half diameter of suction pipe with a minimum of 6 inches.
3. Suction Pipe.
4. Frostproof Casing.
5. Flexible Couplings.
6. O. S. & Y. Gate Valve (See 143k and Notes).
7. Eccentric Reducer.
8. Suction Gauge.
9. Positive Suction Fire Pump.
10. Umbrella Cock or Automatic Air Release.
11. Discharge Gauge.
12. Reducing Tee.
13. Discharge Check Valve.
14. Relief Valve (if required).
15. Discharge Pipe.
16. Drain Valve or Ball Drip.
17. Hose Valve Manifold with Hose Valves.
18. Pipe Supports.

Fig. 5.3.   Suction tank and fire pump arrangement. (From NFPA 20, *Standard for the Installation of Centrifugal Fire Pumps*)

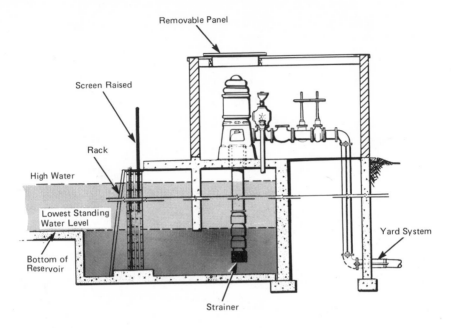

Fig. 5.4.   Fire pump arrangement from water reservoir. (From NFPA 20, *Standard for the Installation of Centrifugal Fire Pumps*)

type of fire pump to be utilized as a water supply to the automatic sprinkler system with a water reservoir suction situation.

Fire Department Connections

Fire department connections are always considered as secondary water supply sources to sprinkler systems. These essential connections should be provided on all automatic sprinkler systems. Fire department connections relative to standpipe systems were examined in Chapter 1, "Standpipe and Hose Systems," and in function with automatic sprinkler systems in Chapter 2, "Fire Department Procedures for Automatic Sprinkler Systems."

The indicating O.S. & Y. or P.I.V. type valves that control the water supply to the sprinkler system are most important. It is critical to provide supervision of the sprinkler system control valves that will activate a signal on the premises or at a supervising office.

Also, as indicated in Chapter 3, "Introduction to Automatic Sprinkler Systems," some insurance organizations now recommend the locking of sprinkler system water control valves in the open position. The *NFPA Sprinkler System Standard* recommends the following procedures relative to the water control valves:*

---

*From NFPA 13, *Standard for the Installation of Sprinkler Systems*, NFPA, Boston, 1975, p. 42.

Valves controlling sprinkler systems, except underground gate valves with roadway boxes, shall be supervised open by one of the following methods:

1. Central station, proprietary, or remote station alarm service.

2. Local alarm service which will cause the sounding of an audible signal at a constantly attended point.

3. Locking valves open.

4. Sealing of valves and approved weekly recorded inspection when valves are located within fenced enclosures under the control of the owner.

Water supply arrangements are seldom provided solely for automatic sprinkler systems or standpipe systems unless such systems are the only water-type fire protection systems in an occupancy. Therefore, with industrial, manufacturing, or storage occupancies, the water supply will usually be connected to all the fire protection systems, including the private water main system on the property. The private water main system with hydrants and hose houses is often referred to as a "yard system." These extensive industrial yard systems almost always employ both a primary and a secondary water supply source, with both systems interconnected. If one system is inoperative or placed out of service, water for fire protection can be maintained from the alternative water supply source.

Figure 5.5 illustrates an industrial occupancy with the water supply sources of a fire pump in combination with a suction tank or a gravity tank, a connection to a public water main, and a fire department connection. Note the looped main arrangement on the yard system providing water to the hydrants and the sprinkler system risers throughout the plant.

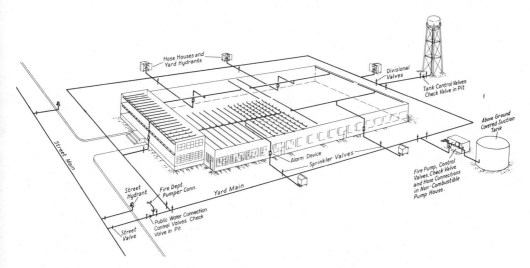

Fig. 5.5. Typical water supply arrangement for sprinkler system and hydrants at an industrial plant. Note the connection to the public water main and fire department connection and the fire pump suction tank or gravity tank water supply. (Factory Mutual System)

PIGING MATERIALS

The traditional piping material used in automatic sprinkler systems has been the welded and seamless steel pipe, black and hot dipped zinc coated, manufactured under the ASTM A 120 specifications. However, copper tubing has been approved in both the *NFPA Sprinkler System Standard* and the *BOCA Life Safety Fire Suppression System standard.*[3]

Plastic Pipe

In an undergraduate paper prepared in 1967 and titled "A Study of the Effects of a Selected Fire Exposure on a Simulated Plastic Automatic Sprinkler Piping System," Dick A. Decker reported for Light Hazard Occupancies, P.V.C. schedule 80 plastic piping seems suitable for application with wet pipe sprinkler systems where water is maintained in the piping system.[4] In 1969 in an undergraduate paper titled "A Study of the Effects of a Fire Situation on the Reliability of a Simulated Plastic Automatic Sprinkler Piping System," Thomas M. Czarnecki investigated the problems of internal pressure generation and possible plastic pipe rupture under thermal exposure.[5] In a Palo Alto Fire Department research report prepared in 1972 and titled "Residential Life Safety Systems Test Utilizing Plastic Pipe," Gerald E. Marks and Donald C. Shaw reported on their investigation of the use of an automatic sprinkler system with plastic pipe for a Light Hazard Occupancy in Palo Alto, California.[6] In a 1975 paper titled "A Unique Sprinkler System Design for a Hi-Rise Building," David M. Banworth described the successful utilization of plastic pipe in an automatic sprinkler system in a 12-story high-rise apartment building in Prince George's County, Maryland.[7] The system described by Banwarth utilized the conventional steel standpipe system as the sprinkler risers at both ends of the building to supply the water to the sprinkler systems. At each floor beyond the water control valve and the water flow indicator, the sprinkler system was installed entirely of P.V.C. schedule 40 plastic pipe, varying in size from 1 to 2 inches in diameter. The plastic pipe was installed in every situation in this building behind one layer of $\frac{1}{2}$-inch fire-rated gypsum wallboard. Figure 5.6 shows the connection between the plastic sprinkler piping and the standpipe in the high-rise apartment building.

Both the *NFPA Sprinkler System Standard* and the *BOCA Life Safety Fire Suppression System Standard* provide for the utilization of any suitable material in the piping system that has been investigated, approved, and listed for sprinkler system service by a nationally recognized testing and inspection agency laboratory. However, plastic pipe is not presently recognized or listed by any nationally recognized laboratory for utilization in automatic sprinkler systems.

Because plastic pipe can resist corrosive environments, reduce the overall cost of a sprinkler system, and be easily installed, it is expected that in the future it will be utilized to an even greater extent. Presently, it would seem that plastic pipe systems are most suitable for wet pipe sprinkler system installa-

Fig. 5.6. Connection between standpipe system and plastic piping of sprinkler system in high-rise apartment house. Note the water flow indicator located on the metal supply pipe at the right, for this floor of the building.

tions in Light Hazard Occupancies. Therefore, it is expected that such systems may be found in occupancies such as schools, office buildings, apartment buildings, and health care facilities.

In an undergraduate paper titled "The Hydraulic Characteristics of Plastic Pipe," Stanley Budzynski described the superior flow capabilities of the plastic pipe, and the Hazen-Williams coefficients of 150 to 169 he obtained in his studies in 1973.[8] Due to the ease of installation and the economic advantages of the material, it is expected that in the future plastic pipe will be utilized more extensively in sprinkler systems for mobile homes and single-family residences.

In a *Fire Journal* article titled "21 Ways to Better Sprinkler System Design,"[9] Rolf Jensen stated:*

> ... However, one may not realize that NFPA 13 officially recognized copper tube in the late 1950s, and that some copper tube systems using UL-approved materials were installed in 1938, the year of UL's first copper tube listings. It also seems important to note that NFPA 13 permits the use of plastic pipe and *has* done so for some years. It only remains for someone to develop a suitable material and to prove it by test.

*From Jensen, Rolf, "21 Ways to Better Sprinkler System Design," *Fire Journal,* Vol. 68, No. 1, Jan. 1974, pp. 47–48, 57.

PIPING ARRANGEMENT AND SCHEDULES

The typical piping arrangement for an automatic sprinkler system involves an underground supply main to the vertical sprinkler riser (see Figure 5.1). The riser usually extends the full vertical height of the building; however, in high-rise buildings, the systems are normally separated into vertical zones. The water supply mains from the risers are identified as cross mains, and the cross mains supply the branch lines. The sprinkler heads are placed on the branch lines, with the spacing of the sprinkler heads in the pipe schedule type of design varying the discharge density of the system.

The traditional piping arrangements have been identified relative to the location of the sprinkler riser and the cross mains as: (a) center central feed, (b) side central feed, (c) central end feed, and (d) side end feed. Upon examination of the diagrams of these piping arrangements, it is apparent that the flow (or feed) of water to the sprinkler heads is in a single direction. Figure 5.7 illustrates the piping arrangement for the traditionally designed pipe schedule systems.

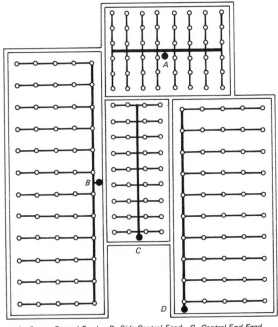

Fig. 5.7. Automatic sprinkler system piping arrangement. (From "Installing Sprinkler Equipment," Factory Mutual System)

A- Center Central Feed   B- Side Central Feed   C- Central End Feed
D- Side End Feed

In 1968 Bernard J. Shelley conducted studies demonstrating the hydraulic advantages of using a looped arrangement for the cross main from the sprinkler riser.[10] In 1971 Richard P. Thornberry developed a computer program which simplified the design procedures for the utilization of a looped sprinkler cross main to supply the sprinkler system with the branch lines connected into the looped feed main.[11] Thus, the sprinkler system became a water supply grid system for each floor. Obviously, such a system is more advantageous in terms of the cost of installation in occupancies with limited compartmentation involving large open areas.

The Georgia-Pacific building in Portland, Oregon, was one of the first examples of the successful application of the principle of a looped cross main system to the sprinklers on each floor. The sprinkler system piping arrangement utilized lateral feed pipes off the looped cross main on each floor of the 27-story building (see Figure 5.8).

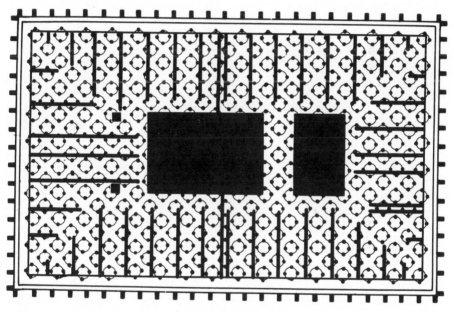

Fig. 5.8.   Looped feed main with lateral feed mains on automatic sprinkler system. (From "The Georgia-Pacific Building, A High-Rise Fire Resistive Office Building with Automatic Sprinklers Installed Throughout," by Richard E. Ritz)

The building was designed as an office occupancy, and had provisions for five-foot modules for the electrical and mechanical utilities. This design arrangement permitted the use of moveable interior partitions, thus providing a means for the movement of the sprinkler head locations. A mechanical swing joint was designed to connect the branch line piping to the lateral piping, allowing the sprinkler head to be located in any of five possible modular locations. The swing joint piping arrangement utilized in the Georgia-Pacific building is illustrated in Figure 5.9.

The building provided 16,000 square feet of floor space on each floor, and the sprinkler system was hydraulically designed to provide a sprinkler discharge density of .15 gpm/sq ft. Richard E. Ritz, in "The Georgia-Pacific Building,

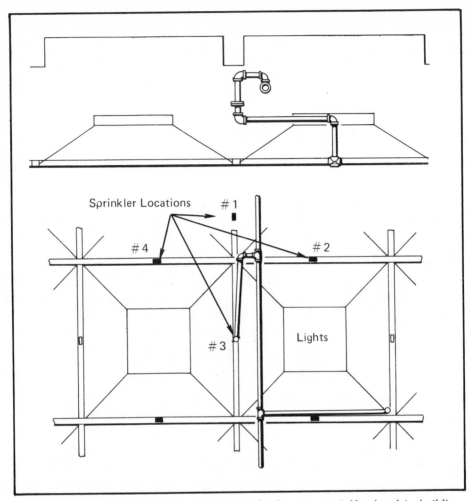

Fig. 5.9. Swing joint piping arrangement for locating sprinkler head in building with modular interior design with moveable partitions. (From "Georgia-Pacific Building," National Automatic Sprinkler and Fire Control Association, Inc.)

A High-Rise Fire Resistive Office Building with Automatic Sprinklers Installed Throughout," explained the sprinkler system was vertically divided into two systems.[12] The initial (or low) system supplied water from the public water main at normal city pressure to the first seven floors. The second (or high) system provided water from the public water main with a fire pump as the primary supply and a pressure tank with 6,000 gallons of water located on the top floor. Both the high and the low systems were provided with separate fire department connections.

The *NFPA Sprinkler System Standard* recommends the following maximum floor area to be protected by a single automatic sprinkler system:

Light Hazard	52,000 sq ft
Ordinary Hazard	52,000 sq ft

Solid piled storage in excess of 15 ft
  in height or palletized or rack stor-
  age in excess of 12 ft in height.   40,000 sq ft
Extra Hazard   25,000 sq ft

Piping Schedules

Piping schedules are not used when the automatic sprinkler system is hydraulically designed, since the pipe is sized according to the pressure and flow requirements of the sprinkler system. These requirements are determined by the sprinkler discharge density desired for the hazard and the size of the fire area, or the estimated maximum area of application.

The sprinkler pipe schedule for Light Hazard Occupancies, as provided in the *NFPA Sprinkler System Standard*, is presented in Table 5.1. Note that Table 5.1 includes the pipe schedule for installations where sprinklers are provided above and below a ceiling from the same branch line. The usual procedure for providing coverage above and below a ceiling is shown in Figure 5.10, with the branch line above or below the ceiling.

The pipe schedule for Ordinary Hazard Occupancies is presented in Table 5.2, and includes the schedule of sprinkler heads allowed for installations where coverage is required above and below a ceiling. Table 5.3 shows the pipe schedule for Extra Hazard Occupancies, as provided in the *NFPA Sprinkler System Standard*.

Table 5.1.   Light Hazard Sprinkler Pipe Schedule*

STEEL		COPPER	
1- in. pipe	2 sprinklers	1- in. tube	2 sprinklers
$1\frac{1}{4}$-in. pipe	3 sprinklers	$1\frac{1}{4}$-in. tube	3 sprinklers
$1\frac{1}{2}$-in. pipe	5 sprinklers	$1\frac{1}{2}$-in. tube	5 sprinklers
2- in. pipe	10 sprinklers	2- in. tube	12 sprinklers
$2\frac{1}{2}$-in. pipe	30 sprinklers	$2\frac{1}{2}$-in. tube	40 sprinklers
3- in. pipe	60 sprinklers	3- in. tube	65 sprinklers
$3\frac{1}{2}$-in. pipe	100 sprinklers	$3\frac{1}{2}$-in. tube	115 sprinklers
4- in. pipe	See below	4- in. tube	See below

The area supplied by any one 4-inch pipe or tube size on any one floor shall not exceed 52,000 square feet.

NUMBER OF SPRINKLERS ABOVE AND BELOW			
STEEL		COPPER	
1- in	2 sprinklers	1- in	2 sprinklers
$1\frac{1}{4}$-in	4 sprinklers	$1\frac{1}{4}$-in	4 sprinklers
$1\frac{1}{2}$-in	7 sprinklers	$1\frac{1}{2}$-in	7 sprinklers
2- in	15 sprinklers	2- in	18 sprinklers
$2\frac{1}{2}$-in	50 sprinklers	$2\frac{1}{2}$-in	65 sprinklers

*From NFPA 13, *Standard for The Installation of Sprinkler Systems.*

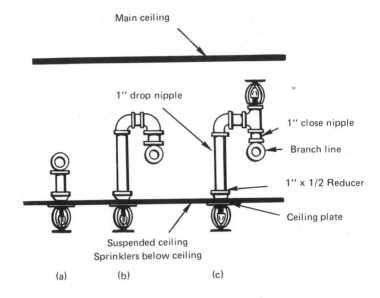

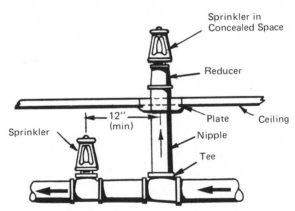

Fig. 5.10. *Typical procedure for providing sprinkler coverage above and below ceiling from same branch line.* (From NFPA 13, *Standard for the Installation of Sprinkler Systems*)

Table 5.2. Ordinary Hazard Sprinkler Pipe Schedule*

STEEL		COPPER	
1- in. pipe..................	2 sprinklers	1- in. tube..................	2 sprinklers
1¼-in. pipe..................	3 sprinklers	1¼-in. tube..................	3 sprinklers
1½-in. pipe..................	5 sprinklers	1½-in. tube..................	5 sprinklers
2- in. pipe..................	10 sprinklers	2- in. tube..................	12 sprinklers
2½-in. pipe..................	20 sprinklers	2½-in. tube..................	25 sprinklers
3- in. pipe..................	40 sprinklers	3- in. tube..................	45 sprinklers
3½-in. pipe..................	65 sprinklers	3½-in. tube..................	75 sprinklers

*From NFPA 13, *Standard for the Installation of Sprinkler Systems*.

Table 5.2 (*Continued*)

4- in. pipe.................100 sprinklers	4- in. tube.................115 sprinklers
5- in. pipe.................160 sprinklers	5- in. tube.................180 sprinklers
6- in. pipe.................275 sprinklers	6- in. tube.................300 sprinklers
8- in. pipe..................... See below	8- in. tube..................... See below

The area supplied by any one 8-in. pipe or tube size on any one floor shall not exceed 52,000 sq ft except that for solid piled storage in excess of 15 ft in height or palletized or rack storage in excess of 12 ft the area served by any one 8-in. pipe or tube size shall not exceed 40,000 sq ft. Where single systems serve both such storage and ordinary hazard areas, storage area covered shall not exceed 40,000 sq ft and total area covered shall not exceed 52,000 sq ft.

When the distance between sprinklers on the branch lines exceeds 12 ft or the distance between the branch lines exceeds 12 ft, the number of sprinklers shall be as follows for given sizes of pipe:

NFPA 13, Table 3–5.3

STEEL	COPPER
$2\frac{1}{2}$-in. pipe................. 15 sprinklers	$2\frac{1}{2}$-in. tube................. 20 sprinklers
3- in. pipe................. 30 sprinklers	3- in. tube................. 35 sprinklers
$3\frac{1}{2}$-in. pipe................. 60 sprinklers	$3\frac{1}{2}$-in. tube................. 65 sprinklers

NUMBER OF SPRINKLERS ABOVE AND BELOW

STEEL	COPPER
1- in............................ 2 sprinklers	1- in............................ 2 sprinklers
$1\frac{1}{4}$-in............................ 4 sprinklers	$1\frac{1}{4}$-in............................ 4 sprinklers
$1\frac{1}{2}$-in............................ 7 sprinklers	$1\frac{1}{2}$-in............................ 7 sprinklers
2- in............................15 sprinklers	2- in............................18 sprinklers
$2\frac{1}{2}$-in............................30 sprinklers	$2\frac{1}{2}$-in............................40 sprinklers
3- in............................60 sprinklers	3- in............................65 sprinklers

Table 5.3. Extra Hazard Sprinkler Pipe Schedule*

STEEL	COPPER
1- in. pipe................. 1 sprinkler	1- in. tube................. 1 sprinkler
$1\frac{1}{4}$-in. pipe................. 2 sprinklers	$1\frac{1}{4}$-in. tube................. 2 sprinklers
$1\frac{1}{2}$-in. pipe................. 5 sprinklers	$1\frac{1}{2}$-in. tube................. 5 sprinklers
2- in. pipe................. 8 sprinklers	2- in. tube................. 8 sprinklers
$2\frac{1}{2}$-in. pipe................. 15 sprinklers	$2\frac{1}{2}$-in. tube................. 20 sprinklers
3- in. pipe................. 27 sprinklers	3- in. tube................. 30 sprinklers
$3\frac{1}{2}$-in. pipe................. 40 sprinklers	$3\frac{1}{2}$-in. tube................. 45 sprinklers
4- in. pipe................. 55 sprinklers	4- in. tube................. 65 sprinklers
5- in. pipe................. 90 sprinklers	5- in. tube.................100 sprinklers
6- in. pipe.................150 sprinklers	6- in. tube.................170 sprinklers
8- in. pipe..................... See below	8- in. tube..................... See below

The area served by any one 8-in. pipe or tube size on any one floor shall not exceed 25,000 sq ft.

*From NFPA 13, *Standard for the Installation of Sprinkler Systems.*

According to the pipe schedule for Light Hazard Occupancies, a 4½-inch pipe is allowed to supply the maximum area to be covered by a single sprinkler system consisting of 52,000 square feet. Also, in Light Hazard Occupancies, when a fire area exceeds 100 sprinkler heads that have no subdividing partitions, the pipe schedule for Ordinary Hazard Occupancies should be used to determine the size of risers and cross mains.

Eight-inch pipe is the largest pipe in the pipe schedule for Ordinary Hazard Occupancies. A maximum area of 52,000 square feet is allowed for the 8-inch pipe, with the reduction to 40,000 square feet in solid storage areas over 15 feet in height and in palletized or racked storage areas over 12 feet in height. The pipe schedules for Light and Ordinary Hazard Occupancies presented in Tables 5.1 and 5.2 provide a maximum of eight sprinkler heads allowed on the branch line on either side of a cross main. However, the pipe schedule for Extra Hazard Occupancies presented in Table 5.3 is predicated on the provision of six sprinkler heads on either side of a cross main on the branch line. The maximum area to be covered by the 8-inch pipe in the Extra Hazard Occupancy is, of course, 25,000 square feet as the maximum area for a single sprinkler system. The pipe schedule for Extra Hazard Occupancies, as published by the NFPA, is a guide, and the specific standard for the occupancy involved or the materials constituting the primary hazard should always be considered.

Spacing of Branch Lines and Sprinkler Heads

For both Light and Ordinary Hazard Occupancies, the maximum distance allowed between the branch lines by the *NFPA Sprinkler System Standard* is 15 feet. Where the sprinkler system is protecting solid storage of materials above 15 feet or palletized or racked storage of materials above 12 feet, the maximum distance specified between the branch lines is 12 feet. The *NFPA Sprinkler System Standard* also requires a maximum distance between branch lines of 12 feet in Extra Hazard Occupancies.

Light Hazard Occupancies with smooth ceilings or beam and girder construction may utilize a protection area per sprinkler head up to a maximum of 200 square feet. When the Light Hazard Occupancy involves a ceiling construction of open wood joists, the maximum protection area permitted is 130 square feet. For all the other types of ceilings in Light Hazard Occupancies, the protection area is limited to 168 square feet.

Ordinary Hazard Occupancies are limited to a coverage area of 130 square feet per sprinkler head, regardless of the ceiling construction. However, as would be expected due to the unique fire problems presented in the storage occupancies, the maximum protection area for the solid storage or rack storage area is 100 square feet per sprinkler head.

The maximum coverage allowed per sprinkler head in extra hazard occupancies is 90 square feet, regardless of the ceiling construction. For the various requirements relative to the specific location of sprinkler heads under different occupancies and roof conditions, Chapters 3 and 4 of the *NFPA Sprinkler System Standard* should be consulted in detail.

Sidewall sprinkler heads are primarily limited to Light and some Ordinary Hazard Occupancies, although they may be used in special situations where a directional discharge of water is required. The spacing requirements for the sidewall sprinkler head in Light Hazard and Ordinary Hazard Occupancies varies with the ceiling construction in a manner similar to the requirements for the standard sprinkler heads. The characteristics of the sidewall sprinkler head will be examined in detail with the other sprinkler heads in Chapter 6, "The Automatic Sprinkler Head." For Light Hazard Occupancies, the distance between sidewall sprinkler heads on the branch line should not exceed 14 feet. When there are noncombustible smooth ceilings in Light Hazard Occupancies, the maximum coverage area is 196 square feet. Special sidewall sprinklers are listed that have area coverages as great as 300 square feet with increased pressures. However, with ceiling construction of combustible materials consisting of plasterboard, metal, or wood lath and plaster, the maximum allowable coverage area per sprinkler head is 168 square feet. When the smooth ceiling is constructed of a wood, fiberboard, or other combustible material, the maximum coverage area per sprinkler head is 120 square feet.

To install sidewall sprinkler heads in Ordinary Hazard Occupancies, the maximum distance between the heads on the branch line is 10 feet. When the ceilings are of noncombustible materials, the maximum coverage area per sidewall sprinkler head is 100 square feet. But when the ceiling construction is of combustible materials, the maximum coverage area per sprinkler head is 80 square feet.

Existing NFPA codes recognize the principle of having complete automatic sprinkler system protection throughout the building. However, there are specifications and local requirements which allow and even mandate the installation of partial sprinkler systems. Typically, many cities have enacted ordinances for installing automatic sprinkler systems in basements of mercantile buildings when the area exceeds certain established limits. Basically, automatic sprinkler protection requires installation throughout the entire structure or building. Such protection must be extended to all areas of the building, including the underside of storage shelves, working tables, stairways, and other similar areas which may obstruct the distribution of the water discharge from the sprinkler heads located at the ceiling. Figure 5.11 illustrates an automatic sprinkler installation drawing for a manufacturing occupancy. Note the distance between the branch lines, the heads on the branch lines, and the spacing of the sprinkler heads as indicated.

HYDRAULIC SPRINKLER SYSTEM DESIGN

In 1886 Ferguson developed the concept of the pipe schedule approach to automatic sprinkler system design with his rules for perforated pipe systems. Ferguson indicated the sum of the orifices should not exceed the diameter of the feed main. The present pipe schedules reviewed in this chapter are based on a similar concept of the diameter of the pipe relative to the $\frac{1}{2}$-inch orifice area of the standard sprinkler head.

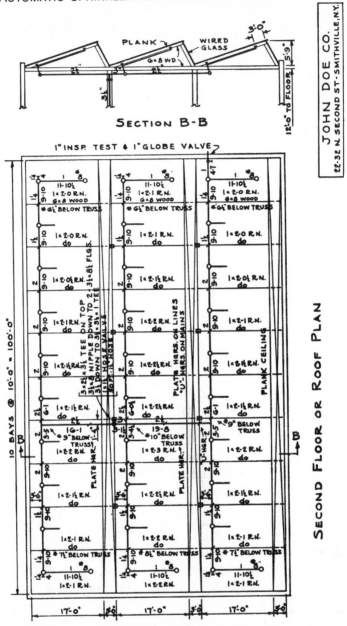

Fig. 5.11. Sprinkler system installation drawing. (From NFPA 13, *Standard for the Installation of Sprinkler Systems*)

In an article titled "Hydraulic Performance of Sprinkler Installations,"[13] Hoyle and Bray indicated the following two principal disadvantages to the pipe schedule process of sprinkler system layout and design.*

---

*From Hoyle, H. and Bray, G., "Hydraulic Performance of Sprinkler Installations," *Fire Technology,* Vol. 3, No. 4, Nov. 1967, p. 291.

1. The larger the size of the system, the lower the rate of discharge from any given number of the most unfavorably placed sprinklers in operation with a given water pressure at the valves, and

2. If pipe sizes are adequate to supply water to sprinklers at the highest level of a multistory building, the pipes for the sprinklers at the lower levels are larger than required, which unnecessarily increases system cost.

Thus, Hoyle and Bray suggested a modified hydraulic approach to the design and layout of automatic sprinkler systems. They recommend a minimum sprinkler discharge density of 0.06 gpm/sq ft for Extra Light Hazard Occupancies consisting of office or residential structures. A sprinkler discharge density for Ordinary Hazard Occupancies consisting of industrial structures varying from a minimum of 0.084 to 0.12 gpm/sq ft. Hoyle and Bray consider that other occupancies require higher sprinkler discharge densities, with the greatest density of 0.72 gpm/sq ft recommended for high-piled and highly combustible storage areas. Therefore, the allowable spacing of the sprinkler heads depends on the discharge characteristics of the head, which is primarily a consideration of the design of the deflector, and the sprinkler orifice size, which determines the flow with the pressure available at the sprinkler head. Hoyle and Bray also recommend the use of $\frac{17}{32}$-inch diameter sprinkler heads whenever a sprinkler discharge density above 0.24 gpm/sq ft is selected.

In "Hydraulic Sprinkler Systems Design—A Computer Approach," Joseph Merry and Earl Schiffhauer described the digital computer as a necessary asset in the hydraulic design procedures for automatic sprinkler systems.[14] Thus, the availability and development of computer programs for the hydraulic design of sprinkler systems have provided the means for the general use of hydraulic design procedures for sprinkler systems. Merry and Schiffhauer also reported at the Eastman Kodak Company their program required the following input data for hydraulic calculations by the computer:*

- The required flow out of the most remote sprinkler, which is dependent on sprinkler spacing and necessary discharge density.

- The lengths and diameters of all pipe sections in the system.

- Orifice coefficients which depend on sprinkler size.

- Appropriate "C" factor–pipe roughness coefficient.

- Certain control numbers which depend on the system layout.

The computer program provided the hydraulic data needed to determine the compatibility of the sprinkler system design with the water supply available. Essentially, the flow required for the area of application and the residual pressure required for the sprinkler discharge density determined the total flow. Figure 5.12 illustrates an isometric sketch of a sprinkler system design with the hydraulic data calculated by a computer program for the 27 design points in the system.

---

*From Merry, Joseph T. and Schiffhauer, Earl J., "Hydraulic Sprinkler Systems Design—A Computer Approach," *Fire Technology,* Vol. 2, No. 2, May 1966, p. 98.

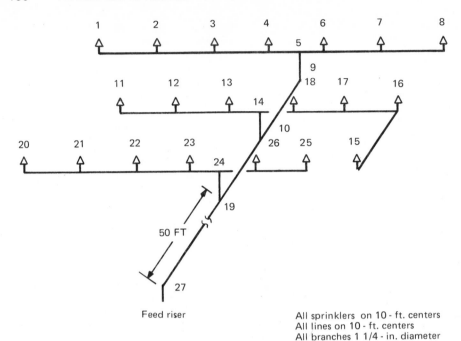

Feed riser

All sprinklers on 10 - ft. centers
All lines on 10 - ft. centers
All branches 1 1/4 - in. diameter
All mains 3 - in. diameter

Hydr. Calc. (.25 gpm/sq ft) Sample System Dwg. No.... Engr...... Date....						
Junction	1	2	3	4	5	6	7
Total Pressure	19.91	20.43	22.28	26.25	34.56	25.45	26.10
Net Pressure	19.91	19.65	20.51	23.04	32.41	25.45	25.11
Junction Flow	25.00	24.83	25.37	26.90	187.17	28.28	28.09
Upstream Flow	25.00	49.83	75.21	102.11	187.17	28.28	56.37
Pipe Diameter	1.38	1.38	1.38	1.38	2.06	1.38	1.38
Junction	8	9	10	11	12	13	14
Total Pressure	28.43	38.39	39.05	25.78	26.44	28.80	34.80
Net Pressure	26.18	37.95	37.27	25.78	25.44	26.51	32.62
Junction Flow	28.68	187.17	188.50	28.46	28.27	28.86	188.50
Upstream Flow	85.05	187.17	375.68	28.46	56.73	85.60	188.50
Pipe Diameter	1.38	3.06	3.06	1.38	1.38	1.38	2.06
Junction	15	16	17	18	19	20	21
Total Pressure	20.11	20.79	22.66	26.69	41.46	22.00	22.57
Net Pressure	20.11	20.00	20.87	23.42	37.66	22.00	21.71
Junction Flow	25.12	25.05	25.59	27.12	171.98	26.28	26.11
Upstream Flow	25.12	50.17	75.77	102.90	547.66	26.28	52.39
Pipe Diameter	1.38	1.38	1.38	1.38	3.06	1.38	1.38
Junction	22	23	24	25	26	27	
Total Pressure	24.60	28.95	38.07	33.45	34.29	57.41	
Net Pressure	22.65	25.41	36.25	33.45	32.98	57.41	
Junction Flow	26.67	28.25	171.98	32.44	32.21	547.66	
Upstream Flow	79.06	107.32	171.98	32.44	64.65	547.66	
Pipe Diameter	1.38	1.38	2.06	1.38	1.38	6.065	

Fig. 5.12. Isometric sketch and computer output for hydraulically designed sprinkler system. (From "Hydraulic Sprinkler System Design—A Computer Approach," by Merry and Schiffhauer)

Merry and Schiffhauer outlined the following advantages to the utilization of the computer approach in the hydraulic design of a sprinkler system as opposed to the dependence on the pipe schedule design approach:

> Ideally, all sprinkler systems should be hydraulically designed to assure adequate coverage. The computer approach gives you this assurance; the pipe schedule does not.
>
> The computer program is simple and versatile enough to allow the designer to customize the sprinkler system to the particular building layout, available water supply, and available water pressure. The pipe schedule approach being general in nature, cannot take into account specific building characteristics, and the result may be over- or under-design.
>
> A hydraulic approach allows the designer to supercede existing pipe schedule rules specified in NFPA 13. In a hydraulically designed system, the number of sprinklers on a branch line or the number of branch lines on a cross main is flexible. Many alternate designs, therefore, are available. The computer approach allows us to investigate all practical alternatives to arrive at the most economical design.
>
> The computer approach provides a means of investigating any special problems, such as space limitations or interferences, and of arriving at an assured coverage design.
>
> Whenever calculations are required, the computer approach greatly reduces engineering time. It also eliminates the human error factor present in lengthy, repetitive hand calculations.
>
> Reasonably uniform water distribution can be achieved with the computer approach. This, of course, results in the more efficient use of existing water supplies. Also, adequate coverage is maintained with the smallest practical water supply. This is especially useful in systems fed by gravity tanks.
>
> The computer approach produces an easy to read, detailed print-out. Review of the system is then possible at any time.
>
> The computer approach yields reliable results, especially in areas where judgement is weak.

In an article titled "Water Net—A Computerized Design Aid," A. K. Rosenhan reported on the development of a computer program which can be used in the analysis and design of many types of water systems varying from water main grids for public systems to automatic sprinkler systems.[15] The computer program can provide a print-out consisting of the flow and pressure losses for each section of pipe and will give the pressure at each pipe junction if an input is given consisting of the following: available combination of constant or variable head input and output flows; pump capacities with flow and pressure; storage or pressure tanks; flow devices; length of pipes with pipe sizes; fittings; and pipe roughness "C" coefficients. The computer program is based on the theoretical concepts of the Hazen-Williams flow relationships, a modified Bernoulli equation, and the Hardy Cross method of pipe network analysis.

Peter J. Chicarello, in his study titled "Analytical Methods for Calculating Sprinkler Discharge,"[16] reported that accuracy and simplicity can be improved by estimating the flows and pressures in sprinkler system piping if the hydraulic calculations are based on the use of the total pressure and the total pressure discharge constant. He also indicated until the total pressure constants become available, the normal pressure and the normal pressure constant will not introduce significant error into the hydraulic calculations, and this procedure is a significant improvement over the use of velocity pressure.

In "Velocity Pressure Effect on Sprinkler System Discharge," H. B. Kirkman and L. E. Campbell described an experimental study utilizing a branch line with seven sprinkler heads, with pipe sizes varying from 1 to 2 inches in diameter.[17] They reported the determination of the discharge coefficient K for the sprinkler head is usually considered to be constant when it is not a constant. Apparently, most manufacturers determine the value of K as an average of readings taken from the actual discharge measured with pressure at the sprinkler head varying from 5 to 60 psi. Therefore, the actual recorded value of the discharge coefficient for the automatic sprinkler head (K) is a computed average value which Kirkman and Campbell indicated may vary as much as 15 percent between the higher and lower pressures.

Kirkman and Campbell also reported the indicated discharges were deficient in percentages, varying from 6 to 14 percent when velocity pressure was considered in the computations. They concluded the use of the Hazen-Williams formula for flow determination, and the use of the roughness coefficient C of 120 for black steel pipe, will indicate excessive pressure requirements for sprinkler system pipes less than 30 years old. Since the sprinkler discharge coefficient is an average of obtained values for hydraulically calculated and designed sprinkler systems, Kirkman and Campbell recommended that velocity pressure should not be utilized in the calculations. The system junction points should be balanced within the limits of 2 psi for branch lines and 4 psi for feed mains with the pressure averaged within the limits.

In "Friction Loss in Sprinkler Piping," P. H. Merdinyan described six samples of sprinkler piping, four from wet systems and two from dry systems, which had been in service for a period of time varying from 10 to 63 years.[18] Seven-foot lengths of the pipe samples in diameters of 1, $1\frac{1}{2}$, and 2 inches were tested relative to the pressure losses through the pipe samples to determine the adequacy of the Hazen-Williams C coefficient of 120 for black steel pipe. The samples, as obtained from the field, were tested in the Grinnell sprinkler system laboratories, and the plot of the flow and pressure relationships relative to the friction loss are shown in Figure 5.13.

Merdinyan determined the C coefficient of 120 seems to be adequate, as based on the six pipe samples tested in his study, and recommended the continued use of the coefficient. It should be kept in mind that Kirkman and Campbell ("Velocity Pressure Effect on Sprinkler System Discharge"[17]) recommended the use of the Hazen-Williams coefficient of 120 only for pipe in service less than 30 years; for older pipes they preferred a form of Fanning's formula to determine the friction loss in sprinkler piping.

The friction loss characteristics of copper tubing have been examined by John M. Foehl in "Flow Characteristics—Copper Sprinkler Conductors."[19] Foehl indicated the Hazen-Williams C coefficient for copper tubing of 140

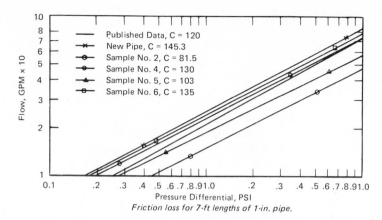

Friction loss for 7-ft lengths of 1-in. pipe.

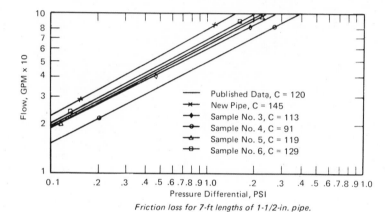

Friction loss for 7-ft lengths of 1-1/2-in. pipe.

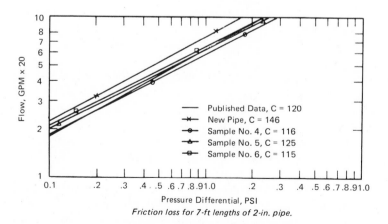

Friction loss for 7-ft lengths of 2-in. pipe.

Fig. 5.13. Friction loss for used sprinkler pipes as compared to new pipe with "C" coefficient values. (From "Friction Loss in Sprinkler Piping," by P. H. Merdinyan)

appeared to be a conservative coefficient. He stated he believed the Hazen-Williams C coefficient of 150 would be more accurate for copper tubing as utilized in automatic sprinkler systems. However, he favors the friction loss factors for copper tubing based on either the D'Arcy-Weisbach, or the Fair, Whipple, and Hsaio formulas. These friction loss formulas are presented in Table 5.4. The hydraulic design of sprinkler systems utilizing the Hazen-Williams factor of 140, the D'Arcy-Weisbach, and the Fair, Whipple, and Hsaio friction loss factors under identical NFPA and FM design criteria are illustrated in Figure 5.14. A discharge density of 0.30 gpm/sq ft applied over an area of 4,000 sq ft is illustrated, with an available pressure of 75 psi at the top of the riser.

Table 5.4. Comparative Friction Loss Formulas*

FORMULA	NOMENCLATURE
Hazen-Williams $$P = \frac{4.52 \, Q^{1.85}}{C^{1.85} \, d^{4.87}}$$	P = friction loss, psi/ft of tube Q = flow, gpm d = average inside tube diameter, in C = dimensionless constant
Fair, Whipple, and Hsaio* $$H = \frac{0.000307L \; V^{1.754}}{D^{1.246}}$$	H = friction loss, feet of head L = length of tube in ft = 1 ft V = velocity of flow, fps D = average inside tube diameter, in
D'Arcy-Weisbach $$P = \frac{0.08078 V^2 \, F}{D}$$	P = friction loss, psi/ft of tube F = friction factor, dimensionless V = velocity of flow, fps D = average inside tube diameter, in

*The loss of head, H, must be converted to friction loss in psi by multiplying H by 0.433. Velocity must be converted to flow in gpm.

With the comparison of designs as indicated in Figure 5.13, the branch line sizes were identical regardless of the friction loss formulas utilized. However, Foehl found variations in the pressure calculations at the junctions of the branch lines with the cross mains. These pressure differences resulted in reductions in the diameters of the tubing employed for the cross mains when applying the D'Arcy-Weisbach or the Fair, Whipple, and Hsaio friction loss factors in place of the Hazen-Williams C factor of 140.

When the NFPA criteria requiring the opening of all the sprinkler heads on both sides of the cross main in the design area of 4,000 sq ft was applied, reductions in the amount of 6-inch cross main required were achieved with the D'Arcy-Weisbach and the Fair, Whipple, and Hsaio friction loss factors.

*From "Flow Characteristics—Copper Sprinkler Conductors," by John M. Foehl.

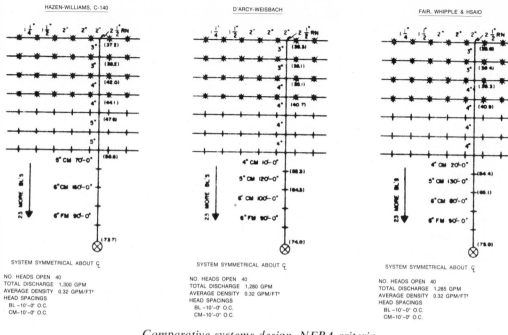

Comparative systems design, NFPA criteria

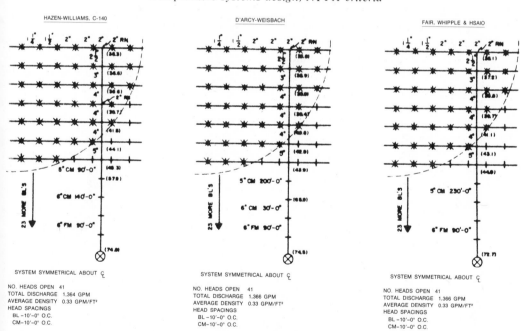

Comparative systems design, Factory Mutual Engineering Corporation criteria

Fig. 5.14. *Comparison of hydraulic sprinkler system design with various friction loss formulas with the NFPA and Factory Mutual criteria.* (From "Flow Characteristics—Copper Sprinkler Conductors," by John M. Foehl)

The Factory Mutual criteria was applied utilizing a point of origin for the fire in the corner adjacent to the most remote sprinkler head. This sprinkler design area had a radius of 71.5 feet to achieve the required 4,000 square feet. The result was that the 6-inch cross main was eliminated with the Fair, Whipple, and Hsaio friction loss factors, and reduced to a total length of 30 feet with the D'Arcy-Weisbach factors.

National Fire Protection Association Procedure

In 1966 the NFPA initially introduced the procedures for the hydraulic design of sprinkler systems in the *NFPA Sprinkler System Standard*, and in 1972 prescribed water demand design curves for hydraulically designed systems. It should be understood the hydraulically designed sprinkler system involves the initial selection of a desired sprinkler discharge density in terms of gpm/sq ft distributed uniformly over a selected and assumed area of application. The sprinkler discharge density and the assumed area of application are determined primarily by the combustible characteristics of the fuel, the geometric arrangement of the fuel, to a minor extent by the rating of the sprinkler heads, and by the type of sprinkler system.

The *NFPA Sprinkler System Standard* provides procedures for hydraulically designed systems relative to the maximum area to be covered by a single sprinkler system (the area of coverage per sprinkler head) with the other general installation requirements. However, the procedures relative to the pipe schedules exhibited in Tables 5.1, 5.2, and 5.3 are not utilized. Remember, the number of sprinklers per branch line, the number of branch lines, and the size of pipe are all determined by the design characteristics of the sprinkler discharge density, area of application, and the available water supply.

There are some variations in the *NFPA Sprinkler System Standard* relative to hydraulically designed systems. A minimum residual pressure of 7 psi is permitted on any sprinkler head, while a minimum pressure of 15 psi is recommended on the pipe schedule designed systems. The area of coverage per sprinkler head or the protection area limitations are modified to allow a maximum coverage of 225 square feet per head for Light Hazard Occupancies with smooth ceilings and beam and girder construction, while the pipe schedule systems are limited to 200 square feet. For the Extra Hazard Occupancies, the limit on the protection area per sprinkler head is 90 square feet with the pipe schedule systems; this limit is increased to a maximum of 100 square feet per sprinkler head for the hydraulically designed sprinkler system. Chapter 8 of the *NFPA Sprinkler System Standard* allows the utilization of protection areas beyond these limitations for sprinkler heads in high-rise buildings with Light Hazard Occupancies. However, the installation must be in conformance with the listings and approvals of a nationally recognized testing laboratory.

The NFPA has developed procedural guidelines for the design and calculations of hydraulically designed sprinkler systems, as presented in Table 5.5.

The hydraulically designed sprinkler system is required to have a nameplate containing the basic design information located at the water control sprinkler system valve. The basic information required on the nameplate includes the location, number of sprinklers in the hydraulically designed section, and the basis of the design. The basis of the design includes the sprinkler discharge density over the designed area of application, including the gpm and residual pressure requirement at the base of the sprinkler system riser. Figure 5.15 illustrates the sprinkler system nameplate.

Table 5.5. NFPA Calculating Procedure for Hydraulically Designed Automatic Sprinkler Systems*

7-4.3.1 In order to maintain consistency in calculating a sprinkler system, manually or by computer, the following rules shall be followed:

(a)* The design area shall be the hydraulically most remote area and usually includes sprinklers on both sides of the cross main.

(b) Each sprinkler in the design area shall discharge at a flow rate at least equal to the stipulated minimum water application rate (density). Begin calculations at the sprinkler hydraulically farthest from the supply connection. With common system configurations this will be the end sprinkler on the end branch line.

(c) Calculate pipe friction loss in accordance with Hazen and Williams formula for a "C" value of 100 for black steel pipe in dry pipe systems, C-120 for black steel pipe in other than dry pipe systems, C-140 for copper tube and cement-lined cast-iron pipe, and C-100 for unlined cast-iron pipe. The authority having jurisdiction may recommend other "C" values.

(d) The density shall be calculated on the basis of the floor area.

(e) Include pipe, fittings and devices such as valves, meters and strainers and calculate elevation changes which affect the sprinkler discharge.

(f) Calculate the loss for a tee or a cross where flow direction change occurs based on the equivalent pipe length of the piping segment in which the fitting is included. The tee at the top of a riser nipple shall be included in the branch line; the tee at the base of a riser nipple shall be included in the riser nipple; and the tee or cross at a cross main—feed main junction shall be included in the cross main. Do not include fitting loss for straight thru flow in a tee or cross.

(g) Calculate the loss of reducing elbows based on the equivalent feet value of the smallest outlet. Use the equivalent feet value for the "standard elbow" on any abrupt ninety-degree turn, such as the screw-type pattern. Use the equivalent feet value for the "long turn elbow" on any sweeping ninety-degree turn, such as a flanged, welded or mechanical joint-elbow type.

(h) Friction loss shall be excluded for tapered reducers, for reducing elbows serving a sprinkler at the end of a branch line, and for all fittings directly supplying a sprinkler.

(i) Orifice plates or sprinklers of different orifice sizes shall not be used for balancing the system, except for special use such as exposure protection, small rooms or enclosures or directional discharge. (See 4-4.20 for definition of small rooms.)

(j) Feed mains, cross mains and branch lines within the same system may be looped or gridded to divide the total water flowing to the design area.

(k) The water allowances for inside hose and for outside hydrants may be combined and added to the system requirement at the system connection to the underground main. The total water requirement shall be calculated through the underground main to the point of supply.

7-4.3.2 Minimum operating pressure of any sprinkler shall be 7 psi.

*From *NFPA 13, Standard for the Installation of Sprinkler Systems.*

This system as shown on......................company

print no.............................. dated........................

for..

at.. contract no................

is designed to discharge at a rate of.................gpm

per square foot of floor area over a maximum area

of.......................... square feet when supplied with

water at a rate of ....gpm at.................psi

at the base of the riser.

Fig. 5.15. *Nameplate for hydraulically designed sprinkler system.* (From NFPA 13, *Standard for the Installation of Sprinkler Systems*)

Building Officials and Code Administrators International Procedure

It should be kept in mind the *BOCA Life Safety Fire Suppression System Standard* for the design and installation of the fire suppression system for life safety is predicated solely for hydraulically designed systems. As indicated in Table 4.2 of Chapter 4, "Water Specifications for Automatic Sprinkler Systems," the fire suppression system for life safety is designed and intended for occupancies with low and moderate fire loads, with the sprinkler discharge densities determined from the fire load of the occupancy.

The area of application for the discharge density is limited to the compartment size. This system is primarily intended for occupancies such as office, residential, apartment, school, and health care facilities. For larger compartment sizes, the area of application for sprinkler design purposes is determined on the basis of the ceiling heights in the area. The student should review the *BOCA Life Safety Fire Suppression System Standard* information presented in Chapter 4 of this text, including Table 4.2. The BOCA Standard also requires electrical supervision or monitoring of the water control valves to the system.

BUILDING MODIFICATIONS FOR SPRINKLER SYSTEMS

One important consideration relative to the basic design procedures for automatic sprinkler systems involves modifications to the structure to facilitate the proper operation of the sprinkler system. If the automatic sprinkler system is to be installed in a new building, and the building is to be fully covered with an automatic sprinkler system, there are usually economic advantages involved in the proper design of the building to complement the automatic sprinkler system. (See Chapter 3, "Introduction to Automatic Sprinkler Systems," for a detailed analysis of economic tradeoff features.)

When installing sprinkler systems in high bay or high ceiling areas, it is important to consider the installation of draft curtains or barriers. Draft curtains contain the thermal column from the fire occurrence within the area immediately above the fire, thus facilitating the operation of the sprinkler heads in the fire area and preventing the operation of heads in adjacent areas not over the area of fire involvement. Figure 5.16 shows draft curtains installed in a high bay storage area.

The vertical openings in buildings should be protected and enclosed to prevent and retard the spread of fire from floor to floor. It is most important to enclose vertical shafts and to protect concealed spaces so that heated gases and flames cannot enter. It should be kept in mind that the basic concept and design of the automatic sprinkler system is predicated on the system's operating on the fire before the fire has spread from the initial area or compartment of origin. The sprinkler system is not intended to operate and control the fire

Fig. 5.16. Draft curtains in truss area of high-ceiling storage building.

occurrence which initiates or spreads very rapidly to an area greater than a single floor. Therefore, enclosing and protecting vertical opernings is most important, in order to prevent the spread of heated gases to upper floors where sprinkler heads beyond the fire area would open, reducing the available water without providing suppressing action in the fire area.

Concealed spaces above ceilings, below floors, and in other inaccessible areas, if not filled with noncombustible insulation or constructed with noncombustible material, should also be protected with automatic sprinklers. One of the most common areas is the rather extensive area above ceilings which is utilized for plenums with communication, electrical, and mechanical services of the building. (See Figure 5.10 earlier in this Chapter for typical means of providing sprinkler protection above and below ceiling from the same sprinkler system branch line.)

In order for a building to be effectively protected by means of an automatic sprinkler system, it is essential the system be installed throughout the building. Partial sprinkler systems are based on two rather hazardous assumptions: (1) the fire will initiate in an area protected by the automatic sprinkler system, and (2) if the fire initiates in an adjacent area, the sprinkler system will prevent the spread of fire throughout the building.

The fire experience relative to partial sprinkler systems indicates the system is effective when the area it covers is completely and effectively protected by fire-rated construction from the remainder of the building, and the fire originates in the sprinklered area of the building. Generally, sprinkler systems are not able to provide effective control from the fire propagation that results from a rapidly spreading fire situation. A similar situation is the inadequacy of interior sprinklers when the exterior of the building is constructed of combustible materials. Such situations, when exterior exposures pose a threat of fire ignition, require the installation of an exposure sprinkler system on the exterior of the building.

Thorough fire protection engineering is required in industrial or manufacturing buildings that are equipped with both sprinkler systems and automatic roof vents. Various authorities recommend designing the venting system to include thermal or smoke detectors that operate the vents. Thus, in most situations the smoke vents will be in operation prior to the sprinkler system. The operation of the sprinklers adjacent to the smoke vents can hamper and seriously limit the effectiveness of the smoke vents.

In an article titled "Sprinkler Heads Nullify Smoke Vents in Warehouse Fire," Alvin Koog described a fire in a rolled paper storage occupancy in which the operation of the sprinkler head located directly in the center of the smoke vent area prevented the effective venting of the smoke from the area.[20] Thus, the placement of sprinkler heads in the center of the vent opening for automatic smoke vents would seem to preclude their effective operation. The integrated design and operation of all of the fire protection systems in a building must always be a consideration for effective fire protection design.

Automatic sprinkler systems in buildings in areas that are subject to earthquakes require special consideration for the bracing of the sprinkler piping and the prevention of piping rupture. Also, it is essential to provide adequate clearance around the sprinkler system piping where it passes through floors. Such clearance spaces should be packed with noncombustible materials in order to prevent the passage of smoke and heat through openings. Sway bracings and flange joints are necessary for sprinkler systems in earthquake prone areas. Figure 5.17 illustrates sway bracing procedures and a typical flange joint installation for a sprinkler system in an earthquake-prone area.

The design of the sprinkler system and the modification of the building construction feature is important in new buildings, and is usually critical for existing buildings.

SUMMARY

The basic design of automatic sprinkler systems involves the provision of a water supply by means of a system of pipes and sprinkler heads that distribute the water to the fire area while the fire is in the initiating stages. In order for an

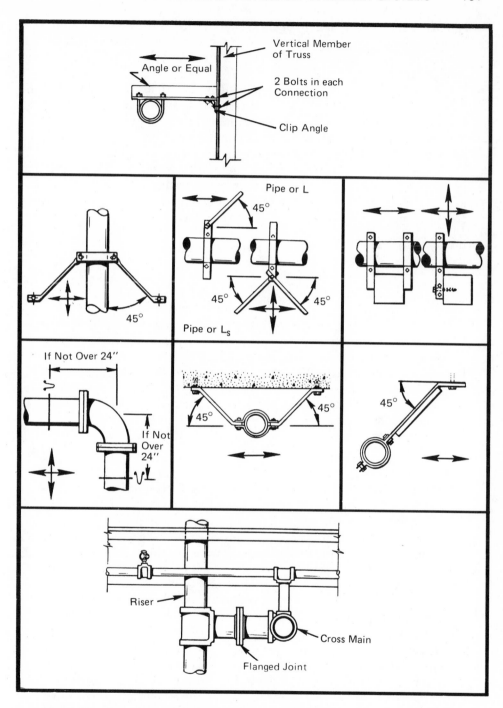

Fig. 5.17. Sway bracing procedures and a typical flange joint installation for a sprinkler system in earthquake-prone areas. (From NFPA 13, *Standard for the Installation of Sprinkler Systems*)

automatic sprinkler system to be most effective, complete sprinkler protection of the building is required. The design of the sprinkler system must be developed so the system can provide adequate water in proper form to suppress and control the fire.

From the time of the initial perforated pipe systems, procedures for the arrangement of sprinkler piping systems have been taken from pipe schedules in rule-book fashion. However, since they are economical and because of their effectiveness in delivering a uniform sprinkler discharge density in the fire area, hydraulically designed systems are being used more extensively. Designing sprinkler systems in combination with standpipe systems with use of the standpipes for sprinkler system risers is increasing, as is the utilization of looped and gridded systems. The use of different piping materials, including copper tubing, is increasing, and the widespread use of plastic pipe is imminent.

SI Units

The following conversion factors are given as a convenience in converting to SI units the English units used in this chapter:

1 square foot =	0.0929 m²
1 inch	= 25.400 mm
1 foot	= 0.305 m
1 psi	= 6.895 kPa
1 gallon	= 3.785 litres
1 gpm	= 3.785 litres/minute

ACTIVITIES

1. When the connection to the public or private water system is unable to provide adequate pressure for the requirements of the automatic sprinkler system, a fire pump is usually provided. To be considered suitable for the primary water supply arrangement, how must a fire pump be arranged when installed on the water main supplying a sprinkler system?
2. Write one-paragraph descriptions for each of the following, including in your descriptions their functions as water supply sources for sprinkler systems:
   a. Pressure tanks   b. Gravity tanks   c. Suction tanks
3. Write a brief summary explaining how the *NFPA Sprinkler System Standard* and the *BOCA Life Safety Fire Suppression System Standard* affect automatic sprinkler system design.
4. A new high-rise apartment has been constructed in your city. You have been put in charge of designing an automatic sprinkler system to protect the building. List several of the factors you must consider when installing the system. Then write a brief description explaining how you would go about accounting for each factor on your list.
5. Both the *NFPA Sprinkler System Standard* and the *BOCA Life Safety Fire Suppression System Standard* contain provisions for the use of any

suitable material for sprinkler system service if that material has been investigated, approved, and listed for such service by a nationally recognized testing and inspection agency laboratory. What is the present status of plastic pipe with such laboratories? What are some of the characteristics of plastic pipe that might bring about a change in its status in the future? Presently, what seems to be its most suitable use? In what occupancies might such systems be found in the future?

6. What are two basic reasons why hydraulically designed sprinkler systems are being used more often? Within the next ten years, why will most automatic sprinkler systems be hydraulically designed systems?

7. Explain how the basic concept and design of the automatic sprinkler system can be affected in buildings in which concealed spaces aren't protected and vertical shafts aren't enclosed.

8. What requirements of the *NFPA Sprinkler System Standard* do not apply to hydraulically designed systems? Why?

9. How can a hydraulically designed automatic sprinkler system be identified? For the fire department, what is the critical information relative to such a system? How may this information be obtained?

10. One of your first duties as a newly appointed assistant to your city's building inspector is to personally inspect and evaluate each building in your assigned area in regard to fire protection features. One building has a partial sprinkler system primarily protecting an area used for storing plastics and gasoline. Most of the building, however, is not protected. When you question the building manager about the reasons for the installation of a partial system, he answers, "We don't need a complete sprinkler system because the rest of the building is composed of noncombustible materials." Based on what you have learned in this Chapter, what is your answer to him? Would you recommend a change? In a written statement, defend your reasoning by explaining why or why not.

SUGGESTED READINGS

"Installing Sprinkler Equipment," Factory Mutual System, Norwood, Mass., 1963.

Jensen, Rolf, "21 Ways to Better Sprinkler System Design," *Fire Journal*, Vol. 48, No. 1, Jan. 1974, pp. 47–48, 57.

Merry, Joseph T. and Schiffhauer, Earl J., "Hydraulic Sprinkler System Design—A Computer Approach," *Fire Technology*, Vol. 2, No. 2, May 1966, pp. 95–107.

Ritz, Richard E., "The Georgia Pacific Building, A High-Rise Fire Resistive Office Building with Automatic Sprinklers Installed Throughout," *Fire Journal*, Vol. 63, No. 5, Sept. 1969, pp. 5–9.

BIBLIOGRAPHY

[1]"Installing Sprinkler Equipment," Factory Mutual System, Norwood, Mass., 1963.

[2]NFPA 13, *Standard for the Installation of Sprinkler Systems,* NFPA, Boston, 1975.

[3]*Standard for the Design and Installation of the Fire Suppression System for Life Safety,* 1st ed., 1974, Building Officials and Code Administrators International, Inc., Chicago.

[4]Decker, Dick A., "A Study of the Effects of a Selected Fire Exposure on a Simulated Plastic Automatic Sprinkler Piping System," Undergraduate paper, 1967, Fire Protection Curriculum, University of Maryland, College Park.

[5]Czarnecki, Thomas, "A Study of the Effects of a Fire Situation on the Reliability of a Simulated Plastic Automatic Sprinkler Piping System," Undergraduate paper, 1969, Fire Protection Curriculum, University of Maryland, College Park.

[6]Marks, Gerald E. and Shaw, Donald D., "Residential Life Safety Systems Test Utilizing Plastic Pipe," Research Report, July 1972, Palo Alto Fire Department, Palo Alto, Calif.

[7]Banwarth, David M., "A Unique Sprinkler System Design for a Hi-Rise Building," 1975, Fire Protection Curriculum, University of Maryland, College Park, Md., Second International Fire Protection Engineering Institute.

[8]Budzynski, Stanley G., "The Hydraulic Characteristics of Plastic Pipe," Undergraduate paper, 1973, Fire Protection Curriculum, University of Maryland, College Park.

[9]Jensen, Rolf, "21 Ways to Better Sprinkler System Design," *Fire Journal,* Vol. 68, No. 1, Jan. 1974, pp. 47–48, 57.

[10]Shelley, Bernard J., "A Study of the Flow Characteristics of a Looped Branch Line Piping Arrangement for an Automatic Sprinkler System as Compared with a Conventional Branch Line Piping Arrangement for Automatic Sprinkler Systems," Undergraduate paper, 1968, Fire Protection Curriculum, University of Maryland, College Park.

[11]Thornberry, Richard P., "A Computer Analysis of Flow Networks for Looped Automatic Sprinkler Systems," Undergraduate paper, 1971, Fire Protection Curriculum, University of Maryland, College Park.

[12]Ritz, Richard E., "The Georgia-Pacific Building, A High-Rise Fire Resistive Office Building with Automatic Sprinklers Installed Throughout," *Fire Journal,* Vol. 63, No. 5, Sept. 1969, pp. 5–9.

[13]Hoyle, H. and Bray, G., "Hydraulic Performance of Sprinkler Installations," *Fire Technology,* Vol. 3, No. 4, Nov. 1967, pp. 291–305.

[14]Merry, Joseph T. and Schiffhauer, Earl J., "Hydraulic Sprinkler Systems Design—A Computer Approach," *Fire Technology,* Vol. 2, No. 2, May 1966, pp. 95–107.

[15]Rosenhan, A. K., "Water Net—A Computerized Design Aid," *Fire Technology,* Vol. 4, No. 3, Aug. 1968, pp. 179–184.

[16]Chicarello, Peter Joseph, Jr., "Analytical Methods for Calculating Sprinkler Discharge," *Fire Technology,* Vol. 3, No. 1, Feb. 1972, pp. 45–52.

[17]Kirkman, Hugh B. and Campbell, Layard E., "Velocity Pressure Effect on Sprinkler System Discharge," *Fire Technology,* Vol. 6, No. 1, Feb. 1970, pp. 68–72.

[18]Merdinyan, P. H., "Friction Loss in Sprinkler Piping," *Fire Technology*, Vol. 4, No. 4, Nov. 1968, pp. 304–309.

[19]Foehl, John M., "Flow Characteristics—Copper Sprinkler Conductors," *Fire Technology*, Vol. 4, No. 3, Aug. 1968, pp. 169–178.

[20]Koog, Alvin, "Sprinkler Heads Nullify Smoke Vents in Warehouse Fire," *Fire Engineering*, Vol. 125, No. 1, Jan. 1972, pp. 40–41.

Chapter **6**

The Automatic
Sprinkler Head

THE DEVELOPMENT OF AUTOMATIC SPRINKLER HEADS

When Henry S. Parmelee of New Haven, Connecticut developed the first practical and operational automatic sprinkler head in 1874, the foundation was laid for the evolution of the modern automatic sprinkler system. Every significant improvement in the automatic sprinkler head has involved the introduction of new sprinkler system concepts or modifications to existing sprinkler systems.

As previously indicated in Chapter 3, "Introduction to Automatic Sprinkler Systems," the Parmelee sprinkler head made possible the present concept of an automatic sprinkler system. In *Automatic Sprinkler Protection*, Dana states that Parmelee developed five sprinkler heads, and the fifth head developed in 1787 was the most successful.[1] This fifth sprinkler head had a brass cap over the distributor, and was similar in appearance to the Parmelee Number 3 and Parmelee Number 4 sprinkler heads (see Figure 3.3 in Chapter 3). The head was modified with the use of a rotating slotted turbine for the distributor, resulting in improved water distribution and less clogging from sediment. When the sprinkler head was heated to the melting point of the solder, the water pressure dislodged the brass cap and the flow of water was dispersed by the revolving turbine. According to Dana, approximately 200,000 of these rather effective Parmelee heads were installed.

The development of automatic sprinkler heads continued during the latter part of the 19th century, and the Grinnell sprinkler head Types B, C, and D were developed from 1884 to 1888 and solved the recurring problem of leakage. Dana reported these Grinnell sprinkler heads quickly became the most

166

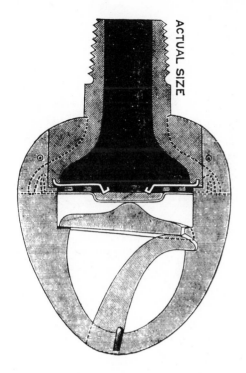

ACTUAL SIZE

Fig. 6.1. Grinnell Type B sprinkler head, 1884. (From Automatic Sprinkler Protection, by Gorham Dana)

widely used sprinkler heads in the United States. Figure 6.1 illustrates this early type of Grinnell sprinkler head.

From early in the 1900s until early in the 1950s, improvements continued to be made in the basic design of the automatic sprinkler head. During this period the Grinnell "Quartzoid" sprinkler head was introduced. This sprinkler head was initially developed in 1924 using the frangible bulb as the activation element. The Globe Automatic Sprinkler Company produced the "Saveall" automatic sprinkler head, a sprinkler head activated by the melting of an organic pellet rather than a fusible metal element. The Grinnell "Duraspeed" sprinkler head developed in the early 1930s can be considered to be the first quick-response sprinkler head since the heat collector configuration of its fusible metal element creates a large surface area to mass ratio, thus reducing the response time of the head. Figure 6.2 illustrates the Grinnell "Quartzoid" sprinkler heads, the Globe "Saveall" head, and the Grinnell "Duraspeed" sprinkler head. The Grinnell "Quartzoid" sprinkler head with the pintle on top of the deflector is a small orifice sprinkler. Both small and large orifice sprinklers with the standard $\frac{1}{2}$-in. pipe thread must have the pintle.

The Standard Sprinkler Head

Primarily as the result of fire department utilization of spray and fog nozzles following the development of such nozzles by the Coast Guard and the Navy during World War II, interest became centered on the development of spray

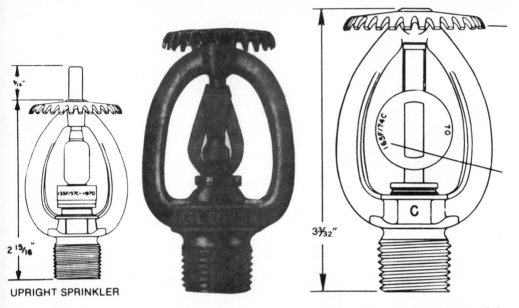

UPRIGHT SPRINKLER

Fig. 6.2. Automatic sprinkler heads. Left to right: Grinnell "Quartzoid", Globe "Saveall", and Grinnell "Duraspeed". (Grinnell Fire Protection Systems Company, Inc., and Globe Automatic Sprinkler Co.)

streams from automatic sprinkler heads. In an *NFPA Quarterly* article titled "New Developments in Upright Sprinklers," Norman J. Thompson reported the Factory Mutual System initiated experimental work in 1947 on the concept of an automatic spray sprinkler head.[2] These studies resulted in the development of specifications for the manufacture of an upright automatic sprinkler head. Several manufacturers initially developed experimental spray sprinkler heads.

By 1952 the Factory Mutual System had conducted sufficient experiments to indicate that in both fire suppression capabilities and water distribution range, upright spray sprinkler heads were superior to the old style sprinkler heads which existed at that time. The spray sprinkler head achieved both an improved distribution of the water into the fire area and a more uniform distribution of the water spray. Thompson explained the following experimental variables resulted in the improved fire suppression characteristics of the spray sprinkler head:*

> As the result of all our experimental work with sprinklers and spray nozzles, it is obvious that the most important single factor is to discharge water at the maximum available rate onto the burning material.
>
> It is also essential that we accomplish as much as is practical in cooling the products of combustion above the fire, especially under the ceiling and around other structural elements so as to reduce the temperature exposure.

---

*From Thompson, Norman J., "New Developments in Upright Sprinklers," *NFPA Quarterly*, Vol. 46, No. 1, July 1952, pp. 7–8.

With good distribution, an improvement in protection is obtained with better atomization.

A final factor of obvious importance, judging from our test work, is the concentration of water vapor and subsequent reduction in oxygen at or near ceiling levels.

The graph in Figure 6.3 shows a comparison of the uniformity of water distribution from the spray sprinkler head and the old style sprinkler head. As indicated in Figure 6.3, the feet away from the head represents the lateral distance from the head.

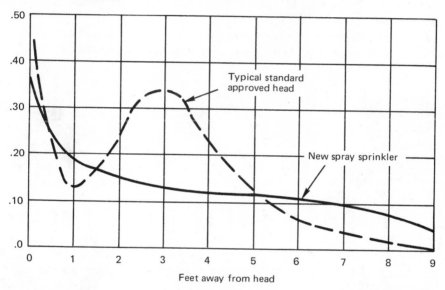

Feet away from head

Fig. 6.3. Distribution of water from spray sprinkler and standard sprinkler heads in Factory Mutual studies with a 30 gpm flow measured 3½ feet below deflector and flow of 30 gpm. (From "New Developments in Upright Sprinklers," by Norman J. Thompson)

One series of Thompson's experiments were concerned with a gasoline spray and wood crib fire. In the experiment the spray sprinklers operated at pressures of 5, 15, and 33 psi eight feet below the sprinkler head, and produced the following results (as compared to identical conditions with old-style sprinklers): destroyed ceiling areas were reduced from 14 square feet to 0 square feet; crib weight loss reduced from 160 lbs to 43 pounds; ceiling temperature reduced from about 1200°F to 800°F or less; and total water flow reduced from 252 gpm to 192 gpm.

In a later article titled "Proving Spray Sprinkler Efficiency," Thompson presented the results of comparative tests of standard and spray sprinklers involving a wood crib. The wood crib was located 3 feet below sprinklers that were installed under a wood joist ceiling construction.[3] The standard sprinklers were spaced 8 × 10 feet, while the spray sprinklers were spaced 9 × 10 feet. The results indicated with less than 50 percent of the spray sprinklers operating, the spray sprinklers lowered the ceiling temperatures and reduced the ceiling and crib damage.

The spray sprinkler head, originally developed in the upright configuration from the experiments at Factory Mutual, was also developed as a pendent sprinkler head. In 1955 this spray type sprinkler head was designated by the NFPA as the standard sprinkler head. The previous standard head was designated as the old style sprinkler head. Old style heads may still be found in service on some sprinkler systems, and since the basic difference in the standard sprinkler head and the old style sprinkler head is the size and design of the sprinkler head deflector, old style heads are still being manufactured. Standard and old style sprinkler heads are shown in Figure 6.4.

Since 1955 the standard sprinkler head has been required in all new sprinkler system installations. The standard sprinkler head should be utilized only in the

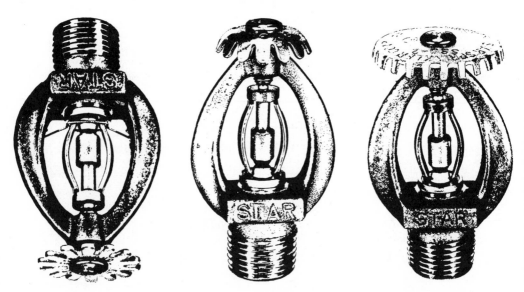

Fig. 6.4. Standard and old style sprinkler heads. Left to right: standard pendent, old style head, standard upright (Star Sprinkler Corporation)

position for which it was designed (as an upright or pendent sprinkler), since all of the water discharges in a single direction. When discharging water at 15 psi, the standard sprinkler head provides a uniform distribution 4 feet below the deflector throughout a 16-foot diameter circle.

Figure 6.5 illustrates water discharge patterns from standard and old style sprinkler heads. (Note, as indicated in Figure 6.5, the old style heads distribute the bulk of the water to the ceiling area where it is deflected downward in relatively large drops. This ceiling deflection process is represented by the hump in the water distribution graph shown in Figure 6.3 earlier in this chapter.) In his book *Fire Behavior and Sprinklers*, Norman J. Thompson stated

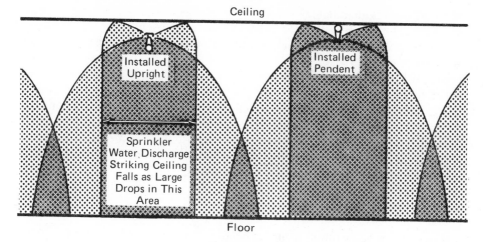

Old Style Sprinkler Heads

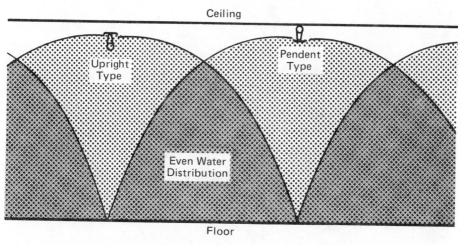

Standard Sprinkler Heads

Fig. 6.5. Sprinkler discharge diagrams for the old style and standard sprinkler heads. (From *Fire Protection Handbook,* NFPA)

with the old style sprinkler head approximately 66 percent of the water is discharged upward.[4]

In 1955 the NFPA's recognition of the spray sprinkler head as the standard sprinkler head resulted in changes in the *NFPA Sprinkler System Standard.* These changes provided for greater coverage per sprinkler head in the pipe schedules, as necessitated by the improved water distribution characteristics of the standard sprinkler head. Standard sprinkler heads in the upright position cannot be interchanged with standard sprinkler heads in the pendent position, and each is identifiable by deflector markings: (1) SSU for Standard Sprinkler Upright, and (2) SSP for Standard Sprinkler Pendent. Currently, old style sprinklers are permitted to be used only in systems where these heads were installed prior to 1953.

Figure 6.6 shows a standard pendent sprinkler head in operation. (In Figure 6.6 the water discharge from the standard pendent sprinkler head is being distributed outward from the deflector in an umbrella-like configuration rather than upward towards the ceiling as with the old style sprinkler head.) Special hazard situations in which water distribution is required above the sprinkler heads (thus necessitating the installation of pendent sprinklers in the upright position) will be discussed later in this chapter.

The On-Off Sprinkler Head

The concept of the on-off sprinkler head—a head that could not only activate, but could also stop the discharge of water upon establishing effective fire

Fig. 6.6. Standard pendent sprinkler head in operation. (International Fire Service Training Association, Oklahoma State University)

control—had been considered for many years. In an article titled "A Fully Approved On-Off Sprinkler," Philip H. Merdinyan reported that in 1878 Joseph Miller designed an on-off sprinkler head that was never commercially produced.[5] In 1884 F. H. Prentice utilized the design principle of the expansion of ether in a closed container for his patent of an on-off sprinkler head. Prentice's design was used only for a few heads. In his report *An Attempt to*

Develop an Automatic, Individually-operated Reusable and Self-resetting Sprinkler Head, John E. Luley reported on the 1969 experimental development of a sprinkler head operated by a heat detector.[6] In a *Fire Journal* article titled "Revolutionary New Sprinkler," Raymond E. Shea describes the 1969 experimental work on the first commercially applicable on-off type of sprinkler head.[7] This sprinkler head was finally commercially developed by the Grinnell Fire Protection Systems Company, Inc., and in October, 1971, was submitted to both the Underwriters Laboratories and the Factory Mutual Laboratories. In 1972 laboratory approval was obtained for the on-off sprinkler head.

The Grinnell "Aquamatic" on-off sprinkler head is shown in Figure 6.7. This sprinkler head is currently available only as a standard pendent sprinkler head or as a recessed pendent sprinkler head.

Fig. 6.7. Standard pendent on-off sprinkler head. (Grinnell Fire Protection Systems Company, Inc.)

Hidden Type of Pendent Sprinkler Head

Architects and building owners sometimes object to the installation of automatic sprinkler systems for aesthetic reasons. This objection has also been made concerning recessed pendent sprinkler heads. As a result of such objections, a hidden pendent type sprinkler head was developed. The hidden pendent type sprinkler head does not protrude through highly decorated ceiling areas, and is approved by both the Factory Mutual and the Underwriters Laboratories. Although completely hidden by a cover plate, the hidden pendent type sprinkler head is capable of providing adequate and effective performance. The cover plate for the sprinkler head is available in colors and patterns to match many of the current decorative ceiling assemblies. The Star Sprinkler Corporation was one of the first manufacturers to provide an approved and listed sprinkler head of the hidden pendent type, and their sprinkler head was trade-named the "Unspoiler." The cover plate for this hidden pendent type sprinkler head acts as a heat collector. The heat from the 4-inch diameter cover plate is conducted directly to the fusible element by metallic strips, or fins, as illustrated by the Star Sprinkler Corp. version of the hidden type automatic sprinkler head shown in Figure 6.8. The cover plate is held to the sprinkler head assembly by fusible metallic hook assemblies that drop away immediately prior to operation of the sprinkler head. The Grinnell Corp. also provides a hidden pendent sprinkler head trade-named the "Cleanline" automatic sprinkler head.

Sprinkler Head Actuators

As previously mentioned, the on-off sprinkler head was developed about 1969, and approved by the Underwriters and Factory Mutual Laboratories in 1972. Due to the life safety situation in certain occupancies, the need for a smoke detector operated sprinkler head has been considered in facilities such as patient rooms in hospitals, nursing homes, and other related types of health

Fig. 6.8. Hidden type of standard pendent sprinkler head. (Star Sprinkler Corp.)

care facilities. Thus, electronic actuators have been developed recently that will activate a standard fusible link in the automatic sprinkler head when the actuator is electrically energized by the operation of a smoke or heat detector. These actuators can also be attached to other fusible link operated devices, including fire dampers in ducts and fire doors in walls. The smoke detector operation of sprinkler heads is currently being considered on an experimental basis in selected high life hazard situations. The smoke detectors should be located within the water discharge area of the sprinkler head. However, the possibility of sprinkler head operation in areas having heavy smoke concentration but no fire development may be a problem. This is especially true in occupancies not having the extensive degree of compartmentation usually found in health care facilities. Actuators that initiate the operation of the automatic sprinkler head are available for the adaptation of existing installations to smoke detector operation for selected sprinkler heads.

In an article titled "The Potential of Actuators in Fire Engineering," L. E. Medlock reported on the use of explosively operated actuators that include a piston or bellows type of actuator which may be used to activate the automatic sprinkler head.[8] Actuators of the electronic squib type and the electrical-chemical type which are attached to, or adjacent to, the fusible element of the automatic sprinkler head have been developed in the United States. Figure 6.9 shows an electrical-chemical type actuator utilized with a pendent type sprinkler head. This is one type of actuator which has been considered for use with smoke detectors for the protection of patients in health care facilities.

Actuators may also be utilized with thermal operated detectors, as well as with any other desired electrical switching devices. The type of smoke detector used may be of any desired operational mode such as ionized chamber, photoelectric, or combustion gas type. Actuators are available that are listed and approved by both Factory Mutual and Underwriters Laboratories. The devices utilized in the United States are generally of low electrical voltage with an operating range from 6 to 30 volts. The unit illustrated in Figure 6.9 is suitable

Fig. 6.9. Standard pendent sprinkler head operated by electrical-chemical actuator initiated by a smoke detector. (S R Products, Inc.)

for use with automatic sprinkler heads with operating temperatures ranging from 135 to 212°F.

The Quick-Response Automatic Sprinkler Head

For life safety purposes, the need for a quick-response sprinkler head has developed simultaneously with the installation of automatic sprinkler systems in many occupancies. Typical examples include the installation of automatic sprinkler systems in health care facilities, high-rise office buildings, and apartment buildings. Some of the testing and approval authorities have recognized that a great deal of variation exists in the operating time of sprinkler heads with identical temperature ratings. Norman J. Thompson, (*Fire Behavior and Sprinklers*[4]) explained these observations as follows:*

> The operating times of sprinklers, even of the same temperature rating, vary considerably depending on the design of the sprinklers. This variation, even under identical fire conditions, is due to the difference in thermal lag of the temperature sensitive element. The amount of heat which must be transferred to the element depends upon its mass whereas the rate of heat transfer is governed principally by the surface area of the element, since the transfer of heat is primarily by convection and conduction with radiation a factor of minor importance, except possibly in fast fires of high intensity. Consequently, the sprinkler with the highest ratio of surface area to mass in the temperature sensitive element will operate the soonest under the same fire conditions.

Thompson also prepared a chart of the operating times of four sprinkler heads plotted against the atmospheric temperature from data collected under fire test conditions. The sprinkler heads were located 10 to 12 feet above the fire, and approximately 7 feet laterally from the center of the fire. (It should be noted there is a significant difference in the operating time between the slowest and the fastest sprinklers in the 165°F rating.) The differences between all of the sprinkler heads is most pronounced when the velocity of temperature rise in the fire area is below 150°F per minute. Figure 6.10 presents Thompson's data indicating the operating times of four standard automatic sprinkler heads.

Once the fire-induced temperature rise exceeds 40°F per minute, it will be noticed that sprinkler B, a "fast" 212°F rated sprinkler, will operate faster than sprinkler A, a "slow" 165°F rated sprinkler. Thus, we have the established experimental evidence of 165°F rated sprinkler heads when exposed to a temperature rise of 50°F per minute, which operate at 2.8 minutes for a "fast" head (curve C), and 3.5 minutes for a "slow" head (curve A).

The quick-response sprinkler heads have been designed to reduce the operating time of the sprinkler head by providing increased surface to mass areas for the operating elements, or they have utilized actuators. Figure 6.11 shows a quick-response sprinkler head designed to provide a large surface to mass ratio

---

*From Thompson, Norman J., *Fire Behavior and Sprinklers*, National Fire Protection Association, Boston, 1964, p. 91.

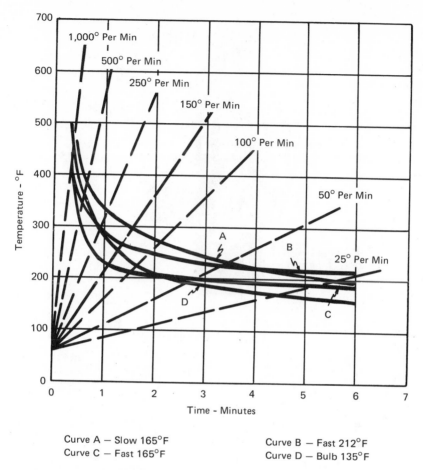

Curve A — Slow 165°F   Curve B — Fast 212°F
Curve C — Fast 165°F   Curve D — Bulb 135°F

Fig. 6.10. Operating times of standard sprinklers. (From *Fire Behavior and Sprinklers,* by Norman J. Thompson)

for the operating elements with the use of metallic vane type heat collectors.

Historically, the Grinnell "Duraspeed" sprinkler with the circular metallic heat collector type of activating element is considered to be a "fast" sprinkler head (see Figure 6.2).

Grinnell has modified their "Duraspeed" sprinkler in the ordinary 165°F rating into a quick-response sprinkler by the utilization of a self-contained electronic squib quick-response attachment. The attachment consists of a printed circuit board, resistor, capacitor, battery, thermostat, and squib. It is contained in a heat-resistant ceramic material, and is attached to the sprinkler with a mounting bracket. This quick-response sprinkler is available in both the standard upright and pendent sprinkler head, and also as a sidewall head. The thermostat operates at 135°F, completing the electronic circuit and allowing energy from the battery to activate the squib. The squib discharges molten metal directly into the heat-responsive element of the "Duraspeed" sprinkler

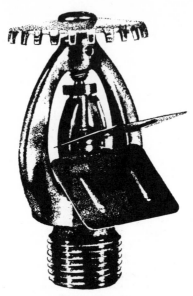

*Fig. 6.11.   Metallic vane type
of quick response sprinkler head.*
(Star Sprinkler Corp.)

which has an operating temperature of 165°F. The molten metal causes immediate operation of the sprinkler head initiating the water discharge. The quick-response attachment on the "Duraspeed" sprinkler head is reported to reduce the operating time of the sprinkler head in one standard test condition from 115 to 30 seconds. Figure 6.12 shows the electronic squib quick-response attachment on a standard pendent sprinkler head.

Underwriters Laboratories have just begun to list the quick-response sprinkler head in the last few years. At this writing these sprinkler heads reflect the most recent developments in sprinkler heads; currently they have been further improved to include the concept of providing extended discharge coverages from the sprinkler head. These performance improvements are being made in response to the need for sprinkler heads that can be used primarily in life safety situations including fires that occur in health care facilities and apartment occupancies.

VARIOUS TYPES OF AUTOMATIC SPRINKLER HEADS

In the previous material concerning the development of automatic sprinkler heads, some of the principal improvements to automatic sprinkler heads have been described. In the following material, the various types of sprinkler heads will be examined relative to their components, function, and design principles.

Presently in the United States, sprinkler heads are of metallic construction—generally steel or bronze alloy. However, Boris Laiming (in "New Sprinklers—Russian Style"[9]) reported on Russia's development of a nonmetallic sprinkler head. The nonmetallic sprinkler head described by Laiming was designed with

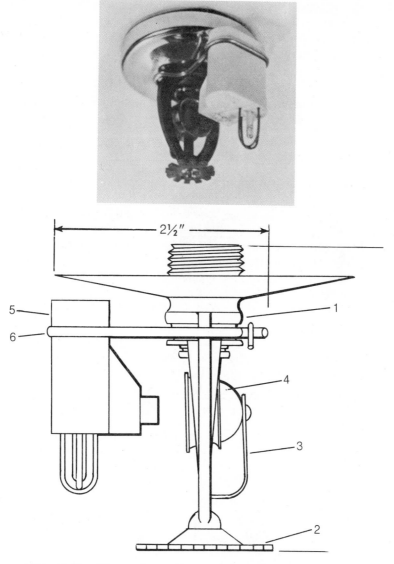

Fig. 6.12. Electronic squib type of quick response sprinkler heads. (Grinnell Fire Protection Systems Company, Inc.)

the sprinkler assembly constructed of a plastic cone and nipple. The cone has a spherical chamber, a cylindrical opening at the top, and four slits on the side. Thus a central stream and four tangential streams enter the chamber through the openings; the impact of the five streams impinge to produce the spray from the sprinkler head. The cylindrical opening in the nipple is sealed by a plastic cap which is held against a seat by two levers. The levers fit into slots in the nipple; the other ends are tied together by the fusible link assembly. When the fusible link actuates, the water pressure forces the cap and both levers clear of the sprinkler. The Russian plastic centrifugal sprinkler is illustrated in Figure 6.13.

Reportedly, use of the plastic centrifugal sprinkler head results in improved water distribution and operates from seven to eleven times faster than the standard Russian sprinkler head. It was originally developed for installation on the plastic pipe systems that have been recommended for sprinkler systems in Russia since 1960. The plastic centrifugal sprinkler head is highly recommended due to the savings in bronze and copper that result from its use.

Components of the Automatic Sprinkler Head

By 1900 the basic design features of the principal components of the automatic sprinkler head had been established. Since that time, any major changes have been related to the sprinkler head deflector and the activating element of the head. Figure 6.14 illustrates the component parts of a standard upright sprinkler head with a link and lever fusible activating element. The link and lever, the triangular strut, the frangible bulb, and the organic or fusible pellet design are the principal activating element arrangements presently utilized in standard sprinkler heads.

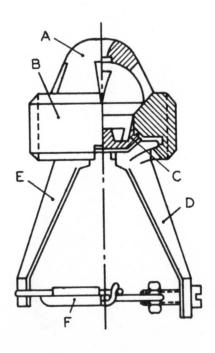

A — Cone   D — Right lever
B — Nipple   E — Left lever
C — Plastic cap F — Fusible link

Fig. 6.13. Russian plastic centrifugal sprinkler head. (From "New Sprinklers-Russian Style," by Boris Laiming)

Figure 6.15 illustrates the component parts of the link and lever fusible element upright standard sprinkler head in the completely assembled mode prior to operation, and Figure 6.16 illustrates the component parts of the upright standard sprinkler head with a link and lever fusible element as the parts separate with the operation of the sprinkler head. In Figure 6.16, the fusible element is shown to have separated vertically in the proper activating procedure. The roller contained between the two link plates accelerates the separation of the plates upon release of the solder.

All of the automatic sprinkler heads listed by Underwriters or Factory Mutual Laboratories will have located upon them the year of manufacture, the

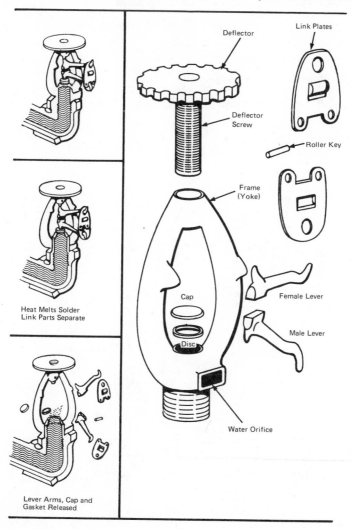

Fig. 6.14. Principal components and operation of an upright sprinkler head of the link and lever design. (Insurance Services Office)

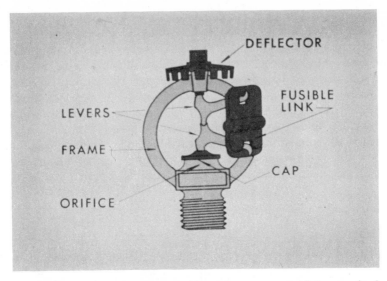

Fig. 6.15. Identification of the component parts of the standard upright sprinkler head of link and lever design. (International Fire Service Training Association, Oklahoma State University)

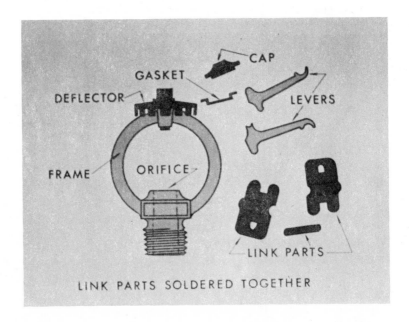

Fig. 6.16. Illustration of the separation of the component parts of the link and lever design sprinkler head following operation. (International Fire Service Training Association, Oklahoma State University)

temperature rating of the activating element, and the name of the manufacturer. If the sprinkler head is a standard sprinkler and not an old style head, it will also carry on the deflector the designation of SSU (Standard Sprinkler Upright) or SSP (Standard Sprinkler Pendent). Some manufacturers instead of using these abbreviations, place the complete wording—"Upright Sprinkler" or "Pendent Sprinkler"—on their deflectors. Old style sprinkler heads may carry only the temperature rating of the activating element and the date of manufacture, and not the name of the manufacturer. However, the manufacturer can usually be identified by the sprinkler head's design features. Figure 6.17 illustrates the usual locations of the required information for the various types and styles of automatic sprinkler heads. In the upper left part of Figure 6.17, the standard upright sprinkler head with the heat collector type of fusible element was previously identified by the manufacturer's designation as the "Duraspeed" sprinkler head. The sprinkler in the upper right part of Figure 6.17 is an old style sprinkler head. Note particularly the difference in the design of the deflectors of the standard upright sprinkler head and the old style sprinkler head.

The center left sprinkler head in Figure 6.17 is a frangible bulb type standard upright sprinkler head, previously referred to by the manufacturer's designation as the "Quartzoid" sprinkler head. The sprinkler head in the center right is a standard upright head with a fusible pellet type activating element. Both of the sprinkler heads shown at the bottom of Figure 6.17 are of the link and lever fusible element design with the sprinkler head on the left being a standard upright, and the head on the right being a standard pendent sprinkler.

Figure 6.18 is an illustration of the old style sprinkler head of the link and lever design. Again, there is a dramatic difference in the design of the deflectors of the standard upright sprinkler head and the old style sprinkler head. Also, the rest of the sprinkler head for a link and lever design fusible element sprinkler head is identical. Thus, the old style head could be easily converted to a standard upright sprinkler head merely by changing the deflector on the frame of the sprinkler head.

In a 1970 study titled "Modification of Sprinkler Heads to Obtain Discharge Above and Below the Deflector in Discrete Percentages," Frank J. Fabin found that deflectors could be modified to discharge a desired amount of water upward onto the ceiling, and a desired amount directly downward.[10] Fabin modified a standard upright sprinkler head by making holes or slits in the deflectors and changing their angles. Thus, Fabin was able to modify the deflector relative to the desired percentage of water discharged above the head by changing the angle of the deflector after adding slits in it. Fabin could modify the deflector to discharge from 10 to 90 percent of the water above the deflector. As a result of this study, he also found the deflector was susceptible to damage and minor deflection if dropped from heights as low as six feet. He therefore recommended the deflector construction be strengthened to prevent accidental deflection, and thought be given to designing a sprinkler head with a one-sided frame (rather than the completely circular frame as presently utilized).

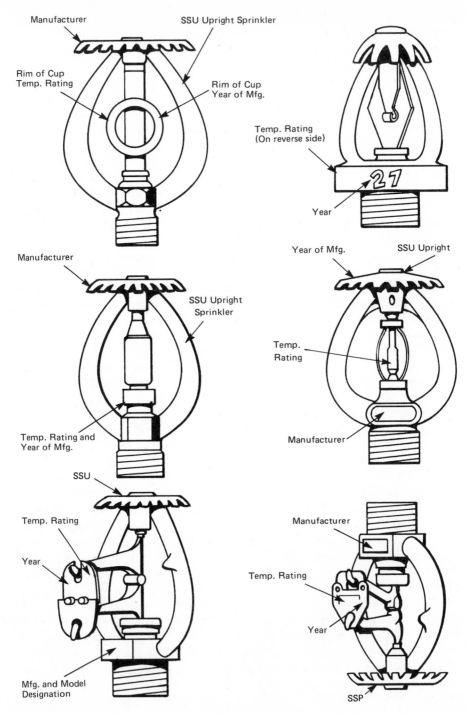

Fig. 6.17. Sprinkler head styles and types with location of identification information on the sprinkler head. (Insurance Services Office)

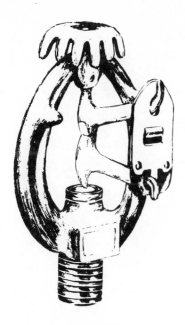

Fig. 6.18. An old style sprinkler head of the link and lever fusible element design. (Insurance Services Office)

Functional Types of Automatic Sprinkler Heads

The types of automatic sprinkler heads considered in this section will be classified according to their functions.

It should be remembered the previously discussed old style sprinkler head can be utilized in the upright or pendent position, while the standard sprinkler head cannot be interchanged between the upright and the pendent positions because of differences in deflector design. Figure 6.19 shows standard upright and pendent sprinkler heads from the same manufacturer; both have a link and lever fusible element design. Note the pendent deflector on the right has extensive slots, and the upright deflector on the left has vertical penetrations, or lips, extending downward from the edge of the deflector which is a solid shield.

When the pendent sprinkler head is utilized on a dry pipe sprinkler system, the head must be a dry pendent sprinkler head. This head prevents water from entering the standoff from the sprinkler branch line until the sprinkler head is in operation. Thus, this feature prevents the manual draining of several hundred pendent sprinkler heads for every operation of the dry pipe valve. Figure 6.20 shows the design of the dry pendent sprinkler head. Fortunately, dry pendent sprinkler heads are not generally installed in large areas and, while these heads are often found in the office areas of unheated structures (such as warehouses), the areas protected and the number of heads involved are usually small.

Sidewall Sprinkler Heads

Although vertical sidewall sprinkler heads have been utilized for many years, only recently has the interest in the sidewall sprinkler head shown a dramatic

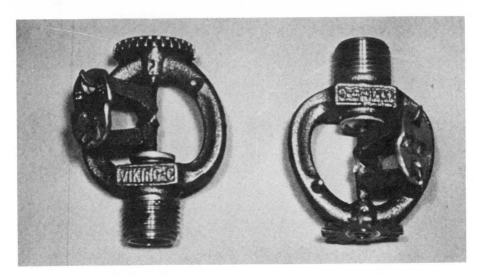

Fig. 6.19.   Standard upright and pendent sprinkler heads with link and lever fusible element design. (International Fire Service Training Association, Oklahoma State University)

increase; this increase has been due to the application and installation of automatic sprinkler systems in existing health care and housing occupancies for the elderly. Figure 6.21 illustrates a vertical sidewall sprinkler head of the standard sprinkler design in both the upright and pendent models. In certain situations, the sidewall sprinkler has been found to have some economic advantages relative to the installation of sprinkler systems in existing and new occupancies with many compartments or rooms.

Recently, most of the sprinkler manufacturers have developed a horizontal type of sidewall sprinkler head. This head is also available as a quick-response type sprinkler head with an electronic squib attachment (see Figure 6.12 earlier in this chapter). At this writing, the horizontal sidewall sprinkler head is the latest type of sidewall sprinkler head. The horizontal sidewall sprinkler head is shown in Figure 6.22 in both the standard configuration and the quick-response type of sprinkler with the electronic squib type quick-response attachment.

The horizontal sidewall sprinkler head is also available with the deflector on the head modified to provide for extended coverage of the water spray. The extended coverage horizontal sidewall sprinkler head has been developed to provide from 336 square feet of water discharge coverage to as high as 392 square feet with a large orifice head, as compared to the approximately 196 square feet of water discharge coverage provided by the standard sidewall sprinkler. Note the differences in the deflector design for the standard horizontal sidewall sprinkler head (shown in Figure 6.22) and the extended coverage horizontal sidewall sprinkler head (shown in Figure 6.23).

The horizontal sidewall type of sprinkler head allows considerable variation in providing for the specific needs of the occupancy hazard as determined by the sprinkler system designer. Where necessary, the horizontal sidewall head may be utilized with the quick-response and the extended-coverage features.

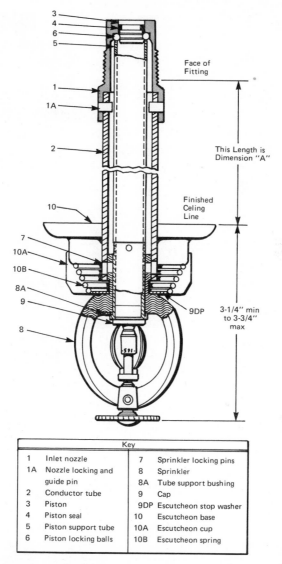

		Key		
1	Inlet nozzle		7	Sprinkler locking pins
1A	Nozzle locking and		8	Sprinkler
	guide pin		8A	Tube support bushing
2	Conductor tube		9	Cap
3	Piston		9DP	Escutcheon stop washer
4	Piston seal		10	Escutcheon base
5	Piston support tube		10A	Escutcheon cup
6	Piston locking balls		10B	Escutcheon spring

Fig. 6.20. Dry pendent sprinkler head. (From "Automatic Sprinkler Protection—Further Developments," by Marshall E. Peterson)[11]

The sidewall sprinkler head, and especially the horizontal sidewall sprinkler head, is one of the most versatile sprinkler heads in use today; in the future, the application of this type of head with its quick-response and extended-coverage modifications will be found more in health care and residential occupancies, including high-rise apartment occupancies.

Decorative and Coated Sprinkler Heads

The recessed and semirecessed type of pendent sprinkler heads have been available (even in the old style sprinkler heads) for installations where the ap-

Fig. 6.21. Vertical sidewall sprinkler heads of the standard upright and pendent types. (Star Sprinkler Corp.)

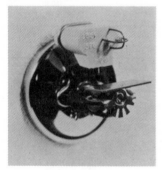

• Horizontal sidewall
  Quick response

• Horizontal sidewall

Fig. 6.22. Horizontal sidewall sprinkler head of the standard type, and same head equipped with quick response electronic squib attachment. (Grinnell Fire Protection Systems Company, Inc.)

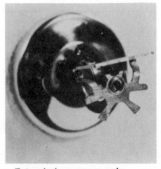

• Quick response
  extended coverage

• Extended coverage only

Fig. 6.23. Horizontal sidewall sprinkler head of the extended coverage type, and same extended coverage head equipped with quick response attachment of the electronic squib type. (Grinnell Fire Protection Systems Company, Inc.)

pearance of the sprinkler heads is an important consideration. As previously indicated in this chapter, the design of the hidden type of pendent sprinkler head is the solution to preserving the aesthetic and decorative aspects of ceilings in sprinklered areas. (See Figure 6.8 earlier in this chapter.) In addition to the recessed and semirecessed sprinkler heads, chrome sprinkler heads of the pendent type are often utilized to provide a more finished appearance in many public occupancies such as offices, department stores, and restaurants. (See Figure 6.21 earlier in this chapter for an example of a semirecessed type of sidewall pendent sprinkler head.)

In areas exposed to corrosive atmospheres or fumes, the metallic elements of sprinkler heads are usually protected with coatings of lead or wax. Figure 6.24 shows a selection of some of the types of decorative and coated sprinkler heads. The top row of sprinkler heads are all standard upright heads, with the first head being a sidewall head of the vertical upright type. The second head is a "Quartzoid" upright sidewall head. The third head in the top row is a "Quartzoid" standard upright sprinkler head with the frangible bulb activating element. The fourth head is an old style head with a protective lead coating and fusible element of the triangular strut type. The final sprinkler head on the right in the top row is a standard upright sprinkler with a protective wax coating, and a fusible element of the link and lever design.

All of the sprinkler heads in the bottom row of Figure 6.24 are standard pendent sprinkler heads. The first sprinkler head on the left is a "Quartzoid" sprinkler head of the type found in recessed and semirecessed fixtures. All of the remaining heads have link and level fusible elements, and the second sprinkler head on the left is of a semirecessed design. The third sprinkler head is of a recessed design, and the sprinkler head on the right is a chrome-plated standard pendent sprinkler head.

Fig. 6.24. Selected decorative and coated sprinkler heads. (International Fire Service Training Association, Oklahoma State University)

The On-Off Sprinkler Head

As previously indicated, a completely automatic on-off sprinkler head received approval from both Underwriters and Factory Mutual Laboratories in 1972. The Grinnell "Aquamatic" on-off automatic pendent sprinkler head is shown in Figure 6.7 earlier in this chapter. This head is presently available as a pendent sprinkler head, and also, as shown in Figure 6.25, as a pendent recessed model.

The on-off sprinkler has a 165°F temperature rating. Merdinyan ("A Fully Approved On-Off Sprinkler"[5]) described the operation of the on-off sprinkler head as follows:*

It is the bimetallic disc that is the key element to successful functioning of the sprinkler.

Besides the snap disc, the on-off sprinkler consists of a body, an inlet, a deflector, a pilot valve, a piston, a spring, a restriction, and 0 rings.

In service, the aquamatic sprinkler is held closed by pressure in the piston chamber. When exposed to a temperature of 165°F, the disc responds by opening the pilot valve to drain water from the piston chamber faster than it can be replenished through the restriction.

Water pressure then forces the piston down, thus permitting unobstructed flow of water through the main body passage. After the fire has been extinguished and the air around the snap disc has cooled, the snap disc responds by closing the pilot valve. Water flowing through the restriction quickly fills the piston chamber, forcing the piston up toward the sprinkler inlet until it shuts off.

Figure 6.26 is a diagrammatic presentation of the operating components of the Grinnell "Aquamatic" on-off sprinkler head, with an indication of the configuration of parts in both the open and closed positions.

Merdinyan's article indicated the following advantages for the on-off sprinkler head:**

1. It makes the best use of available water by directing it where it is needed—on the fire.

2. It eliminates the need to turn off the main valve for inspection or to replace sprinklers. This reduces the potential risk of catastrophic loss due to valve shutoff.

3. In some instances it may reduce the cost of a new installation through savings resulting from smaller pumps, pipes, and reservoirs.

4. It minimizes water damage.

An indication of the potential water minimization effects of the on-off sprinkler head was reported in Merdinyan's article relative to the evaluation fire test conducted at the Factory Mutual Sprinkler Test Facility and observed by Underwriters Laboratories personnel. The test situation involved two separate,

*From Merdinyan, Philip H., "A Fully Approved On-Off Sprinkler," *Fire Journal*, Vol. 67, No. 1, Jan. 1973, p. 11.
**From *Ibid.*, p. 15.

Component Parts:

1. Body
2. Adjusting Collar
3. Orifice Cover
4. Flat Head Screw
5. Cover Plate
6. Aquamatic Sprinkler

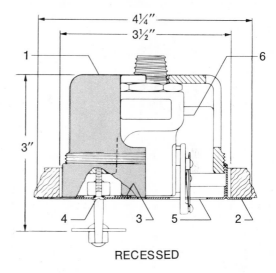

RECESSED

Fig. 6.25. Recessed on-off pendent sprinkler head. (Grinnell Fire Protection Systems Company, Inc.)

but identical, fire environments for sixteen on-off sprinklers compared with sixteen standard sprinklers. The standard sprinklers utilized a total discharge of 11,148 gallons of water. The on-off sprinkler heads utilized only 6,524 gallons, yet achieved better weight loss control of the wood crib used in the fire test. The standard sprinklers had a weight loss on the crib of 55 percent, while the on-off sprinklers obtained a weight loss of 46.5 percent. The difference in water utilization for the identical fire conditions between the standard sprinkler heads and the on-off sprinkler heads occurred because the on-off sprinkler heads shut off when the water discharge was no longer needed to control fire or to reduce temperatures. Only eight of the on-off sprinklers were open for 50 percent or

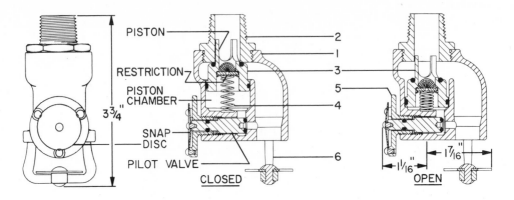

Component Parts: 1. body, 2. inlet, 3. piston, 4. spring, 5. pilot valve assembly, 6. yoke & deflector assembly

Fig. 6.26. Component parts of the on-off sprinkler head in both the closed and open positions. (Grinnell Fire Protection Systems Company, Inc.)

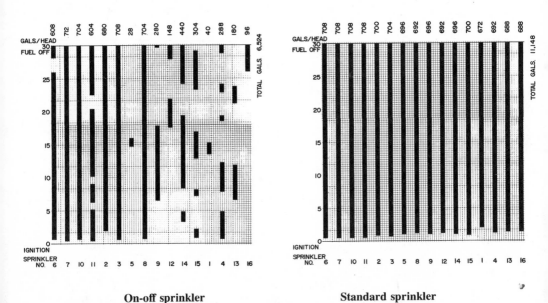

On-off sprinkler   Standard sprinkler

Fig. 6.27. Water flow charts for comparison fire tests conducted at Factory Mutual for both on-off and standard sprinklers. (From "A Fully Approved On-Off Sprinkler," by Philip H. Merdinyan)

more of the 30-minute test, and the remaining eight were open less than 50 percent of the time. The cycling effect of the operation of eleven of the on-off sprinkler heads is shown in the water flow charts presented in Figure 6.27.

EVALUATION OF AUTOMATIC SPRINKLER HEADS

The evaluation of automatic sprinkler heads is presently conducted by the Underwriters Laboratories and the Factory Mutual System. Standards for the evaluation of automatic sprinkler heads have been developed by the respective laboratories based on the experience of laboratory personnel, the interest of advisory groups, and the areas of concern reported by representatives of the sprinkler or fire protection industry.

Underwriters Laboratories Evaluation

In 1919 the Underwriters Laboratories published *The Standard for Automatic Sprinklers,* the first standard relative to the evaluation of automatic sprinkler heads.[12] The present standard, published on May 15, 1974, and titled *Automatic Sprinklers for Fire Protection Service,* is in its fifth edition.[13] The Underwriters Laboratories' evaluation procedure takes into consideration the characteristics of sprinkler heads in terms of basic construction, including orifice sizes, thread sizes, temperature ratings of the heads, and coatings or platings that may be applied.

The sprinkler heads are classified relative to the size of discharge orifices in accordance with the criteria presented in Table 6.1. Also, the operating temperatures of the sprinkler heads must conform to the temperature rating classifications shown in Table 6.1.

The temperature ratings are presented with the maximum ceiling temperatures in the area where the sprinkler head is to be utilized. Temperature rated sprinkler heads of Intermediate, High, and Extra-High ratings will be discussed later in this chapter relative to their use in some industrial occupancy situations where they are utilized to conserve the discharge of water and to reduce the number of heads opening on a sprinkler system. The color coding of sprinkler heads to indicate activating temperatures is now standard procedure throughout the automatic sprinkler industry, with the color being applied either to all or a portion of the frame of the standard sprinkler head. For sprinkler heads with lead or wax coatings, a color dot is placed in the center of the deflector; color coding is not required for the chrome or decorative heads.

The Underwriters Laboratories' Standard, *Automatic Sprinklers for Fire Protection Service,* defines the old style sprinkler head and the standard sprinkler head relative to their water discharge characteristics as follows:*

>   **Old Style Sprinkler**—A sprinkler intended for installation in the upright
> or pendent position, designed to distribute water so that approximately 40

---

*From *Automatic Sprinklers for Fire-Protection Service,* U. L. 199, 5th ed., Underwriters Laboratories, Chicago, Ill., May 15, 1974, pp. 5–6.

Table 6.1. Orifice Size and Temperature Ratings of Automatic
Sprinkler Heads*

NOMINAL ORIFICE SIZE		
ORIFICE TYPE	NOMINAL ORIFICE SIZE Inch mm	DISCHARGE COEFFICIENT "K"
Small Orifice	$\frac{1}{4}$ 6.4	1.3–1.5
Small Orifice	$\frac{5}{16}$ 7.9	1.8–2.0
Small Orifice	$\frac{3}{8}$ 9.5	2.5–2.9
Small Orifice	$\frac{7}{16}$ 11.1	4.0–4.4
Standard Orifice	$\frac{1}{2}$ 12.7	5.3–5.8
Large Orifice	$\frac{17}{32}$ 13.5	7.4–8.2

TEMPERATURE RATINGS

TEMPERATURE RATING	OPERATING TEMPERATURE		COLOR	MAXIMUM CEILING TEMPERATURE	
	Degrees F	Degrees C		Degrees F	Degrees C
Ordinary	135–170	57.2– 76.7	Uncolored[1]	100	38
Intermediate	175–225	79.4–107	White	150	66
High	250–300	121 –149	Blue	225	107
Extra High	325–375	163 –191	Red	300	149
Very Extra High	400–475	204 –246	Green	375	191
Ultra High	500–575	260 –302	Orange	475	246

[1]Sprinklers with an operating temperature of 135°F (57.2°C) may be colored black.

percent of the discharge is directed upward and 60 percent is directed downward. When installed in the upright position, this discharge will cover 10 ft (3.05m) diameter circle, 10 ft (3.05m) below the sprinkler, when the sprinkler is discharging water at the rate of 15 gal per min (0.95 l/s).

Standard Sprinkler—A sprinkler intended for installation in the upright or pendent position, designed to distribute water downward in an umbrella shaped pattern. The discharge from a standard sprinkler having a standard orifice will cover a circle 16 ft (4.88m) in diameter, 4 ft (1.22m) below the sprinkler, when the sprinkler is discharging water at the rate of 15 gal per min (0.95 l/s).

The various performance tests applied by the Underwriters Laboratories are abstracted from *Automatic Sprinklers for Fire Protection Service* and presented in Table 6.2. These performance tests consist of: The Load and Strength of the Heat Responsive Element Test; The Leakage, Hydrostatic Strength, and Leak Point Test; The 30-Day Leakage Test; The Water Hammer Test; The Operating Temperature (Bath) Test; The Operation-Air Oven Test; The Pressure Operation Test; The Cold Soldering Test; The High Temperature Test;

*From *Automatic Sprinklers for Fire-Protection Service*, Underwriters Laboratories.

The Assembly Load Test; The Strength of Frame Test; The Calibration Test (which determines the average discharge coefficient K as to the suitability of the discharge coefficient relative to the nozzle orifice size within the criteria previously shown in Table 6.1); The 10 Pan Distribution Test; The 16 Pan Distribution Test; The 100 Pan Distribution Test; The Sidewall Distribution Test; The Fire Test; The 10 Day Corrosion Test; The 30 day Corrosion Test; The Vibration Test; and the Wax Evaporation Test. The automatic sprinkler heads of the coated or sidewall types have variations of these standard tests. It should be noted The 100 Pan Distribution Test is utilized only for old style sprinkler heads. Additional special conditions may be specified by the testing laboratories when unique or specially developed sprinkler heads are being tested. For example, Merdinyan's "A Fully Approved On-Off Sprinkler" article reported the on-off sprinkler head was subjected to the special test conditions of a comparison fire test, a cycling test, a clogging test, and a temperature cycling test.

Table 6.2.
Underwriters Laboratories Performance Tests*

11. Load on Heat Responsive Element Test

11.1   The average and maximum loads exerted on the heat responsive element, and the overall load tolerance based on the design load for the assembly, shall be determined.

12. Strength of Heat Responsive Element Test

12.1   A heat responsive element in the Ordinary temperature rating shall be designed to (1) sustain a load of 15 times its maximum design load for a period of 100 hours or (2) demonstrate the ability to sustain the maximum design load when tested in accordance with paragraph 12.2.

12.2   Compliance with item 2 in paragraph 12.1 is to be determined by subjecting sample heat responsive elements to loads in excess of the design maximum which will produce failure at times up to 1000 hours.

13 Leakage, Hydrostatic Strength, and Leak Point Test

13.1   An automatic sprinkler shall not leak at or below 500 psi (3.45 MPa) when subjected to an internal hydrostatic pressure for a period of 1 minute.

13.2   An automatic sprinkler shall withstand, without rupture, an internal hydrostatic pressure of 875 psi (6.03 MPa).

14. 30 Day Leakage Test

14.1   An automatic sprinkler shall not leak when subjected to a hydrostatic pressure of 300 psi (2.07 MPa) for a period of 30 days, There shall be no distortion or other physical damage, as evidenced by a reduction in the leak point to 500 psi (3.45 MPa) or less.

15. Water Hammer Test

15.1   An automatic sprinkler shall be capable of withstanding, without leakage,

---

*From *Automatic Sprinklers for Fire-Protection Service*, Underwriters Laboratories.

Table 6.2. *(Continued)*

3000 applications of a pressure surge, increasing rapidly from 50 to 500 psi (0.34 to 3.45 MPa). There shall be no distortion or other physical damage, as evidenced by a reduction in the leak point to 500 psi, or less.

16. Operating Temperature (Bath) Test

16.1 A group of ten automatic sprinklers of the same type and marked temperature rating shall, when bath tested, operate within a range having a maximum temperature not in excess of 10°F (5.5°C) or 107 percent of the minimum temperature of the range, whichever is greater. For the purpose of this determination, the marked temperature rating is to be included as one of the ranged values, making a total of eleven values in the range. When tested, each sprinkler shall operate with a positive action.

17. Operation—Air Oven Test

17.1 An automatic sprinkler of other than the dry-type with a temperature rating of 375°F (191°C) or less shall operate with a sharp, positive action under zero gauge pressure when exposed to air within the temperature and time limits shown in Table 17.1.

17.2 The operating temperatures of each of a similarly marked group of a given model and type of automatic sprinkler shall, when air oven tested, be within a range having maximum and minimum values within 20°F (11°C) for any individual test.

18. Operation Test

18.1 An automatic sprinkler of other than the upright dry-type shall operate at service pressures of 5–175 psi (0.034–1.21 MPa). All moving parts shall release with sharp, positive action and shall be thrown clear of the sprinkler frame and deflector so as to not impair distribution.

19. Operation—Cold Soldering Test

19.1 An automatic sprinkler shall not, when discharging water at a service pressure of 100 psi (689 kPa), prevent the operation of a second ordinary-temperature rating automatic sprinkler having frame arms parallel to the first and located 6 feet (1.83 m) distant on an adjacent pipe line in the same horizontal plane.

20. High Temperature—Uncoated Test

20.1 An uncoated automatic sprinkler shall withstand, for a period of 90 days without evidence of weakness or failure, an exposure to a high-ambient temperature in accordance with Table 20.1, or 20°F (11°C) below the rated operating temperature of the samples (whichever is the lower temperature), but in any case the test temperature shall be not less than 120°F (48.9°C). Following the exposure, each sprinkler shall be tested for conformance with the requirements for leakage, paragraph 13.1. Sprinklers of other than the dry-type with a temperature rating of 375°F (191°C) or less are then subjected to the Operation—Air Oven Test, paragraph 17.1. Sprinklers of other than the dry-type with a temperature rating over 375°F (191°C) are then subjected to the Operating Temperature (Bath) Test, paragraph 16.1. Dry-type sprinklers shall be subjected to the plunge test described in paragraph 30.3.

Table 6.2. *(Continued)*

TABLE 20.1
HIGH-TEMPERATURE EXPOSURE TEST
CONDITIONS

Temperature Rating		Test Temperature	
Degrees F	Degrees C	Degrees F	Degrees C
135–140	57.2– 60.0	120	48.9
145–170	62.8– 76.7	125	51.7
175–225	79.4–107.2	175	79.4
250–300	121 –149	250	121
325–375	163 –191	300	149
400–475	204 –246	375	191
500–575	260 –302	475	246

21. High Temperature—Coated Test

21.1   A coated automatic sprinkler shall withstand, for a period of 90 days without evidence of weakness or failure, an exposure to a high-ambient temperature in accordance with Table 5.1, or 20°F (11°C) below the rated operating temperature of the samples (whichever is the lower temperature), but in any case the test temperature shall be not less than 120°F (48.9°C). Following the exposure, the coating shall not show evidence of deterioration as defined in paragraph 8.4. Sprinklers of other than the dry-type with a temperature rating of 375°F (191°C) or less are then subjected to the Operation—Air Oven Test, paragraph 17.1. Sprinklers of other than the dry-type with a temperature rating over 375°F (191°C) are then subjected to the Operating Temperature (Bath) Test, paragraph 16.1. Dry-type sprinklers shall be subjected to the plunge test described in paragraph 30.3.

22. Assembly Load Test

22.1   The load which is impressed on an automatic-sprinkler frame due to assembly of the operating parts into the frame shall be determined.

23. Strength of Frame Test

23.1   Each automatic-sprinkler frame used in the Assembly Load Test shall not show permanent distortion in excess of 0.002 inch (0.05 mm) when subjected to a test loading of two times its assembly load.

24. Calibration Test

24.1   The discharge coefficient "K" of a sprinkler shall be determined and shall conform with the requirements of Section 6.

25. 10 Pan Distribution Test

25.1   The distribution of water from a sprinkler, flowing at a rate of 15 gallons per minute (0.95 l/s) for standard- and small-orifice sprinklers and at a rate of 21 gallons per minute (1.32 l/s) for large-orifice sprinklers, shall be measured by the 10 pan method illustrated in Figure 25.1.

25.2   The water-distribution pattern from a standard sprinkler shall not exceed a 16 foot (4.88 m) diameter circular area located in a horizontal plane 4 feet (1.22 m) below the sprinkler deflector.

26. 16 Pan Distribution Test

26.1   Four small- or standard-orifice sprinklers, flowing at the rate of 15 gallons per minute (0.95 l/s) per sprinkler and tested by the 16 pan method (pans between four sprinklers), shall discharge water at an average density of not less than 0.15 gallons per minute per square foot (0.102 l/s per m²) for the 16 pans and not less than 0.12 gallons per minute

Table 6.2. *(Continued)*

per square foot (0.082 l/s per m²) for any individual pan.

27. 100 Pan Distribution Test

27.1 The distribution of water from an "old style" sprinkler shall be in a uniform, solid pattern. With a single sprinkler flowing at the rate of 15 gallons per minute (0.95 l/s), 90 percent of the water shall be discharged within a 10- by 10-foot (3.05- by 3.05-m) area.

28. Sidewall Distribution Test

28.1 The water distribution from two sidewall sprinklers flowing at the rate of 15 gallons per minute (0.95 l/s) each shall average a minimum of 0.050 gallons per minute per square foot (0.034 l/s per m²) for the floor area covered by the measuring pans, as illustrated in Figure 28.1. The minimum amount of water distributed to any individual measuring pan shall be not less than 0.030 gallons per minute (0.002 l/s).

29. Fire Test

General

29.1 An open sprinkler when tested under the conditions described below and discharging at a rate of 15 and also at 25 gallons per minute (0.95 and 1.58 l/s) per sprinkler for a standard-orifice, and at a rate of 21 and also at 35 gallons per minute (1.33 and 2.22 l/s) per sprinkler for a large-orifice, shall limit the loss in weight of the wood crib to not more than 20 percent. Following application of the water, the sprinkler discharge shall cause the ceiling temperature to be reduced to some level less than 530°F (295°C) above ambient within 5 minutes after start of

water discharge. From the time the temperature first falls below 530°F (295°C) above ambient to the end of the test, the ceiling temperature shall not exceed this amount for more than three consecutive minutes and the average temperature for this period shall not exceed 530°F (295°C) above ambient.

29.2 Small orifice and sidewall-type sprinklers, intended for use in light hazard occupancies only, are not subjected to the Fire Test.

30. 10 Day Corrosion Test

30.1 An automatic sprinkler shall withstand an exposure to salt spray, hydrogen sulphide, and sulphur dioxide—carbon dioxide atmospheres in accordance with paragraphs 31.3–31.10 for a period of 10 days each. Following the exposure, each test sample shall be operable when subjected to the Operation—Air Oven Test, and the average temperature of operation shall not show an increase in excess of 10 percent when compared to the average temperature of operation of samples not subjected to the 10 Day Corrosion Test, and no individual sample shall exceed the operation temperature limits of the Operation—Air Oven Test.

31. 30 Day Corrosion Test

General

31.1 An automatic sprinkler having a corrosion-resistant coating or plating shall withstand an exposure to salt spray, hydrogen sulphide, and sulphur dioxide—carbon dioxide atmospheres in accordance with paragraphs 31.3–31.10 for a period of 30 days. Following the exposure, each test sample shall be operable

Table 6.2. *(Continued)*

when subjected to the Operation—Air Oven Test, and the average temperature of operation shall not show an increase in excess of 10 percent when compared to the average temperature of samples not subjected to the 30 Day Corrosion Test, and no individual sample shall exceed the operation temperature limits of the Operation—Air Oven Test.

inch (0.21 mm) displacement, for a period of 120 hours. Following the exposure, a sprinkler shall not show a reduction in average leak point in excess of 20 percent, nor leak at a pressure of less than 500 psi (3.45 MPa), and shall operate normally when tested in the Operation—Air Oven Test. See paragraph 30.2 for operation test of a dry-type sprinkler.

32. Vibration Test

32.1   An automatic sprinkler shall withstand an exposure to vertical vibration at a rate of 35 cycles per second, with an amplitude of 0.04 inch (0.10 mm), 0.08

33. Wax Evaporation Test

33.1   A wax used for coating a sprinkler shall not contain sufficient volatiles to cause shrinking, hardening, cracking, or flaking of the applied coating.

The evaluation criteria for the Air Oven Test, as conducted by the Underwriters Laboratories, are shown in Figure 6.28, relative to the temperature exposure curve within the oven, and the maximum operating temperature and operating time for the various temperature ratings of the automatic sprinkler heads. It should be noted that very Extra-High and the Ultra-High rated sprinkler heads are not subjected to this test. The two temperature curves for the oven should be noted with the separate curve for the Ordinary rated sprinkler head.

Currently, the performance tests of the Underwriters Laboratories shown in Table 6.2 are undergoing continuing review; significant changes are expected in the Heat Responsive Element Test, the Distribution Tests, and the Sidewall Tests.

Factory Mutual Evaluation Procedures

Generally, the Factory Mutual System's procedures for the evaluation of automatic sprinkler heads are similar to the Underwriters Laboratories' standards. However, there are some minor differences relative to the implementation of specific test criteria.

Figure 6.29 shows the air oven utilized by Factory Mutual for conducting automatic sprinkler head tests. The air oven is specially constructed to expose the sprinkler head to an air environment in which the temperature rises following the prescribed time-temperature relationship shown in Figure 6.28. The velocity and movement of the air stream are controlled and relatively uniform in the vicinity of the sprinkler head. In Figure 6.28, the sprinkler head is

(a)

AIR OVEN–OPERATING TEMPERATURE RANGE

Temperature Ratings,		Maximum Actual Operating Temperature,		Maximum Actual Operating Time,
Degrees F	Degrees C	Degrees F	Degrees C	Min.:Sec.
135 — 170	57.2 — 76.7	290	143	6:30
175 — 225	79.4 — 107	445	229	2:30
250 — 300	121 — 149	475	246	4:00
325 — 375	163 — 191	505	263	6:30

(b)

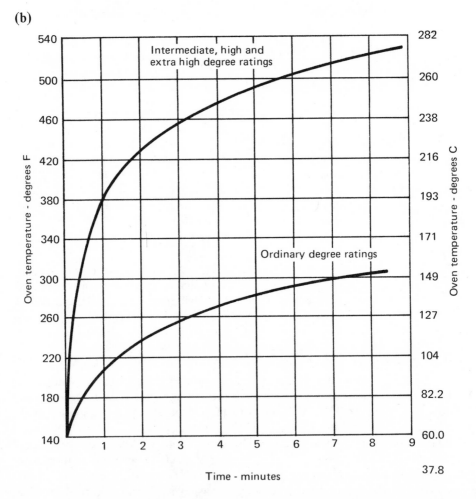

Fig. 6.28. *Time-temperature curve and test criteria for the Operation-Air Oven Test as conducted by Underwriters Laboratories.* (Underwriters Laboratories)

Fig. 6.29. Automatic sprinkler air oven. (Factory Mutual System)

mounted in the glass chamber.) The metal cylinder at the bottom of the oven is a water-cooled jacket. The burner of the gas-fired oven is located toward the bottom of the water-cooled jacket, shielded from the sprinkler head. The gauge on the right indicates the pressure on the sprinkler head while under test, while the electrical pyrometer for temperature measurement is on the left under the time clock.

The hydrostatic pressure testing of the automatic sprinkler head is required by both Factory Mutual and Underwriters Laboratories. Factory Mutual conducts what is known as a "Weep and Spurt Point Test."[14] The weep and spurt points are defined as follows:*

> The sprinkler shall not leak or weep below 500 psi hydrostatic pressure. This minimum weep pressure, producing the first visible water between the valve seat and gasket, shall be retained throughout the life of sprinklers of all ratings even when subjected to the maximum allowable installation temperature.
>
> The spurt point is the hydraulic pressure producing a fine spray from between the valve seat and gasket.

Figure 6.30 shows the test equipment utilized by Factory Mutual to hydrostatically test the automatic sprinkler head to determine the weep and spurt points. With this equipment, an air-operated piston is utilized to raise the hydrostatic pressure in the closed water pipe system. This procedure provides

*From *Approval Standard, The Mechanical Qualities of Automatic Sprinklers,* Factory Mutual System, July 15, 1970, Norwood, Mass., pp. 2–3.

Fig. 6.30. Hydrostatic test of sprinkler head for weep and spurt point. (Factory Mutual System)

for the continued testing of the sprinkler head at higher pressures until the final determination of spurt point is obtained.

The Factory Mutual's evaluation procedures include water distribution tests that are similar, and in some cases identical, to the Underwriters Laboratories evaluation procedures. Figure 6.31 shows the 10 pan distribution test apparatus for the testing of the water distribution from a single sprinkler head. This equipment consists of an electrically powered turntable collector apparatus which rotates under the single sprinkler head to measure the dispersion and density of spray discharged by the sprinkler head.

The sprinkler head must perform within certain minimum average densities of water distribution at 15 and 30 gpm flow for the standard orifice sprinkler head. The minimum average densities in distance of radius from the sprinkler head for the Factory Mutual procedures are shown in Figure 6.32.

The Factory Mutual's evaluation procedures for water distribution tests do not extend beyond a 16 pan procedure. Figure 6.33 shows the test arrangement of 16 pans for measuring the water distribution density from four sprinkler heads. This test procedure is utilized for four or six sprinkler heads.

In their evaluation tests, both the Underwriters Laboratories and the Factory Mutual System utilize the various pan distribution tests, including the 10

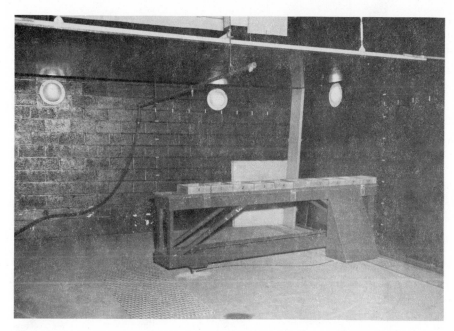

Fig. 6.31.   10 pan distribution test apparatus. (Factory Mutual System)

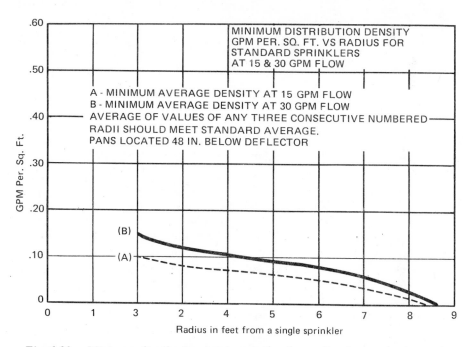

Fig. 6.32.   Minimum distribution density, gpm/sq ft vs radius for standard sprinklers at 15 and 30 gpm. (From *Approval Standard, Distribution and Fire Test Performance of Standard Sprinklers,* Factory Mutual System)

and 16 pan tests illustrated in Figures 6.31 and 6.33, to obtain performance data on the water distribution characteristics of the sprinkler head. However, the 10 pan test is the only test on the single sprinkler head, and the 16 pan test involves multiple sprinkler heads.

Fig. 6.33. 16 pan distribution test arrangement for 4 sprinkler heads. (Factory Mutual System)

In an *NFPA Quarterly* article titled "Eutectic Metals in Fire Protection," O. J. Seeds reported the fusible metal elements in some of the higher temperature rated sprinkler heads (especially the Extra High temperature rated sprinkler heads with a 360°F rating) failed to operate at temperatures above 400°F.[15] Seeds indicated this situation seems to occur when the sprinkler heads have been exposed to elevated temperatures of approximately 300°F over extended periods of time. Seeds also reported it is his belief that such an elevated temperature results in a definite physical change in the fusible eutectic element. He stated it seemed that some of the tin used in the 360 fusible metal alloy migrated into the base metal of the sprinkler head. When some of the copper from the base metal (being either bronze or brass) migrated into the lead and tin of fusible element, the result was a new fusible alloy with higher melting temperatures. In a 1966 undergraduate paper titled "A Study of the Effects of High Temperature on 360°F Solder Link Type Sprinklers," John M. Watts, Jr., attempted to experimentally create the conditions described by Seeds twelve years earlier; Watts observed the failure of the 360°F rated heads to operate at 400°F appears to correlate with the time and temperature of the previous elevated temperature exposure.[16] Watts stated the 360°F rated heads he evaluated would have an operating life expectancy of approximately 50 days when exposed to 300°F temperatures.

In a 1957 study titled *Study of 360°F Solder Type Sprinklers to Determine if Solder is Affected by Continuous Exposure to Heat,* Norman J. Thompson described the oven testing of 228 eutectic fusible metal type sprinkler heads having a 360°F rating.[17] In the oven test, 59 of the heads (25 percent) failed to operate at 400°F. Thompson also reported that a sample of sprinkler heads placed in a constant temperature oven at 300°F for one year were then tested in the air oven; 40 percent of the heads failed to operate, and 30 percent of the heads were delayed in their operation. Thompson concluded that certain physical changes do occur in the fusible eutectic metal elements of the solder of the 360°F rated heads when exposed to elevated temperatures for extended periods of time.

The exposure of sprinkler heads to elevated temperatures for extended periods of time usually results in the premature operation of the head without a fire exposure. This premature operation is known as cold flow, and is the reason for the maximum ceiling temperature ratings presented in Table 6.1 earlier in this chapter.

AUTOMATIC SPRINKLER HEAD INSTALLATION AND OPERATING VARIABLES

The elapsed time of exposure required to operate an automatic sprinkler head when the head is installed in a building and exposed to an accidental fire situation is dependent upon many variables that are related to both the physical environment around the sprinkler head and the design of the sprinkler head. Important physical environmental variables are the distance of the sprinkler head from the ceiling, the surface configuration of the ceiling relative to the heat flow characteristics under the ceiling, the height of the sprinkler heads from the floor, and most importantly, the characteristics of the fuel array, including its configuration. Sprinkler head design variables include the temperature rating, thermal lag, and the actual condition of the head at the time of the fire occurrence, especially relative to previous high temperature exposures, or any loading of the sprinkler head which might delay or invalidate its operation.

Condition of the Automatic Sprinkler Head

Proper maintenance of the automatic sprinkler head is a procedure which requires attention, especially at industrial locations where the heads are exposed to environments that may create loading. Figure 6.34 illustrates a sprinkler head loaded with a fine powder which would tend to insulate the fusible element.

As previously mentioned, the deflector on the standard sprinkler head is important since it can affect the distribution of the water being discharged. Fabin ("Modification of Sprinkler Heads to Obtain Discharge Above and Below the Deflector in Discrete Percentages"[10]) found the standard sprinkler head deflector when dropped could be damaged, and relative minor distortion could affect water discharge. Figure 6.35 shows a sprinkler head with

Fig. 6.34. Sprinkler head loaded with a fine powder, which insulates fusible element. (Gordon M. Betz, Western Electric Co., Inc.)

the deflector severely distorted; such distortion would affect the water distribution from the sprinkler head and would leave some areas uncovered by the discharge.

The loading of sprinkler heads with paint is a problem that occurs in many industrial situations. A standard upright sprinkler head loaded from paint overspray is shown in Figure 6.36. The sprinkler head shown is so overloaded with paint that its operation would be prevented in a fire situation.

The solution to the problem of paint overspray is to cover the sprinkler head with a protective covering when the area around it is being painted. Such protective covering helps to prevent excessive accumulations of paint overspray, and should be changed with every work shift. In some paint-spray situations, a coating of oil or liquid soap on the sprinkler head can be effective if applied at regular intervals during the work day. In an undergraduate paper titled "Effects

Fig. 6.35. Sprinkler head with severely damaged deflector. (Gordon M. Betz, Western Electric Co., Inc.)

Fig. 6.36. Standard upright sprinkler head loaded with paint overspray. (Gordon M. Betz, Western Electric Co. Inc.)

of Protective Materials on the Operation of Sprinkler Heads," Richard E. Rice found the plastic bag to be simple and effective since its effect resulted in relatively little lost time prior to the operation of the sprinkler head.[18] The paper cup, when utilized in Rice's study, actually helped reduce the operation time, since the cup acted as a heat canopy or collector around the sprinkler head.

In "Don't Make Automatic Sprinkler Maintenance Too Routine," Gordon M. Betz indicated the type of sprinkler maintenance needed to ensure head operation requires an understanding of the operation of the various types of automatic sprinklers.[19] Betz also suggested for a complete sprinkler head maintenance program, complete and thorough tests as well as recurring inspections should be made of representative sprinklers selected from various locations in a facility.

Location of the Automatic Sprinkler Head

As indicated earlier, the standard automatic sprinkler head is a relatively slow operating device when compared to many fire detection devices. In a 1962 undergraduate paper, Orville M. Slye, Jr., reported in his study the standard automatic sprinkler head was consistently slower in activation.[20] He stated when exposed to a fast light Class A fire, a slow deep-seated Class A fire, and a fast Class B fire, the standard automatic sprinkler head required a longer response time than a rate-of-rise temperature detector and a 165°F fixed temperature detector. Slye's study also revealed the 165°F rated sprinkler head consistently exhibited a thermal lag that varied in severity from 15°F to as high as 135°F above the rating of the head.

In a *Fire Technology* article titled "A Study of Sprinkler Sensitivity," Layard E. Campbell described the fastest operation under smooth ceiling construction as occurring when the sprinkler head deflector was located 3 inches below the ceiling.[21] As the deflector distance from the ceiling increased beyond the 3-inch point, the operating time for the sprinkler head increased. This 3-inch deflector distance also seemed to prevail when the fire occurred in the bay with a ceiling of beam or girder construction. Campbell found that when the fire occurred at an adjacent or distant bay with a ceiling of beam or girder construction, the deflector distance for fastest operation was, again, 3 inches from the ceiling, with the deflector located below the bottom level of the beams. The location with the slowest operation of the sprinkler head was above the bottom level of the beams, with a deflector distance from the ceiling of 12 inches or more. Campbell also discovered that beamed ceilings tended to increase the sprinkler head operating time relative to the time of operation for smooth ceilings. However, variations in beam spacing and flange width tended to have no significant effect on the sprinkler head operating time.

In "The High Challenge Fireball—A Trademark of the Seventies," Emile W. J. Troup described a high intensity fire test experience at the Factory Mutual research center which indicated the first sprinkler head at the ceiling operated only after the flames impinge on, or approach the ceiling.[22] Thus, in the fire test, the air temperatures measured at the ceiling adjacent to the sprinkler level usually indicated temperatures ranging between 500 and 1,000°F when the initial sprinkler head operated.

In "Effectiveness of Automatic Sprinkler Systems in Exhibition Halls," William A. Webb stated in his studies with fire test situations having a 20 lb/sq ft fuel load and fuel configuration of display booths with the sprinkler heads located 30 feet from the floor, the 160°F standard upright sprinkler head operated at 2 minutes and 56 seconds.[23] The identical fire test situation was recreated, but with the sprinklers located 50 feet above the floor; the first 160°F standard upright sprinkler operated at 5 minutes and 10 seconds. Thus, it would seem that the distance of the sprinkler head from the floor, and the type and configuration of the fuel, are important variables relative to the operating time of the automatic sprinkler head.

In "A Study of the Performance of Automatic Sprinkler Systems," O'Dogherty, Nash, and Young described the studies they conducted in England in 1967.[24] In their studies, the size of the fire area was increased approximately as the square of the height of the sprinkler heads from the floor. They noted the farther the sprinkler is from the center of the fire, the larger the fire will be when the first sprinkler operates. In addition, they indicated the sprinkler head should be mounted as close as possible to the ceiling, allowing a minimum clearance of 1 inch from the ceiling to avoid the area where the hot gases may be cooled by their proximity to the ceiling.

In a 1973 report titled "Experiments with Sprinkler Head Canopies for Fire Protection," DeMonbrun and McCormick explain how the operating times of sprinkler heads can be reduced by installing sprinkler head canopies immediately above the heads.[25] They reported in their experiments the most effective canopies were 10 or 12 inches square, with the edges designed to contain the thermal flow. They also concluded that painting the sprinkler head frames black seemed to further reduce the operating times of the sprinkler heads.

The actual sequence of the operation of a standard upright sprinkler head with the triangular strut fusible element design is shown in Figure 6.37. Note from the illustration how the discharge of water does not extend beyond the sprinkler head's deflector, resulting in the umbrella-shaped water distribution.

The importance of the location of the sprinkler head relative to its operating time has been discussed herein. The location of the sprinkler head is also important in order that it may provide an effective discharge of water into the fire area.

In a *Fire Engineering* article titled "Seattle Tests Show How to Get Better Sprinkler Protection Under Piers," Hugh M. Maguire reported standard upright sprinkler heads, as installed on the undersides of piers in Seattle, were ineffective in controlling the spread of fire across the creosoted understructures of the piers.[26] As a result of the 1969 tests described by Maguire, the Seattle Fire Department now requires that standard pendent sprinkler heads be installed in the upright position under the piers, with the sprinkler head deflectors located 5 to 8 inches below the bottom of the joists. The installations for the underside of the piers must also have the branch lines 8 feet apart, with the heads stagger-spaced on the branch lines.

The graphs in Figure 6.38 show the discharge radii from a standard upright and pendent sprinkler head with $\frac{7}{16}$-inch orifices at pressures of 7, 15, and 30 psi. The graph in Figure 6.39 indicates the gpm discharge from this same standard sprinkler head with a $\frac{7}{16}$-inch orifice and a K factor of 5.5. Factory Mutual has developed a chart that gives the general flow for sprinkler heads of the standard ($\frac{1}{2}$ inch), large orifice ($\frac{17}{32}$ inch), and small orifice ($\frac{3}{8}$ inch) sizes. The

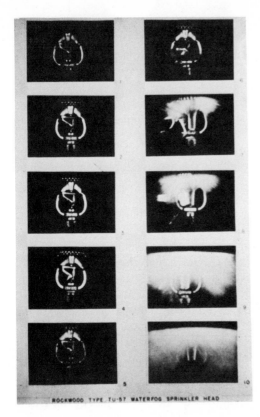

ROCKWOOD TYPE TU-57 WATERFOG SPRINKLER HEAD

Fig. 6.37. Operation of standard upright sprinkler head. (Rockwood Sprinkler Corp.)

Factory Mutual chart is shown herein as Table 6.3, and is useful when specific sprinkler head flow characteristics (as shown in Figure 6.39) are not available. Upon examination of Figures 6.37 and 6.38, it can readily be seen why standard upright sprinkler heads were ineffective when installed in upright positions on the undersides of the Seattle piers.

Relative to the deflector distance for the sprinkler heads under the Seattle piers, it was recommended the pendent heads be installed in the upright position with the deflectors 5 to 8 inches below the bottom of the joists. Norman J. Thompson's book (*Fire Behavior and Sprinklers*[4]) recommends the installation of both old style and standard sprinkler heads at distances of 6 to 8 inches from the ceiling, thus taking into consideration both the speed of operation and the water distribution factors relative to extinguishing effectiveness as measured with wood crib fires.

The NFPA Committee on Automatic Sprinklers states the minimum distance of the sprinkler head deflector below both ceilings and beams is to be 1 inch. The maximum distance varies with the location of the sprinkler head in the bay or under the beam, the type of ceiling area, and the combustible or noncombustible classification of the ceiling area. The NFPA information relative to the maximum distances for sprinkler head deflectors from ceilings is summarized in Table 6.4.

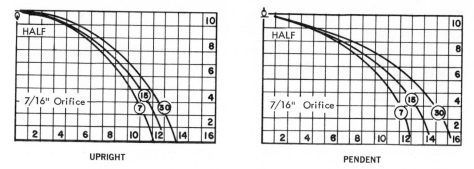

Fig. 6.38.   Discharge radii of standard upright and pendent sprinkler heads at 7, 15, and 30 psi. (Globe Automatic Sprinklers, Norris Industries)

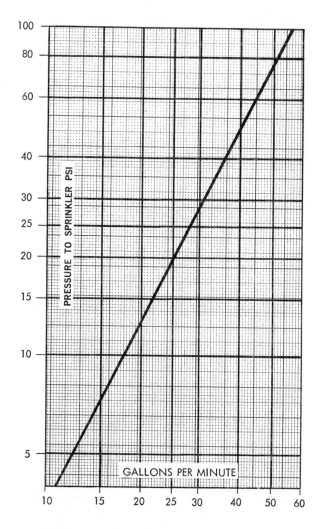

Fig. 6.39. Discharge flow curve characteristics for standard orifice upright and pendent sprinkler heads. (Globe Automatic Sprinklers, Norris Industries)

Table 6.3. General Discharge Flow for Standard, Large, and Small Orifice Sprinkler Heads*

PRESSURE (psi)	FLOW (gpm) $\frac{1}{2}$ in.	$\frac{17}{32}$ in.	$\frac{3}{8}$ in.	PRESSURE (psi)	FLOW (gpm) $\frac{1}{2}$ in.	$\frac{17}{32}$ in.	$\frac{3}{8}$ in.	PRESSURE (psi)	FLOW (gpm) $\frac{1}{2}$ in.	$\frac{17}{32}$ in.	$\frac{3}{8}$ in.
3	9.9	13.9	4.8	36	34.2	48.0	16.8	69	47.3	66.5	23.3
4	11.4	16.0	5.6	37	34.7	48.7	17.0	70	47.7	66.9	23.4
5	12.7	17.9	6.3	38	35.1	49.3	17.3	71	48.0	67.4	23.6
6	14.0	19.6	6.9	39	35.6	50.0	17.5	72	48.3	67.9	23.8
7	15.1	21.2	7.4	40	36.0	50.6	17.7	73	48.7	68.4	23.9
8	16.1	22.6	7.8	41	36.5	51.2	17.9	74	49.0	68.8	24.1
9	17.1	24.0	8.4	42	37.0	51.8	18.1	75	49.4	69.3	24.3
10	18.0	25.3	8.8	43	37.4	52.5	18.4	76	49.7	69.7	24.4
11	18.9	26.5	9.3	44	37.8	53.1	18.6	77	50.0	70.2	24.6
12	19.7	27.7	9.7	45	38.2	53.7	18.8	78	50.3	70.7	24.7
13	20.5	28.8	10.1	46	38.7	54.3	19.0	79	50.7	71.1	24.9
14	21.3	29.9	10.5	47	39.0	54.8	19.2	80	51.0	71.6	25.0
15	22.1	31.0	10.8	48	39.5	55.4	19.4	81	51.3	72.0	25.2
16	22.8	32.0	11.2	49	39.9	56.0	19.6	82	51.6	72.4	25.4
17	23.5	33.0	11.5	50	40.3	56.6	19.8	83	51.9	72.9	25.5
18	24.2	33.9	11.9	51	40.7	57.1	20.0	84	52.2	73.3	25.7
19	24.8	34.9	12.2	52	41.1	57.7	20.2	85	52.5	73.7	25.8
20	25.5	35.8	12.5	53	41.5	58.2	20.4	86	52.9	74.2	26.0
21	26.1	36.7	12.8	54	41.9	58.8	20.6	87	53.2	74.6	26.1
22	26.7	37.5	13.1	55	42.3	59.3	20.8	88	53.5	75.0	26.3
23	27.3	38.4	13.4	56	42.6	59.9	21.0	89	53.8	75.5	26.4
24	27.9	39.2	13.7	57	43.0	60.4	21.1	90	54.1	75.9	26.6
25	28.5	40.0	14.0	58	43.4	60.9	21.3	91	54.4	76.3	26.7
26	29.1	40.8	14.3	59	43.8	61.4	21.5	92	54.7	76.7	26.8
27	29.6	41.6	14.6	60	44.2	62.0	21.7	93	55.0	77.1	27.0
28	30.2	42.3	14.8	61	44.5	62.5	21.9	94	55.3	77.6	27.1
29	30.7	43.1	15.1	62	44.9	63.0	22.0	95	55.5	78.0	27.3
30	31.2	43.8	15.3	63	45.2	63.5	22.2	96	55.8	78.4	27.4
31	31.7	44.5	15.6	64	45.6	64.0	22.4	97	56.1	78.8	27.6
32	32.2	45.3	15.8	65	45.9	64.5	22.6	98	56.4	79.2	27.7
33	32.7	46.0	16.1	66	46.3	65.0	22.7	99	56.7	79.6	27.9
34	33.2	46.6	16.3	67	46.6	65.5	22.9	100	57.0	80.0	28.0
35	33.7	47.3	16.6	68	47.0	66.0	23.1				

The Temperature Rating of the Automatic Sprinkler Head

In "Temperature Rating and Sprinkler Performance," J. M. Rhodes reported in his studies, the use of sprinkler heads in the higher ratings (above the ordinary temperature rating from 135 to 170°F) resulted in fewer sprinkler heads operating during the fire, thus reducing water demands.[27] Factory Mutual has recommended sprinkler heads with an intermediate temperature rating (175 to 225°F), a high temperature rating (250 to 300°F), or an extra high temperature rating (325 to 375°F) for use in various industrial situations. Based on experience gained from a 1962 fire at their facility in Paducah, Kentucky, in which 2,341 sprinkler heads operated, the Atomic Energy Commission re-

*From "Discharge from Sprinklers," *Factory Mutual Engineering Loss Prevention Guide*, Factory Mutual System.

Table 6.4. Maximum Distance of Sprinkler Head
Deflectors Below Ceilings*

MAXIMUM DISTANCE OF DEFLECTORS BELOW CEILINGS
(Inches)

TYPE OF CONSTRUCTION	IN BAYS[1]		UNDER BEAMS	
	COMB.	NONCOMB.	COMB.	NONCOMB.
Smooth Ceiling	10	12	14	16
Beam and Girder	16	16	20	20
Panel up to 300 sq ft	18	18	22	22
Bar Joists	10	12	—	—
Open Wood Joists—center 3 ft or less (see para. 4-3.4 for centers over 3 ft)	6	—	—	—

Minimum below ceiling is 1 in.

Minimum below beams 1 in., maximum 4 in. Do not exceed maximum below ceiling.

[1]Not more than 4 in. below beams where lines run across beams.

commended 286°F temperature rated heads for under-ceiling use in similar situations.[28]

O'Dogherty, Nash, and Young ("A Study of the Performance of Automatic Sprinkler Systems,"[24]) reported the use of higher temperature rated sprinkler heads will result in a fewer number of heads opening. However, they also indicated the advantage of reduced water demand must be considered relative to the large fire which invariably occurs prior to the initial sprinkler head operation. They stated the use of intermediate temperature rated sprinkler heads, as opposed to ordinary temperature rated heads, may result in an approximate 50 percent decrease in the number of sprinklers operating. However, according to the data they obtained from experimental crib fires, the size of the fire at the time of initial sprinkler head operation will also have been increased by approximately 45 percent.

In 1964 Underwriters Laboratories conducted a series of studies for the National Board of Fire Underwriters relative to the utilization of higher temperature rated sprinkler heads.[29] Experiments were made using sprinkler heads with 165, 212, 286, and 360°F temperature ratings in controlled experimental crib fire situations. In the experiments, a gasoline nozzle supplied fuel to the crib in a manner similar to the sprinkler approval fire test conditions previously mentioned. In addition, an evaluation was made of the procedure for using Ordinary temperature rated sprinkler heads with an increased initial water

---

*From NFPA 13, *Standard for the Installation of Sprinkler Systems.*

pressure on the sprinkler heads. The results of these experiments indicated the following conclusions and recommendations:*

It seems apparent that the use of intermediate and higher temperature rated sprinklers in lieu of ordinary ratings is justified for many conditions; particularly under extremes where high hazard, fast-developing fires may be encountered or where the opening of great numbers of sprinklers will overtax weak water supplies. The use of these sprinklers is likely to lessen water damage accompanying a fire, increase the over-all efficiency of water use, and decrease the total water demand. At the same time, some delay in initial sprinkler operation must be anticipated together with the potential of somewhat lessened fire protection of buildings and contents.

It is also apparent that a reduction in the number of sprinklers operating in a given fire situation can be obtained by increasing the initial static pressures on the sprinkler system while using ordinary-degree (165°F) sprinklers. This approach causes a slight increase in water demand indicating a potential for greater water damage, and also indicates an increase in the over-all efficiency of water use. No increased delay in the operating time of the first sprinkler need be anticipated, and somewhat increased fire protection of buildings and contents may be expected.

SUMMARY

The automatic sprinkler head is approximately 100 years old, and a great amount of research effort and activity has been applied toward its improvement and refinement. Relatively recent innovations are the on-off sprinkler head, the quick-response sprinkler head, and the extended coverage sprinkler head: the on-off sprinkler head was approved approximately four years ago, and the quick-response and extended coverage sprinkler heads are only several years old.

The automatic sprinkler system, as originally developed, was intended and primarily designed for the fire protection of buildings, the fire protection of the contents of buildings, and the fire protection of property. Based on the excellent performance record of sprinkler systems and the automatic sprinkler head, current developments and research improvements are designed to further improve the sprinkler head capabilities as a device for assuring life safety. As a result of the current developments and research improvements, significant changes in the present concepts of sprinkler systems and automatic sprinkler heads should develop in the immediate future. These changes should evolve with a more efficient sprinkler head which is constructed of new and non-metallic materials, and which is capable of reduced response time and greater water discharge coverage.

Critical factors that appear to be related to the operation time of a sprinkler head may be: the design of the head (especially the activating element), the thermal lag features of the sprinkler head, and the temperature rating of the

*From *Study of Factors Influencing the Use of High Temperature Sprinklers*, National Board of Fire Underwriters, American Insurance Association, New York, July 29, 1964, p. 8.

head. Variables that affect the operation of the sprinkler head relative to its location include: the height of the sprinkler head from the floor, the distance of the deflector from the ceiling, and the location of the sprinkler head relative to the center or axis of the fire area. Finally, the thermal characteristics of the fire situation will affect the operating time of the sprinkler head. The greater the velocity of the heated air flow of the fire plume, the shorter the sprinkler head response time. Also, the greater the humidity of the heated air flowing around the sprinkler head, the shorter the response time of the sprinkler head.

SI Units

The following conversion factors are given as a convenience in converting to SI units the English units used in this chapter:

1 square foot	=	0.0929 m²
1 foot	=	0.305 m
1 inch	=	25.400 mm
1 pound (mass)	=	0.454 kg
1 psi	=	6.895 kPa
$\frac{5}{9}$ (°F−32)	=	°C
1 gallon	=	3.785 litres
1 gpm	=	3.785 litres/minute

ACTIVITIES

1. The development of the automatic sprinkler head involved many improvements in its design and efficiency. In outline form, trace the history of the development of the automatic sprinkler head. Use as the basis for your outline your reasons why improvements had to be made.

2. Relative to the occupancies protected in your area or community, list the occupancies and types of hazards you believe would have improved protection with the installation of on-off sprinkler heads. Compare and discuss your list with others. Then revise your list to include any constructive comments.

3. Based on your knowledge of present-day automatic sprinkler heads, write a brief explanation describing one major design improvement you would recommend. Explain why.

4. Describe the situations where the use of sprinkler heads with High or Extra-High temperature ratings would be advantageous. Indicate any situations where the use of such sprinkler heads would not be considered. Discuss both types of situations.

5. Consider the installation of standard sprinkler heads relative to their water discharge characteristics as compared with the old style sprinkler heads. In a written statement, explain what you consider to be the principal advantages of standard sprinkler heads over old style heads.

6. List at least six variables that may affect the operational response time of an automatic sprinkler head when the head is installed on a system in a building.

7. The problem of paint overspray on sprinkler heads occurs in many industrial situations. Explain some of the solutions to this problem.
8. In a written statement, summarize what you have learned in this chapter concerning the installation of particular types of sprinkler heads on the underside of piers.

SUGGESTED READINGS

Merdinyan, Philip H., "A Fully Approved On-Off Sprinkler," *Fire Journal*, Vol. 67, No. 1, Jan. 1973, pp. 10–15, 19.

Rhodes, J. M., "Temperature Rating and Sprinkler Performance," *NFPA Quarterly*, Vol. 57, July 1963, pp. 25–29.

Thompson, Norman J., "New Developments in Upright Sprinklers," *NFPA Quarterly*, Vol. 46, No. 1, July 1952, pp. 5–18.

BIBLIOGRAPHY

[1]Dana, Gorham, *Automatic Sprinkler Protection*, 2nd ed., Wiley, New York, 1919.

[2]Thompson, Norman J., "New Developments in Upright Sprinklers," *NFPA Quarterly*, Vol. 46, No. 1, July 1952, pp. 5–18.

[3]_____ "Proving Spray Sprinkler Efficiency," *NFPA Quarterly*, Vol. 47, No. 3, Jan. 1954, pp. 205–212.

[4]_____ *Fire Behavior and Sprinklers*, National Fire Protection Association, Boston, 1964.

[5]Merdinyan, Philip H., "A Fully Approved On-Off Sprinkler," *Fire Journal* Vol. 67, No. 1, Jan. 1973, pp. 10–15, 19.

[6]Luley, John E., *An Attempt to Develop an Automatic, Individually-operated Reusable and Self-resetting Sprinkler Head*, Undergraduate paper, 1969, Fire Protection Curriculum, University of Maryland, College Park.

[7]Shea, Raymond E., "Revolutionary New Sprinkler," *Fire Journal*, Vol. 48, No. 6, Nov. 1969, p. 94.

[8]Medlock, L. E., "The Potential of Actuators in Fire Engineering," *Fire*, May 1974, pp. 619–620.

[9]Laiming, Boris, "New Sprinklers—Russian Style," *Fire Technology*, Vol. 2, No. 2, May 1966, pp. 164–168.

[10]Fabin, Frank J., "Modification of Sprinkler Heads to Obtain Discharge Above and Below the Deflector in Discrete Percentages," Undergraduate paper, 1970, Fire Protection Curriculum, University of Maryland, College Park.

[11]Peterson, Marshall E., "Automatic Sprinkler Protection—Further Developments," *The Building Official and Code Administrator*, Vol. 5, No. 12, Dec. 1971, pp. 5–8, 31.

[12]*Standard for Automatic Sprinklers, (Section III)*, Subject 199, Underwriters Laboratories, Chicago, 1919.

[13]*Automatic Sprinklers for Fire-Protection Service,* U. L. 199, 5th ed., Underwriters Laboratories, Chicago, Ill., May 15, 1974.

[14]*Approval Standard, The Mechanical Qualities of Automatic Sprinklers,* Factory Mutual System, July 15, 1970, Norwood, Mass., pp. 2–3.

[15]Seeds, O. J., "Eutectic Metals in Fire Protection," *NFPA Quarterly,* Vol. 48, No. 2, Oct. 1954, pp. 104–111.

[16]Watts, John M., Jr., "A Study of the Effects of Sustained High Temperature on 360°F Solder Link Type Sprinklers," Undergraduate paper, 1966, Fire Protection Curriculum, University of Maryland, College Park.

[17]Thompson, Norman J., *Study of 360°F Solder Type Sprinklers to Determine if Solder is Affected by Continuous Exposure to Heat,* Factory Mutual System, Norwood, Mass., 1957.

[18]Rice, Richard E., "Effects of Protective Materials on the Operation of Sprinkler Heads," Undergraduate paper, 1966, Fire Protection Curriculum, University of Maryland, College Park.

[19]Betz, Gordon M., "Don't Make Automatic Sprinkler Maintenance Too Routine," *Plant Engineering,* July 1962, pp. 114–116.

[20]Slye, Orville M., Jr., "A Study to Determine the Effect of Fuel Load, Fuel Arrangement, and Fuel Type on Operation Time of Fire Detection Devices," Undergraduate Paper, 1962, Fire Protection Curriculum, University of Maryland, College Park.

[21]Campbell, Layard E., "A Study of Sprinkler Sensitivity," *Fire Technology,* Vol. 5, No. 2, May 1969, pp. 93–99.

[22]Troup, Emile W. J., "The High Challenge Fireball—A Trademark of the Seventies," *Fire Technology,* Vol. 6, No. 3, Aug. 1970, pp. 211–223.

[23]Webb, William A., "Effectiveness of Automatic Sprinkler Systems in Exhibition Halls," *Fire Technology,* Vol. 4, No. 2, May 1968, pp. 115–125.

[24]O'Dogherty, M. J., Nash, P., and Young, R. A., "A Study of the Performance of Automatic Sprinkler Systems," Fire Research Technical Paper No. 17, 1967, Department of Environment, Building Research Establishment, Fire Research Station, Borehamwood, Herts., England.

[25]DeMonbrun, J. R. and McCormick, J. W., "Experiments with Sprinkler Head Canopies for Fire Protection," J-JA-96, Union Carbide Oak Ridge Y-12 Plant, Oak Ridge, Tenn., July 2, 1973.

[26]Maguire, Hugh M., "Seattle Tests Show How to Get Better Sprinkler Protection Under Piers," *Fire Engineering,* Vol. 122, No 7, July 1969, pp. 48–49.

[27]Rhodes, J. M., "Temperature Rating and Sprinkler Performance," *NFPA Quarterly,* Vol. 57, No. 1, July 1963, pp. 25–29.

[28]"Experience Gained by Paducah Fire Indicates Need for Caution in Selecting Sprinkler Heads Ratings," *Serious Accidents Bulletin,* United States Atomic Energy Commission, Washington, D.C., #215, Jan. 17, 1964.

[29]*Study of Factors Influencing the Use of High Temperature Sprinklers,* National Board of Fire Underwriters, American Insurance Association, New York, July 29, 1964.

Chapter **7**

The Wet Pipe
Automatic Sprinkler System

CHARACTERISTICS OF THE WET PIPE AUTOMATIC SPRINKLER SYSTEM

The wet pipe automatic sprinkler system is the oldest and most reliable type of automatic sprinkler system. The automatic sprinkler system installed by Henry Parmelee in his New Haven piano factory in 1875 contained an alarm valve that provided for the operation of a bell or whistle when one or more of the sprinkler heads operated. (See Chapter 3, "Introduction to Automatic Sprinkler Systems.") Thus, Parmelee can be credited with two major accomplishments: (1) he developed the first practical and effective automatic sprinkler head, and (2) he perfected the first effective wet pipe automatic sprinkler system.

Because of its relatively simple design, the wet pipe sprinkler system is the most effective and efficient type of sprinkler system available. The wet pipe sprinkler system consists of a system of water-filled pipes with standard upright, pendent, or sidewall sprinkler heads installed on the pipes. The distribution system of pipes with the sprinkler heads on the branch lines is connected to a suitable water supply of adequate capacity and pressure. In the majority of cases, the water supply is from either a public or private water supply main.

The wet pipe sprinkler system is activated by operation of the automatic sprinkler head; this activation allows water to be discharged into the fire area. In turn, the flow of water from the sprinkler head activates the alarm check valve on the main supply riser to the system. The alarm check valve then initiates the activation of water or electrically operated alarm devices that may be automatically arranged to summon the fire department.

218

Fig. 7.1. Exterior equipment for a wet pipe sprinkler system. (Glen Echo Volunteer Fire Department, Glen Echo, Maryland)

The wet pipe sprinkler system is usually equipped with a fire department connection as a secondary source of water supply. The water control valves should be of the indicating type, should be inspected regularly, and should preferably be kept locked. (Refer back to Figures 2.6 and 2.7 of Chapter 2, "Fire Department Procedures for Automatic Sprinkler Systems," for illustrations showing water control valves: the O.S. & Y., the P.I.V., and the P.I.V.A.)

Some wet pipe sprinkler systems are equipped with a water flow indicator installed on the main riser in place of an alarm check valve. The water flow indicator is a vane-type of device which is also often used on wet pipe sprinkler systems with an alarm check valve on the main riser, and supplemental water flow indicators on the supply mains at each floor. These should be installed on every floor on multistoried buildings so that fire fighting personnel can determine the area of sprinkler system operation and fire involvement immediately upon arrival at the building. (Figure 5.6 in Chapter 5, "Basic Design of Automatic Sprinkler Systems," illustrates the plastic sprinkler piping connection to the standpipe and also shows the water flow indicator.) Design differences and similarities in alarm check valves and water flow indicators will be discussed later in this chapter. Figure 7.1 illustrates the typical features of a wet pipe automatic sprinkler system located on the exterior of the building; a horizontal P.I.V. (post indicator valve), the fire department connection, and the water motor gong.

The total operational features of a wet pipe automatic sprinkler system, with the sprinkler pipes charged with water from a public water main, are illustrated in Figure 7.2. Note that in Figure 7.2 the water control valve is of the O.S. & Y. indicating type, and is located immediately below the alarm check valve on the riser. The alarm check valve is located below the fire department connection to prevent the water from being distributed into the water supply. The sprinkler heads located on the branch lines would appear to be standard upright sprinkler heads since they are located on the top of the sprinkler branch line piping. Figure 7.3 illustrates the total concept of a wet pipe automatic sprinkler system.

Performance Record of the Wet Pipe Automatic Sprinkler System

As discussed at the beginning of this chapter, the wet pipe sprinkler system is the most effective and efficient sprinkler system available primarily because of its relatively simple design. The NFPA has maintained impressive and continuous records of the performance of automatic sprinkler systems since 1925. In 1970 these records indicated an overall efficiency record of 96.2 percent satisfactory performance for all types of automatic sprinkler systems.[1] However, the most generally utilized type of automatic sprinkler system is the wet pipe system.

Typically, the wet pipe sprinkler system will extinguish or control a greater percentage of fire occurrence with fewer sprinkler heads operating, when compared with the performance record of the dry pipe sprinkler system. There are various reasons for the improved performance of the wet pipe system, including the following: (1) increased maintenance is required on dry pipe systems, (2) dry pipe valve operation requires more time, and (3) maintenance of the quick-opening-devices on the dry pipe system, since warehouses and many other unheated industrial occupancies utilize this type of system. Thus, the principal reason for fewer heads operating on the wet pipe system is the fact operation of a sprinkler head on a wet pipe sprinkler system results in water being immediately discharged, since the water is at the sprinkler head at all times. Consequently, the elimination in the wet pipe system of the transport time required for the water to travel from the dry pipe valve to the sprinkler head and the dry pipe valve operation time.

The NFPA's "Automatic Sprinkler Performance Tables, 1970 Edition," reports that 89.4 percent of all the fires were controlled by 10 or less sprinkler heads on wet pipe sprinkler systems, while only 68.5 percent of all the fires were controlled by 10 or fewer sprinkler heads on the dry pipe sprinkler systems. A comparison of the number of sprinkler heads opened by fire for the wet and dry pipe systems is shown in Figure 7.4 for fires activating up to a maximum of 50 sprinkler heads.

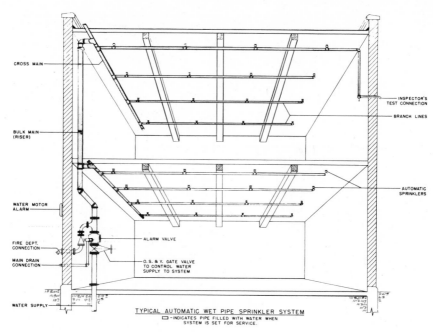

Fig. 7.2. Wet pipe automatic sprinkler system. (Grinnell Fire Protection Systems Company, Inc.)

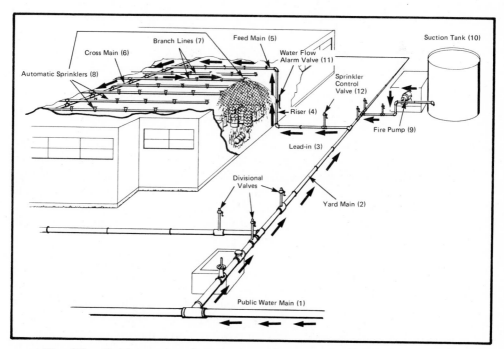

Fig. 7.3. Total concept of the wet pipe automatic sprinkler system. (Factory Mutual System)

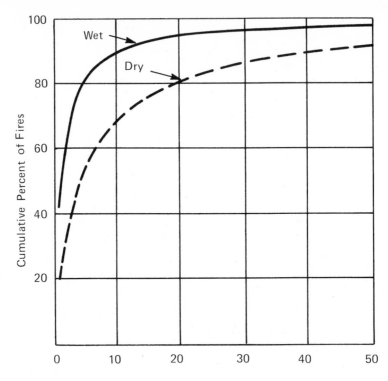

Fig. 7.4. Number of automatic sprinkler heads opened by fire relative to wet pipe and dry pipe sprinkler systems, up to a total of 50 heads, 1925–1969. (From "Automatic Sprinkler Performance Tables, 1970 Edition," NFPA)

The NFPA tabulation for the cumulative number of automatic sprinkler heads operating to control fires for the wet pipe, dry pipe, and all types of sprinkler systems is shown in Table 7.1. Note from Table 7.1 that 37.4 percent of all fires were controlled by the operation of one sprinkler head in sprinklered buildings. However, 42.6 percent of the fires in buildings equipped with wet pipe sprinkler systems were extinguished with one sprinkler head, while only 20.1 percent of the fires with dry pipe systems were extinguished with a single automatic sprinkler head.

Apparently, the principal factor responsible for the greater number of sprinkler heads opening on dry pipe systems involves the longer response time for the dry pipe valve to operate and for the water to travel to the head. During this response time, the fire transmits heat which opens additional sprinkler heads as the fire plume propagates across the ceiling areas.

Only recently has the NFPA begun to analyze data for the various types of sprinkler systems relative to the occupancy or fire occurrence where sprinkler

Table 7.1. Cumulative Number of Sprinkler Heads Operating for Wet Pipe, Dry Pipe, and all Types of Sprinkler Systems, 1925–1969*

NUMBER OF AUTOMATIC SPRINKLERS OPERATING	WET SYSTEMS PER CENT	DRY SYSTEMS PER CENT	UNKNOWN SYSTEMS PER CENT	TOTAL NO. OF FIRES	PERCENT OF TOTAL NO. OF FIRES
1	42.6	20.1	33.1	29,733	37.4
2 or fewer	61.0	32.7	50.0	43,396	54.6
3 or fewer	70.2	41.5	59.8	50,769	63.8
4 or fewer	76.2	48.7	66.7	55,795	70.1
5 or fewer	80.2	53.7	70.9	59,156	73.4
6 or fewer	83.2	57.8	75.0	61,814	77.7
7 or fewer	85.2	61.3	77.7	63,724	80.1
8 or fewer	87.0	64.2	80.3	65,348	82.2
9 or fewer	88.3	66.4	82.0	66,571	83.7
10 or fewer	89.4	68.5	83.5	67,629	85.0
11 or fewer	90.4	70.3	84.5	68,533	86.2
12 or fewer	91.2	72.4	86.1	69,464	87.3
13 or fewer	91.7	73.8	87.0	69,990	88.0
14 or fewer	92.6	75.3	88.0	70,788	89.0
15 or fewer	93.1	76.2	89.9	71,313	89.7
20 or fewer	95.0	81.0	91.5	73,347	92.2
25 or fewer	96.0	84.3	92.9	74,464	93.6
30 or fewer	96.9	86.7	94.2	75,411	94.8
35 or fewer	97.3	88.6	95.0	75,976	95.5
40 or fewer	97.7	90.0	95.8	76,472	96.2
50 or fewer	98.1	91.9	96.7	77,079	96.9
75 or fewer	98.9	94.7	98.0	77,995	98.1
100 or fewer	99.4	96.3	98.5	78,533	98.7
200 or fewer	99.8	99.7	99.9	79,384	99.8
All fires	100.0	100.0	100.0	79,544	100.0
No data or no water				1,881	
Total Number of Fires	54,158	13,217	12,169	81,425	

performance was satisfactory. Satisfactory sprinkler performance is considered to be obtained when the system controls the fire and provides notification of the fire occurrence. Analysis of the data presented in Figure 7.4 shows that an average of 8 wet pipe sprinkler heads opened during each fire in all 16 occupancies, while an average of 23 dry pipe sprinkler heads opened in each fire for all the occupanices. Figure 7.5 shows the average number of wet pipe sprinkler heads that opened in the 16 specific occupancies during the five years from 1965 to 1969.

The highest number of sprinkler heads were opened for fires in industrial occupancies. The rubber industry required the activation of 16 sprinkler heads, the largest average number per fire; the woodworking occupancies activated

*From "Automatic Sprinkler Performance Tables, 1970 Edition," NFPA.

an average of 15 sprinkler heads per fire; and the chemical industry required an average of 11 sprinkler heads per fire. The office and residential occupancies, which are classified as Light Hazard Occupancies according to the *NFPA Sprinkler System Standard*, required an average of three sprinkler heads per fire.[2] This average represented the least number of heads opened per fire from wet pipe sprinkler systems in the five years from 1965 to 1969.

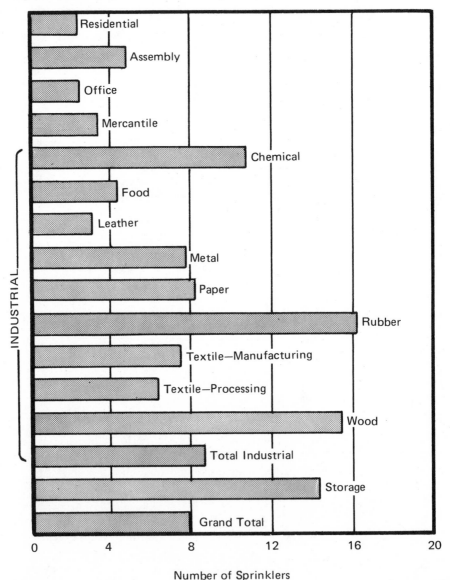

Fig. 7.5. Average number of sprinkler heads opened per fire on wet pipe sprinkler systems, by various occupancies, 1965–1969. (From "Automatic Sprinkler Performance Tables, 1970 Edition," NFPA)

WATERFLOW DEVICES FOR WET PIPE SPRINKLER SYSTEMS

The wet pipe automatic sprinkler system requires the use of a device that indicates when one or more of the automatic sprinkler heads have activated and are discharging water into the fire area. The device which is installed most often is the alarm check valve, often referred to as an alarm valve or a wet pipe valve. The alarm check valve serves both as an alarm valve that indicates when a sprinkler head has operated, and as a check valve that prevents back flow from a sprinkler system. Back flow might be a possible source of contamination when the system is connected to a potable water supply, and may reduce the water pressure when the system has a fire department connection. The fire department connection should always be attached to the riser above the alarm check valve and the water control valve, as shown previously in Figure 7.2. However, should the riser be connected to a manifold supplying other systems, then the fire department connection may be attached to the water manifold piping.

The Alarm Check Valve

The alarm check valves for wet pipe automatic sprinkler systems are approved and evaluated by both the Factory Mutual System and Underwriters Laboratories.[3] There are two principal types of alarm check valves relative to their design and construction; although identical in function, they are identified by the design of the alarm valve clapper arrangement. The two types of alarm check valves are: (1) the pilot valve type, and (2) the divided-seat ring or grooved-seat type. The alarm check valve is essentially a free-swinging clapper valve which is automatically self-resetting and which requires very little maintenance. Thus, it is seldom necessary to manually open an alarm check valve.

Pilot Valve Type of Alarm Check Valve

The pilot valve type of alarm check valve may have a metal-to-metal water seal or a rubber-faced clapper, with a single large clapper hinged at one side. The opposite side of the large clapper has the pilot valve attached to the extreme end of the clapper. This arrangement, a single large clapper valve with the pilot valve, is shown in Figure 7.6.

The pilot valve is arranged with a spring-loaded piston or other mechanism so that the main valve clapper may move a slight distance off its seat without moving the pilot valve off its seat. This arrangement is designed to allow slight surges in water flow to move the main clapper valve, and not cause water flow through the pilot valve to the alarm device. Given the operation of a single sprinkler head, which would create a flow of approximately 21.7 gpm at 15 psi and 10 gpm at 7 psi, the main clapper valve would move off its seat to allow full water flow through the riser, thus moving the pilot valve with it. Water would pass through the pilot valve opening to the retarding chamber, and then to the alarm devices. The alarm devices could be water motor gongs or electric

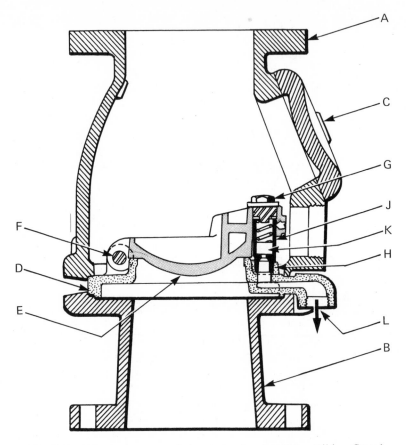

Fig. 7.6.   Pilot valve type of alarm check valve. (The Viking Corp.)

pressure switches that activate electric lights or bells, or automatically transmit the signal to the fire department. The required operational flow range of an alarm valve varies from a minimum flow of 6 gpm to a maximum flow of 24 gpm, depending on the static pressure on the system. As shown in Figure 7.6, when the sprinkler head on the pilot valve type of alarm check valve operates, the flow of water raises the main clapper (E) from its seat (D) which lifts the pilot valve (K) from its seat (H) and allows water to flow through the alarm outlet (L) to the appropriate alarm devices.

Divided-Seat Ring Type of Alarm Check Valve

The divided-seat ring type of alarm check valve refers to a different design and arrangement of the water outlet and the passage for the water to flow out of the alarm check valve, which still consists of a single main clapper hinged at one side of the valve. However, the opening to the alarm device is located at the far side of the seat from the clapper and consists of a number of small holes in a groove in the seat. The clapper always has a rubber or neoprene seat

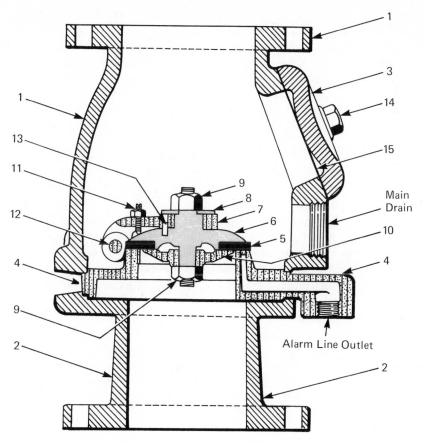

Fig. 7.7. Divided seat ring type of alarm check valve. (The Viking Corp.)

on it, and is a differential type of valve since the alarm check valve side of the clapper is slightly larger than the water supply main side. This ratio in size provides the positive check valve function, and provides more resistance to water surges and minor fluctuations in flow to move the main clapper valve and allow water passage to the alarm devices. The divided-seat ring type of alarm check valve is often found where the valve may be subjected to water surges of a higher pressure. Figure 7.7 illustrates the principal features of the divided-seat ring type of alarm check valve. The main clapper (6) raises upward with the flow of water into the valve housing from its seat (4) with the operation of the sprinkler head, allowing the water to flow into the divided ring openings located under the clapper valve seat gasket (5), and the water then flows from the alarm line outlet to the appropriate alarm device.

Retarding Chambers

The problem of false alarms from water surges can be serious when valves are located on private or public water systems that are close to pumping stations or other sources of severe water surges. Therefore, the alarm outlet line on most alarm check valves is connected to a retarding chamber before the line reaches the water motor gong or electrical pressure switch. Retarding chambers will usually have a capacity from $1\frac{1}{2}$ to $2\frac{1}{2}$ gallons. Figure 7.8 shows an alarm check valve retarding chamber that has a restricted intake and drain with a full $\frac{3}{4}$-inch alarm line outlet at the top.

The retarding chamber is similar to the old surge chamber found on the positive displacement piston pumps used for fire department apparatus and by industry. An industrial steam piston fire pump was previously shown in Figure 1.6 of Chapter 1, "Standpipe and Fire hose Systems." If a water surge is severe enough to allow water to pass through the divided seat or the pilot valve, the water then passes into the retarding chamber through the restricted inlet. If the flow continues and completely fills the chamber and flows out of the outlet at the top, it will then travel through the line and operate the water motor gong and the electrical pressure switch. Intermittent surges or flows into the retarding chamber are drained out of the restricted orifice chamber before the flow fills the chamber and flows out to the water motor gong or pressure switch and activates the alarm devices. Therefore, retarding chambers will be found on alarm check valves of both the divided-seat ring and the pilot valve type. The chamber is intended to prevent water surges and intermittent flows from transmitting water flow alarms.

Alarm check valves do not always require the installation of a retarding chamber since the alarm check valve may be installed on a system not subjected to pressure fluctuations and surges. Thus, a wet pipe sprinkler system that is supplied with water on a system where valves are not subject to sudden closing, and where pumps are not manually or automatically operated on the water supply main to the system, would not require a retarding chamber.

In some industrial situations where the sprinkler system is supplied from a water main with a constant pressure from a gravity tank, the alarm check valve can be installed without a retarding chamber; the alarm check valve would then be referred to as a constant pressure alarm check valve. In areas with excessively high water supply main pressure surges, the retarding chambers may be supplemented with bypass piping arrangements which allow surges to pass from the supply side of the valve to the system side of the valve without moving the main clapper of the alarm check valve off its seat.

Another supplemental feature sometimes used in areas with high pressure and severe surges is to provide an electrically operated excess pressure pump. This pump, which is connected to a bypass piping arrangement, boosts the pressure above the alarm check valve through a restricted orifice to maintain the system pressure higher than the pressure on the supply side of the clapper. However, when a sprinkler head operates, the restricted orifice of the excess pressure pump, as shown in Figure 7.9, prevents the pump from supplying enough water to prevent the alarm valve clapper from operating.

Figure 7.10 illustrates the operation of the alarm check valve with the flow of water through the valve, and the auxiliary flow through the retarding chamber to the electrical pressure switch and the water motor gong.

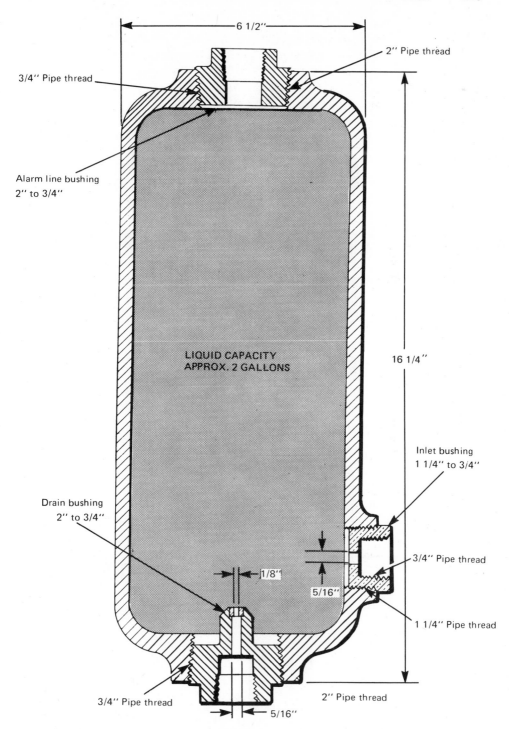

6 1/2"

2" Pipe thread

3/4" Pipe thread

Alarm line bushing
2" to 3/4"

LIQUID CAPACITY
APPROX. 2 GALLONS

16 1/4"

Inlet bushing
1 1/4" to 3/4"

Drain bushing
2" to 3/4"

3/4" Pipe thread

1/8"

5/16"

1 1/4" Pipe thread

3/4" Pipe thread

2" Pipe thread

5/16"

Fig. 7.8. Retarding chamber for alarm check valve. (The Viking Corp.)

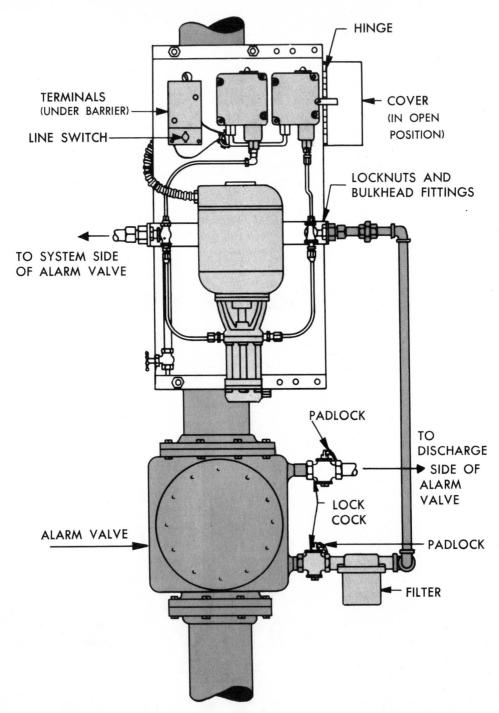

Fig. 7.9. Excess pressure pump arrangement for alarm check valve on a wet pipe sprinkler system. (A.D.T. Security Systems)

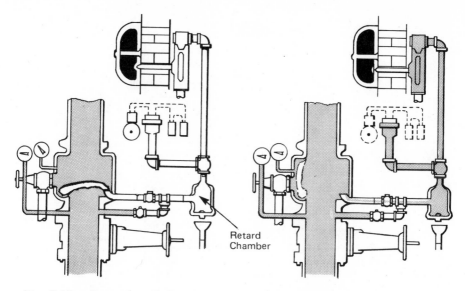

Retard
Chamber

Fig. 7.10. Operation of the alarm check valve, retarding chamber, and alarm devices with activation of sprinkler head. (Insurance Services Office)

The water motor gong, as previously indicated in Chapter 2, "Fire Department Procedures for Automatic Sprinkler Systems," is a local type of alarm device. Automatic sprinkler system water flow alarm devices should always be connected to an appropriate response agency which is a fire department or private central station.

One of the principal reasons for the highly effective sprinkler system performance record in Australia and New Zealand has been the requirement that automatic sprinkler system alarm devices automatically transmit the water flow alarm signals to the fire department.[4]

It is difficult to justify installing a water motor gong on the exterior of a building in order to signify the operation of an emergency response system which indicates there is a fire in the building. Because many sprinklered buildings are located in industrial and manufacturing areas, personnel usually are not available at night or on weekends. Also, there are many psychological and sociological variables which may inhibit an individual from responding to a signal, once alerted by its operation. Therefore, automatic sprinkler water flow alarm devices should always be connected to the responding fire department.

Figure 7.11 shows the construction features of the water motor gong, illustrating how the device obtained its name. The water motor gong shown in Figure 7.11 is a mechanical device, and the striker for the gong is propelled by the hydraulic energy of the water flowing through the water motor gong and out the drain line. There have been objections to the use of water motor gongs because the rust-colored water that drains from them can stain the sides of buildings. In many installations the water flows down the outside surface of the building.

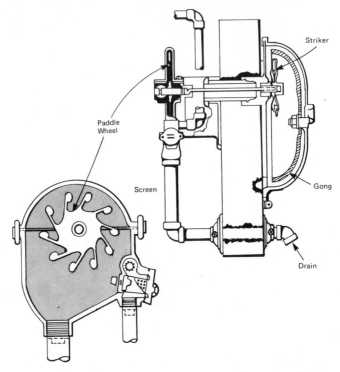

Fig. 7.11.   Construction of the water motor gong. (Insurance Services Office)

The Alarm Check Valve Accessories

The alarm check will have many of the trimmings or accessories previously discussed, and the arrangement of these accessories may vary slightly from valve to valve. However, once familiarity is established with the appearance and the location of the trimmings on one alarm check valve, it is easy to identify the devices, valves, and lines on any other alarm check valve. Every alarm check valve has a main drain valve and a 2-inch diameter drain line (the main drain was previously discussed in Chapter 4, "Water Specifications for Automatic Sprinkler Systems," relative to the main drain test as an indication of water supply availability). The alarm check valve has two water pressure gages; one on the system side of the valve, and the other on the water supply side of the valve. The alarm check valve has a control valve on the line to the alarm devices. This alarm control valve is always an indicating type of valve, and should be sealed in the open position. (If it is closed, water cannot flow to the alarm devices.) The alarm control valve is usually located in the ¾- or 1-inch line leading from the alarm valve outlet to the retarding chamber. The alarm test valve is located on a line coming from the supply side of the alarm check valve to the retarding chamber. The alarm test valve is provided to simulate the operation of the alarm check valve by allowing water to flow into the retarding chamber and the water flow alarm devices. Actual testing of the operation of the alarm check valve should always involve operation of the inspector's test valve which is located at the topmost line of sprinkler heads. A pressure switch that provides electrical alarm signals is located immediately past the retarding chamber on the outlet line to the water motor gong. The water motor gong is usually located at an elevated location above the alarm check valve, but should not be higher than a maximum of 29 feet; the pipe length to the water motor from the alarm valve should not exceed 75 feet. It is essential to allow for complete draining of the water motor gong to prevent freezing in winter conditions. The retarding chamber is provided with a drain line which is usually 1 inch in diameter. Figure 7.12 shows an alarm check valve with the trimmings consisting of bypass piping, retarding chamber, main drain valve and line, alarm control valve, alarm test valve, system and supply water pressure gages, electrical pressure switch, and the line to a water motor gong.

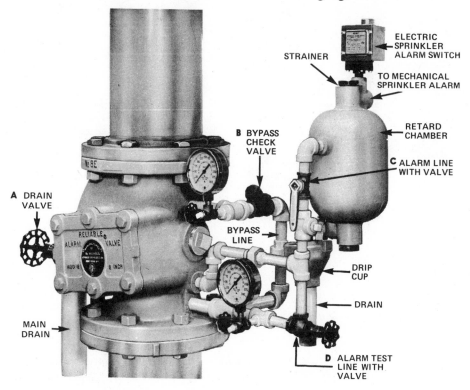

Fig. 7.12. Alarm check valve on wet pipe sprinkler system with typical trimmings. (Reliable Automatic Sprinkler Company, Inc.)

Waterflow Indicators

Waterflow indicators are electrical devices with a metallic or plastic vane that is inserted into the sprinkler piping by a hole drilled in the pipe. Waterflow indicators, when used instead of an alarm check valve on a wet pipe sprinkler system riser, must be used with a check valve below the indicator if the system is connected to a public water system or has a fire department connection into the riser. Where a fire department connection is provided to the sprinkler system riser, the waterflow indicator should be on the system side of the connection.

As previously explained, waterflow indicators are utilized on many wet pipe sprinkler systems to indicate the portion of the system which has been activated. They are available in a variety of sizes, including some for 1-inch pipe; they are also designed for vertical or horizontal installation. Wet pipe sprinklers with waterflow indicators for each floor are often installed in multistoried buildings. Figure 7.13 illustrates a waterflow indicator in both the normal and alarm position.

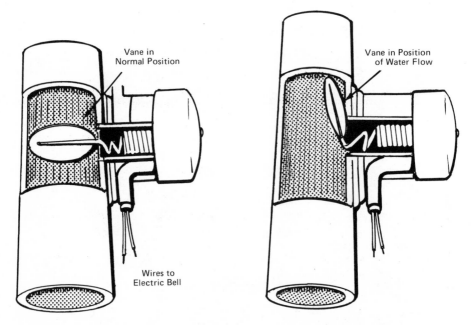

Fig. 7.13. Waterflow indicator in normal and alarm position. (Insurance Services Office)

Since the vane of the waterflow indicator is inserted across the waterway of the sprinkler pipings, the waterflow indicators are provided with a mechanically or electrically operated time-delay feature. When a sprinkler head operates on the wet pipe sprinkler system, the flow of water through the sprinkler piping causes the vane to move in the direction of the waterflow, and the stem of the vane moves until electrical contact is made. The time delay then operates and, if the vane is still in the alarm position following the operation of the time delay, the electrical alarm signal is transmitted from the waterflow indicator to the appropriate response agency or to a local electric alarm bell. Waterflow indicators have an economic advantage relative to the cost for installation of an alarm check valve; however, they may require a separate check valve on the wet pipe sprinkler system at the main riser, depending on the water supply sources and the provision of the fire department connection.

The waterflow indicator is often used to subdivide sprinkler systems into zones or areas to provide immediate and valid information as to the area of operation on the sprinkler system.

Waterflow indicators are also used to provide alarm service for partial sprinkler systems consisting of sprinkler heads in special areas such as laundry chutes, rubbish chutes, and storage rooms. Figure 7.14 shows a waterflow indicator installed in the horizontal position on a combination sprinkler and standpipe riser in a multistoried apartment house.

Fig. 7.14. Waterflow indicator installed in horizontal position on combination sprinkler and standpipe system riser. (Glen Echo Volunteer Fire Department, Glen Echo, Maryland)

Care should always be taken to be certain the waterflow indicator is installed so the movement of the water in the sprinkler piping will activate the device. All indicators have a marking to indicate the proper installation of the device relative to the waterflow in the piping.

Should the indicator be installed backwards, an alarm cannot be received; thus, complete testing of the waterflow indicator with operation of the inspector's test valve is as essential for the waterflow indicator as for the alarm check valve.

INSPECTION AND TESTS OF THE WET PIPE SPRINKLER SYSTEM

The initial tests required on a wet pipe automatic sprinkler system are the hydrostatic tests required on all new systems. A hydrostatic test pressure of not less than 200 psi for two hours should be applied, or 50 psi in excess of the static pressure, when the maximum static pressure exceeds 150 psi. During the hydrostatic test there should be no visible leakage from the sprinkler system piping.

The underground mains leading to the sprinkler system should always be thoroughly flushed before the mains are connected to the system. According to the *NFPA Sprinkler System Standard,* the underground mains which supply a wet pipe automatic sprinkler system should be flushed at a minimum flow of 750 gpm for 6-inch pipe, 1,000 gpm for 8-inch pipe, 1,500 gpm for 10-inch pipe, and 2,000 gpm for 12-inch pipe.

Inspection of the wet pipe sprinkler system at regular intervals is essential. A weekly inspection of all the water control valves and alarm control valves is recommended to be certain they remain open even though they are locked or sealed. As previously indicated, the greatest single cause of sprinkler system failure is closed water control valves. During inspections the pressure indicated on the water gages should always be noted and compared with previous readings. It is not unusual to have a higher pressure on the pressure gage on the system side of the alarm check valve, since water surges of slight pressure and short duration will open the main alarm check valve enough to pass through and cause a higher pressure to be built up on the system side of the valve without activating alarm devices. Complete inspections of wet pipe sprinkler systems should be conducted semiannually using the NFPA recommended inspection form (see Table 3.4 of Chapter 3, "Introduction to Automatic Sprinkler Systems").

In an article "What Fire Fighters Should Know About Fixed Water Systems," Charles W. Bahme recommended that fire department personnel schedule their prefire survey procedures to coincide with the sprinkler system tests made by insurance authorities.[5] These tests include evaluation of the waterflow to the sprinkler system with main drain or hydrant flow tests, and the evaluation of the operation of the alarm check valve and alarm devices with the operation of the inspector's test connection.

In "The Inspector's Test Pipe and Valve," James M. Hammack indicated that a 30- to 40-second delay is acceptable for the activation of the alarm devices following the opening of the inspector's test connection.[6] In addition, Hammack explained that trapped air in the wet pipe sprinkler system piping may cause the alarm devices of the water motor gong type to operate intermittently until the trapped air is exhausted. Such a condition is considered permissible if the alarm device is activated at least 50 percent of the time. Hammack stated the basic function of the inspector's test pipe and valve on the wet pipe sprinkler system is to provide a means of testing the alarm devices and making a positive determination that water will flow through the system to the topmost line of automatic sprinkler heads. Figure 7.15 shows an inspector's test pipe and valve arrangement for a wet pipe sprinkler system.

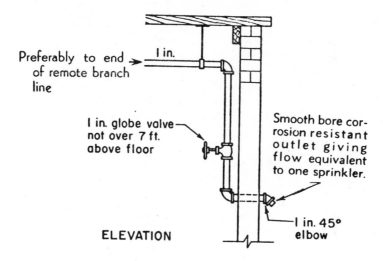

Preferably to end of remote branch line → 1 in.

1 in. globe valve not over 7 ft. above floor →

Smooth bore corrosion resistant outlet giving flow equivalent to one sprinkler.

ELEVATION

1 in. 45° elbow

NOTE: Not less than 4 feet of exposed test pipe in warm room beyond valve when pipe extends through wall to outside

Fig. 7.15. Inspector's test pipe and valve arrangement on a wet pipe automatic sprinkler system. (From "The Inspector's Test Pipe and Valve," by James M. Hammack)

The *NFPA Sprinkler System Standard* recommends the discharge from the inspector's test pipe be at a point where it can be observed, and preferably the discharge be outside the building as indicated in Figure 7.15. However, in many buildings—and especially in multistoried buildings—discharge outside of the building is not effective, efficient, or desirable. Thus, it is permissible to terminate the test pipe inside the building with an approved sight test connection. Note, however, the discharge from the inspector's test valve must have a smooth orifice to approximate the discharge flow from a single sprinkler head; thus at 15 psi, approximately 21 gpm.

The location of the inspector's test valve is important. This valve should always be prominently marked in order to indicate the opening of the valve will activate alarm devices and, on those systems that are automatically connected to fire departments, summon the fire department. False alarms have been caused by building maintenance and custodial personnel operating sprinkler system inspector's test valves which have not been properly marked. Fire department personnel, when conducting prefire surveys, should always check the inspector's test valve to determine if it is properly marked.

Restoration of the Wet Pipe Automatic Sprinkler System After Operation

Upon arrival at the fire scene, the fire department can silence local alarm devices such as electrical bells or water motor gongs by closing the alarm control valve on the 1-inch pipe trimming (usually between the alarm check valve and the retarding chamber, as shown earlier in this chapter in Figure 7.12). Alarm devices having waterflow indicators cannot be easily silenced until the flow of water is stopped. Thus, once the fire is completely under control, the closest water control valve should be closed. However, the water control valve on the main riser should never be closed unless the sprinkler system has no section control valves; no more than one valve should be closed unless it is a looped system; and, in order for the closed water control valve to be immediately reopened in the event of reignition or any other type of fire occurrence, someone should be posted to guard it.

The fused sprinkler heads on the branch lines should be replaced. To replace a head, the water in the branch line should be drained; drainage is accomplished by opening the main drain valve at the alarm check valve, or by using the closest auxiliary drain (if one is provided on the system). Insurance authorities always require that additional sprinkler heads be kept on the premises; such extra heads are usually located adjacent to the alarm check valve at the system's main riser. Many fire departments also carry complete sets of replacement sprinkler heads, while some departments delegate this function to sprinkler service companies. Once the sprinkler heads have been replaced, the water control valve should immediately be reopened in order to refill the system with water. Then the water control valves should be sealed or locked open, and the alarm control valve should be sealed open. It is usually good practice to open the inspector's test valve on the system to be certain that water will flow to the topmost sprinkler head, that the alarm devices are operating, and the system is again completely functional.

Tests for Pipe Clogging on a Wet Pipe Automatic Sprinkler System

When the wet pipe sprinkler system operates, the water flow in the pipes is at a relatively low velocity and flow as compared with the dry pipe system. As previously indicated in Chapter 5, "Basic Design of Automatic Sprinkler Systems." Merdinyan (in "Friction Loss In Sprinkler Piping") stated the friction loss increases with the age of the sprinkler piping. The National Board of Fire Underwriters has found the proportion of sprinkler systems with flow obstructions increased by 100 percent for sprinkler systems over 15 years of service.[7] The accumulation of sediment and foreign objects in sprinkler systems can be prevented by flushing the supply mains to the systems through the hydrants at least semiannually; the pressure, suction, or storage tanks, including gravity tanks, should be cleaned periodically.

The National Board of Fire Underwriters recommended that sprinkler branch line flow tests be conducted from the extremities of the branch lines to determine obstructions in the lines and to clean them of sediment or foreign materials that may be carried into the system by the flow of water. Periodically flushing the branch lines into large drums to determine the extent of accumulations and to discover obstructions to the flow is facilitated with the installation of the line testers at the end of the sprinkler system branch lines, as illustrated in Figure 7.16.

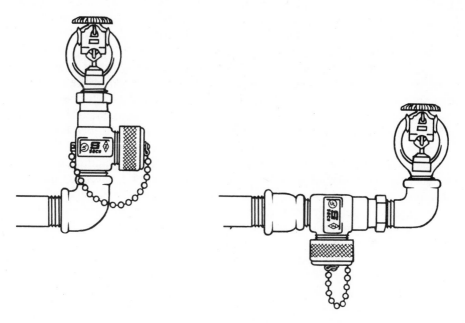

Fig. 7.16. Line testers installed at the end of branch lines on the automatic sprinkler system to facilitate flow testing. (Seco Manufacturing, Inc.)

APPLICATIONS OF WET PIPE SPRINKLER SYSTEMS

The wet pipe automatic sprinkler system is the most generally applied sprinkler system due to the reliability of the system's operation and the provision of these systems in occupied areas. Recent developments relative to the use of wet pipe sprinkler systems have been concentrated on their capabilities for the protection of human lives and the feasibility of providing increased safety to the occupants of varied structures. Another area of increasing application and usefulness is the installation of quick-response and extended-coverage sprinkler heads on wet pipe sprinkler systems for health care and residential facilities in multistoried and high-rise buildings.

In an article titled "Built-in Protection for High-Rise Buildings," Wayne E. Ault indicated the fire test for automatic sprinklers, as conducted by both Underwriters Laboratories and Factory Mutual System, creates a ceiling temperature of approximately 1400°F prior to the initial sprinkler head operation.[8] However, within 5 minutes, ceiling temperatures are reduced to below 600°F, and maintained at this level despite the 350 pound lumber crib and burning spray of n-heptane. Thus, sprinkler systems can reduce the temperature of a severe fire exposure to prevent structural damage and local structural failure even when the fire is a flammable liquid fuel fire that resists complete extinguishment.

In an article titled "Sprinklers Control High-Piled Tire Warehouse Fire," E. N. Proudfoot described the successful control of a high-piled tire warehouse fire having difficult fuel types and fuel configurations.[9] The wet pipe automatic sprinkler system that protected the warehouse was designed to provide a water discharge density of 0.48 gpm/sq ft, over the most remote 2,400 square feet of floor space in the warehouse. The sprinkler heads were $\frac{17}{32}$-inch large orifice standard heads with a temperature rating of 165°F. The pile of tires involved in the fire was stacked four pallets high to a total height of 18.5 feet, with a pile floor area of 26 × 155 feet or a total of 4,030 square feet. The fire operated seven sprinkler heads which, when supplemented by three $1\frac{1}{2}$-inch hose lines, provided a total water density of approximately 0.74 gpm/sq ft in the fire area. This extensive fire was effectively controlled by use of the sprinkler system and fire department personnel. The sprinkler system was kept in operation throughout the duration of the fire, including and during the post fire time when the smoldering tires were removed from the warehouse.

There is often much controversy concerning the installation of sprinkler systems—especially wet pipe automatic sprinkler systems—in areas where electronic equipment is installed or operated. In "Water and Electronics Can Mix," Donald J. Keigher indicated one standard procedure for the cleaning of electronic equipment is to:

(1) spray it with a high-pressure, warm-water detergent solution,
(2) rinse it thoroughly with clean water, and
(3) dry it with warm air.[10]

Keigher's article provided the following data on the wet pipe system installed in the Merchandise Mart Building in Chicago:*

> Built in 1931, the Mart has about 60,000 sprinklers. By 1961 the system reportedly had operated 18 times—9 times to control and extinguish actual fires, 3 times on accidental release of heat (i.e., steam), 5 times because janitors and maintenance people accidentally broke off heads, and once when a berserk employee smashed a head deliberately, actuating the system. That building has had perfect performance for close to 2,000,000 head years experience.

Thus, Keigher recommended the installation of automatic sprinkler systems as the backup or final protection within computer areas, with the installation of

---

*From "Water and Electronics Can Mix," *Fire Journal,* Vol. 64, No. 6, Nov. 1968, p. 69.

ionization-type smoke detectors and portable extinguishers as the initial response protection. Due to the recent serious property losses and the recognized potential for loss of life in high-rise buildings, the use of automatic sprinkler systems in such buildings has been a continual subject for consideration. In recognition of the fact that fires in high-rise buildings are beyond the control of fire department personnel with hose lines, specific requirements for automatic sprinkler systems in many buildings and fire prevention codes for automatic sprinkler systems in high-rise buildings (usually defined as being above 75 or 100 feet) have been formulated. In "Automatic Sprinkler Experience in High-Rise Buildings," W. Robert Powers described a 3½-year study conducted in New York City (from January 1, 1969 to July 1, 1972).[11] During the study period, 115 office buildings experienced 41 fires; all of the fires were satisfactorily controlled or extinguished by sprinkler systems. Thus, 34 fires were controlled with 1 sprinkler, 5 fires with 2 sprinklers, 1 fire by 3 sprinklers, and 1 fire by 4 sprinklers. All of the 115 office buildings were of fire resistive construction, and all of the sprinkler systems were equipped with waterflow devices having central station service which notified the fire department upon activation of the sprinkler system. However, 85 percent of the fires did not require fire department service of even a single hose line.

Powers expanded his study to include sprinkler system performance in 500 additional high-rise buildings, thus making a total sample of 615 high-rise buildings. During the 3½-year study period, 661 fires occurred; 654 of the fires were extinguished or controlled by automatic sprinkler systems. Unsatisfactory sprinkler system performance occurred in only 7 fires, with 5 of these occurrences caused by the water supply being turned off. Thus, the automatic sprinkler systems in the 615 high-rise buildings performed with a 98.9 percent efficiency rating. Most of these sprinkler systems had the waterflow alarm devices installed and supervised by a central station service which provided automatic fire department notification upon receipt of a water flow signal from the sprinkler system. Relative to construction, the ceiling heights were generally about 12 feet, the sprinkler spacing was usually 100 square feet per head, and the buildings were of fire resistive construction. Table 7.2 indicates the number of sprinklers operating and the frequency of fires per floor in the 615 high-rise buildings in Powers' study.

Powers also obtained comparative data relative to the performance of sprinkler systems in about 8,000 sprinklered buildings under 10 stories high. These results indicated satisfactory sprinkler operation in 1,935 instances involving 1,990 fires. Thus, a satisfactory sprinkler performance record of 97.2 percent was compiled. The 55 cases of unsatisfactory performance consisted of the following primary causes: 24 failures due to closed water control valves; 17 occurrences were due to exceptionally severe hazards; 5 were caused by building faults; 5 were related to arson; and 4 failure causes were unknown.

Thus, Powers' study of the New York City experience for the operation of sprinkler systems over a 3½-year period indicated a 100 percent satisfactory performance for high-rise office buildings, a 98.9 percent satisfactory performance for all high-rise buildings, and a 97.2 percent satisfactory perfor-

Table 7.2. Number of Sprinklers Operating Satisfactorily and
Frequency of Fires per Floor in High-Rise Buildings*

NUMBER OF SPRINKLERS OPERATING SATISFACTORILY IN ALL THE HIGH-RISE BUILDINGS		NUMBER OF FIRES PER FLOOR IN HIGH-RISE BUILDINGS			
NUMBER OF SPRINKLERS	NUMBER OF FIRES	FLOOR	NUMBER OF FIRES	FLOOR	NUMBER OF FIRES
		Basement	150	16	9
1	461	1	112	17	13
2	108	2	32	18	3
3	36	2	27	19	3
4	20	4	23	20	6
5	7	5	27	21	1
6	6	6	20	22	3
7	3	7	28	23	3
8	1	8	30	24	2
9	1	9	31	25	1
10	2	10	22	26	1
11	2	11	40	27	7
12	4	12	27	29	1
13	1	13	6	31	1
14	1	14	17	33	1
17	1	15	13	34	1
	Total 654			**Total** 661	

mance in other buildings. Table 7.3 presents Powers' results relative to the number of sprinklers operating, as compared with the NFPA data for 1970 ("Automatic Sprinkler Performance Tables, 1970 Edition"[1]).

Powers explained most of the systems in his study were designed with a minimum pressure of 25 psi at the sprinkler head. Therefore, the New York City performance experience would seem to be primarily due to building construction, prompt fire department response, and a dependable water supply. The water supply was from both public mains and gravity or pressure tanks designed to supply 25 percent of the sprinkler heads in the largest fire area for 20 minutes with 20 gpm flowing per sprinkler head.

In an article titled "Sears Tower: Life Safety and Fire Protection Systems," C. W. Schirmer described the criteria used in the design of the wet pipe automatic sprinkler system for the tallest and largest privately owned office building in the world.[12] This system was designed with a maximum coverage of 225 square feet per sprinkler head, and a minimum 10 psi operating pressure at the end sprinkler.

*From "Automatic Sprinkler Experience in High-Rise Buildings," by W. Robert Powers.

Table 7.3. New York City and National Automatic Sprinkler System
Performance Experience*

COMPARISON OF NEW YORK CITY AND NATIONWIDE EXPERIENCE

PERCENTAGE OF FIRES NO. OF SPRINKLERS OPERATING	NATIONWIDE		NEW YORK	
	ALL OCCUPANCIES	OFFICES	HIGH-RISE	OTHER
1	37.4	82.9	69.7	61.3
2 or Fewer	54.6	95.1	86.1	76.1
3 or Fewer	63.8	97.6	91.5	82.6
4 or Fewer	70.1	100.0	94.6	87.1
5 or Fewer	73.4		96.1	89.7
6 or Fewer	77.7		97.0	91.2
7 or Fewer	80.1		97.4	92.4
8 or Fewer	82.2		97.6	93.5
9 or Fewer	83.7		97.9	93.9
10 or Fewer	85.0		98.2	94.5
11 or Fewer	86.2		98.5	94.6
12 or Fewer	87.3		99.1	95.0
13 or Fewer	88.0		99.2	95.3
14 or Fewer	89.0		99.4	95.4
15 or Fewer	89.7		99.5	95.8
20 or Fewer	92.2		99.7	96.9
25 or Fewer	93.6			97.2
30 or Fewer	94.8			97.5
35 or Fewer	95.5			
40 or Fewer	96.2			97.5
50 or Fewer	96.9		99.8	97.8
75 or Fewer	98.1			98.0
100 or Fewer	99.7			98.2
All Fires (Including Unknown)	100.0	100.0	100.0	100.0

In Evaluation of Multilevel Sprinkler Systems and Container Materials for Fire Protection in High Rack Storage, Butturini, DeMonbrun, McCormick, McLaughlin, and Weathersby presented criteria for the installation of multi-level sprinkler systems in high rack storage situations.[13] Their experiments were conducted with a wet pipe automatic sprinkler system that supplied sprinklers installed in the ceiling 21 feet above the floor, and two levels of intermediate sprinklers installed at 9- and 13-foot heights. A water discharge density of 0.53 gpm/sq ft was provided by the ceiling sprinklers, a discharge density of 0.30 gpm/sq ft was provided at the 13-foot level, and a discharge density of 0.25 gpm/sq ft at the 9-foot level. These studies illustrated the importance of the combustibility of the containers stored in the racks. They also illustrated the need to protect the sprinkler heads at the intermediate levels from the water

*From "Automatic Sprinkler Experience in High-Rise Buildings," by W. Robert Powers.

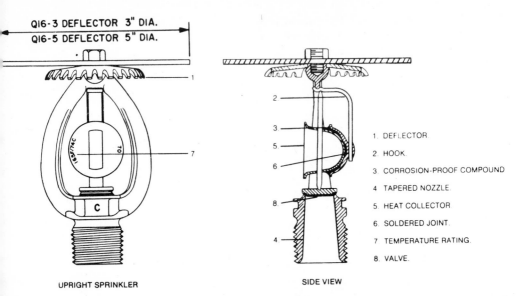

QI6-3 DEFLECTOR 3" DIA.
QI6-5 DEFLECTOR 5" DIA.

1. DEFLECTOR

2. HOOK.

3. CORROSION-PROOF COMPOUND

4. TAPERED NOZZLE.

5. HEAT COLLECTOR

6. SOLDERED JOINT.

7. TEMPERATURE RATING.

8. VALVE.

UPRIGHT SPRINKLER   SIDE VIEW

Fig. 7.17. *Standard upright sprinkler head with baffle for intermediate level installation in rack storage systems.* (Grinnell Fire Protection Systems Company, Inc.)

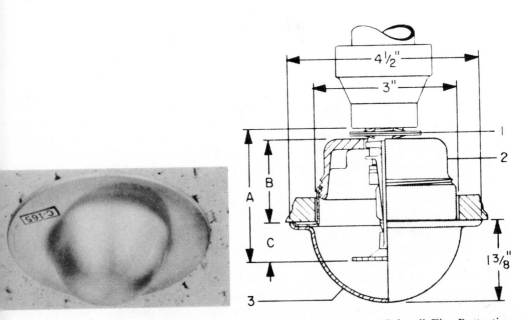

Fig. 7.18. *Recessed pendent sprinkler head dust cap.* (Grinnell Fire Protection Systems Company, Inc.)

discharge of the ceiling sprinklers when the ceiling sprinklers first operated, thus creating a cold soldering condition on the intermediate level sprinklers. As a result of research studies and fire experiences related to the installation of intermediate level sprinklers in high rack storage situations, sprinkler manufacturers provide both the standard upright and pendent sprinkler heads with baffles for intermediate level rack storage situations. Figure 7.17 illustrates a standard upright sprinkler head with provisions for a 3- or 5-inch baffle, for use in rack storage situations.

The wet pipe automatic sprinkler system can be found in locations where the cleanliness of wall and ceiling areas is of utmost importance, such as medical or health laboratories, and electronic equipment assembly areas. Figure 7.18 illustrates a dust cap assembly that is installed over standard pendent recessed sprinklers in such occupancies. The dust cap assembly drops away when the temperature of the cap reaches 165°F, leaving the sprinkler head exposed. The head may be of 135°F or 165°F rating. This assembly is similar to the hidden-pendent sprinkler head discussed in Chapter 6, "The Automatic Sprinkler Head," and shown in Figure 6.8 in that Chapter.

SUMMARY

The wet pipe automatic sprinkler system is the most efficient and effective type of sprinkler system currently available and installed in many occupancies. When compared to the dry pipe automatic sprinkler system, the wet pipe automatic sprinkler system—because of its design with water in the sprinkler pipes —has, with fewer heads operating, continually and consistently extinguished fires. The primary cause of unsatisfactory performance of wet pipe automatic sprinkler systems is the same as with all other types of sprinkler systems: the closing of water supply control valves before the fire occurs, or before the fire is completely extinguished. The two designs for alarm check valves involve the pilot valve and the divided-seat ring type of alarm check valve clappers. The alarm check valves are usually found on wet pipe automatic sprinkler systems without constant pressure sources, and with fire department connections to the riser. Waterflow indicators are utilized instead of alarm check valves in some industrial situations, and are often installed at each floor in many multistoried buildings. The most common types of alarm devices installed on wet pipe automatic sprinkler systems are electrical pressure switches that transmit alarm signals, and water motor gongs.

Wet pipe automatic sprinkler systems require a minimum of maintenance as compared to dry pipe systems. However, in many areas, wet pipe systems more than 15 years old should be checked regularly, especially in the smaller diameter branch lines, for accumulations of sediment and foreign materials.

Wet pipe automatic sprinkler systems—especially systems having quick-response and extended-coverage sprinkler heads—are being installed more frequently in life safety situations. The continued development of improved sprinkler heads for life safety situations, with the use of new pipe materials including the polyvinyl chloride pipe discussed in Chapter 5, "Basic Design of Automatic Sprinkler Systems," should result in increased applications of wet pipe automatic sprinkler systems in apartment buildings, health care facilities, multistoried dwellings, and mobile-home occupancies.

SI Units

The following conversion factors are given as a convenience in converting to SI units the English units used in this Chapter:

$$
\begin{aligned}
1 \text{ square foot} &= 0.0929 \text{ m}^2 \\
1 \text{ foot} &= 0.305 \text{ m} \\
1 \text{ psi} &= 6.895 \text{ kPa} \\
1 \text{ gpm/sq ft} &= 40.746 \text{ litre/min m}^2
\end{aligned}
$$

ACTIVITIES

1. Explain why the wet pipe automatic sprinkler system consistently extinguishes a greater number of fires with fewer sprinkler heads in operation than does the dry pipe sprinkler system.
2. Write your own definition of a wet pipe automatic sprinkler system. Include in your definition a description of the basic characteristics of these systems, and their purpose.
3. Discuss the operation of water flow indicators and their use with wet pipe automatic sprinkler systems. Then complete the following:
   (a) Write an explanation of the operation of the type of valve most often used on wet pipe sprinkler systems when one or more of the sprinkler heads have been activated and are discharging water into the fire area.
   (b) Write a brief statement comparing the design differences and similarities of alarm check valves and water flow indicators.
4. What is the purpose of the retarding chamber? Do you know of any alarm check valves on wet pipe sprinkler systems in your area without retarding chambers?
   (a) Write an explanation of the type of situation not requiring the installation of a retarding chamber on a wet pipe sprinkler system.
   (b) In what type of areas might retarding chambers be supplemented?
   (c) How might they be supplemented?
5. What are some disadvantages concerning the use of water motor gongs? Explain how such disadvantages might be counteracted.
6. Explain why wet pipe automatic sprinkler systems require a minimum of maintenance as compared to dry pipe systems.
7. List the wet pipe automatic sprinkler systems with the occupancy protected in your area, and indicate whether or not the alarm devices for each are arranged for automatic transmission to the fire department.

8. As an authority on wet pipe automatic sprinkler systems, you have been asked to prepare a statement concerning inspection procedures for such systems. Outline a procedure that you believe would best maintain a wet pipe automatic sprinkler system in peak condition.

SUGGESTED READINGS

Keigher, Donald J., "Water and Electronics Can Mix," *Fire Journal*, Vol. 62, No. 6, Nov. 1968, pp. 68–72.

Powers, W. Robert, "Automatic Sprinkler Experience in High-Rise Buildings," *Fire Journal*, Vol. 66, No. 6, Nov. 1972, pp. 47–49.

Schirmer, C. W., "Sears Tower: Life Safety and Fire Protection Systems," *Fire Journal*, Vol. 66, No. 5, Sept. 1972, pp. 6–12.

BIBLIOGRAPHY

[1]"Automatic Sprinkler Performance Tables, 1970 Edition," *Fire Journal*, Vol. 64, No. 4, NFPA, July 1970, pp. 35–39.

[2]NFPA 13, *Standard for the Installation of Sprinkler Systems*, NFPA, Boston, 1975.

[3]*Alarm Valves for Fire Protection Service, U.L. 193*, Underwriters Laboratories, Northbrook, Ill., 1973.

[4]Marryatt, H. W., *Automatic Sprinkler Performance in Australia and New Zealand, 1886–1968*, Australian Fire Protection Association, Melbourne, Australia, 1971.

[5]Bahme, Charles W., "What Fire Fighters Should Know About Fixed Water Systems," *NFPA Firemen*, Vol. 35, No. 4, April 1968, pp. 34–38.

[6]Hammack, James M., "The Inspector's Test Pipe and Valve," *Fire Journal*, Vol. 64, No. 2, March 1970, pp. 28–30.

[7]*Internal Cleaning Sprinkler Piping*, National Board of Fire Underwriters, American Insurance Association, New York, 1959.

[8]Ault, Wayne E., "Built-in Protection for High-Rise Buildings," *Fire Journal*, Vol. 66, No. 4, July 1972, pp. 40–42, 72.

[9]Proudfoot, E. N., "Sprinklers Control High-Piled Tire Warehouse Fire," *Fire Journal*, Vol. 68, No. 2, March 1974, pp. 70–72.

[10]Keigher, Donald J., "Water and Electronics Can Mix," *Fire Journal*, Vol. 62, No. 6, Nov. 1968, pp. 68–72.

[11]Powers, W. Robert, "Automatic Sprinkler Experience in High-Rise Buildings," *Fire Journal*, Vol. 66, No. 6, Nov. 1972, pp. 47–49.

[12]Schirmer, C. W., "Sears Tower: Life Safety and Fire Protection Systems," *Fire Journal*, Vol. 66, No. 5, Sept. 1972, pp. 6–12.

[13]Butturini, W. G., DeMonbrun, R. J., McCormick, W. J., McLaughlin, M. L., and Weathersby, E. W., *Evaluation of Multilevel Sprinkler Systems and Container Materials for Fire Protection in High Rack Storage*, Union Carbide Corp., Y-12 Plant, Oak Ridge, Tenn., 1971.

Chapter 8

The Dry Pipe
Automatic Sprinkler System

CHARACTERISTICS OF THE DRY PIPE AUTOMATIC SPRINKLER SYSTEM

The dry pipe automatic sprinkler system was developed from the wet pipe sprinkler system and was primarily created as a result of the need for automatic sprinkler protection in unheated occupancies and structures. The original dry pipe valves were developed as differential valves, with the first practical valve being the Grinnell Bellows differential valve developed in 1885 (see Figure 3.6 of Chapter 3, "Introduction to Automatic Sprinkler Systems"). A number of mechanical dry pipe valves were also developed, and were widely used until the 1930s.

The dry pipe automatic sprinkler system requires a system of piping similar to the wet pipe system previously shown in Figure 7.2 of Chapter 7, "The Wet Pipe Automatic Sprinkler System." However, the dry pipe sprinkler system does not contain water above the dry pipe valve in the system, but air or nitrogen under pressure. The air pressure restrains the water in the supply main at the clapper arrangement of the dry pipe valve.

The dry pipe sprinkler system is connected to a reliable water supply, usually a public or private water main, and is equipped with an indicating type of water control valve, a fire department connection, and the dry pipe valve. Since the dry pipe system is primarily utilized in unheated occupancies, the dry pipe valve must be protected against freezing. A location in the heated portion of a structure, such as an office in a warehouse occupancy, or in a separate heated enclosure (identified as a dry pipe valve enclosure) protects the valve from freezing.

Compare Figure 8.1 in this chapter with the diagram of a wet pipe automatic sprinkler system shown in Figure 7.2 of Chapter 7. Note that Figure 8.1 illustrates the dry pipe sprinkler system with the fire department connection installed on the supply side of the dry pipe valve, while Figure 7.2 shows the wet pipe sprinkler system with the fire department connection installed on the system side of the alarm check valve above the alarm check valve. When the system has a single riser, the fire department connection, as utilized in both wet and dry pipe sprinkler systems, is always installed above the water control valve in order to enable the fire department to supply the system with water even when the main riser water control valve is closed.

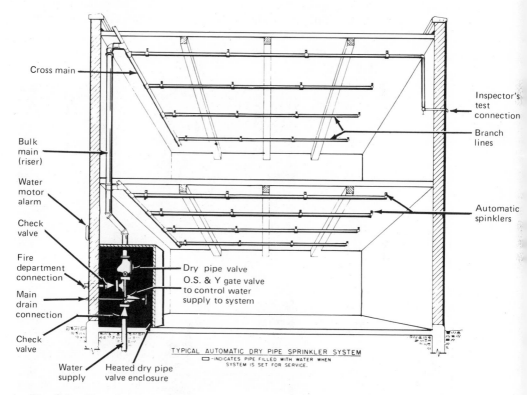

Fig. 8.1. *Dry pipe automatic sprinkler system.* (Grinnell Fire Protection Systems Company, Inc.)

In Figure 8.1, note the heated dry pipe valve enclosure and the installation of the check valve on the supply riser which prevents any water being discharged into the system through the fire department connection from being dissipated throughout the water supply main. The check valve on the fire department connection, the water motor gong, and the inspector's test connection located on the topmost line of sprinkler heads are also standard features.

The wafer type check valve is often used on the riser or the fire department connection since the valve is manufactured for both horizontal and vertical installations. An advantage of using the wafer type check valve includes the space and weight savings with the valve assembly, as compared to the typical swing check valve. These design features make it particularly suited to installation on vertical sprinkler system risers.

Figure 8.2 shows the installation of a wafer type check valve on a combination standpipe and sprinkler system riser, and Figure 8.3 illustrates its component parts. Figure 8.4 shows a conventional swing check valve installed in the horizontal position on the fire department connection to a sprinkler riser.

The check valve assemblies illustrated in Figures 8.2, 8.3, and 8.4 are found on all types of sprinkler or standpipe systems, and are not restricted to installation on dry pipe automatic sprinkler systems. It is important to remember with dry pipe automatic sprinkler systems, the sprinkler piping contains air or nitrogen under pressure above the dry pipe valve.

Fig. 8.2. Wafer type check valve installed in the vertical position on a combination standpipe and sprinkler system riser. (Glen Echo Volunteer Fire Department, Glen Echo, Maryland)

DRY PIPE VALVES

Most of the modern dry pipe valves are of the differential type of design and construction. Due to the design differences in the sizes of the clapper or clappers with the air seat being greater in area than the water seat, a differential advantage is established which enables one pound of air pressure to effectively restrain five to six pounds of water pressure. Most of the dry pipe valves have an intermediate chamber which is located between the air and water seats or

Fig. 8.3. Wafer type check valve showing component parts. (TRW Mission Manufacturing Co.)

Fig. 8.4. Swing check valve installed on the fire department connection to the sprinkler system riser. (Glen Echo Volunteer Fire Department, Glen Echo, Maryland)

adjacent to both the seats, and is normally open to atmospheric pressure. The intermediate chamber has a drain that discharges any water that might entering the chamber by water surges or intermittent flows which pass the water seat, or by leakage of priming water from above the air seat. This drain from the intermediate chamber is usually equipped with a velocity drip valve of the ball check or clapper type design. Figure 8.5 illustrates a modern differential type of dry pipe valve with the valve set for operation on the system, with the priming water and air above the clapper. Figure 8.6 shows a single clapper dry pipe valve with the cover removed and the clapper open (left) and closed (right). In Figure 8.6, left, note the inner circle, the area of the water seat, and the outer circle, the area of the air seat. The space between the two seats is the intermediate chamber.

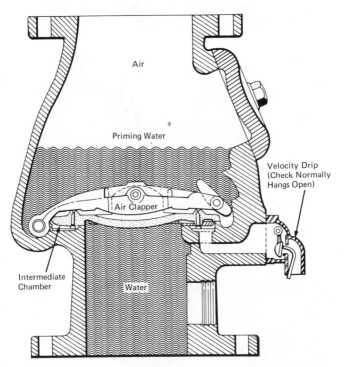

Fig. 8.5. Differential dry pipe valve set for operation on a dry pipe automatic sprinkler system. (Insurance Services Office)

The single clapper arrangement shown in Figure 8.6 is not necessarily a component of all dry pipe valves. Some are designed with two separate, but connected, air and water clappers. Figure 8.7 shows a diagram of a modern differential dry pipe valve of the dual clapper type design. The difference in the size of the water and air clappers with the different configuration and size of the intermediate chamber should be noted. It is of interest to note that the water clapper keeps the intermediate chamber drain check valve (16A, 16B in Figure 8.7) open, and once the valve trips, the clapper moves out of the waterway and

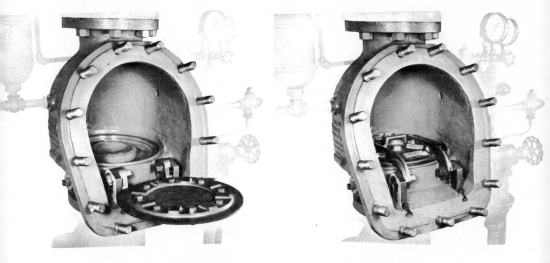

Fig. 8.6. Single clapper design dry pipe valve. (Raisler Sprinkler Co.)

the check valve is allowed to close. Therefore, the water flowing into the intermediate chamber is directed out of the connection to the alarm devices (0–4 in Figure 8.7) from the top of the intermediate chamber drain housing to activate the alarm devices.

The operation of the dry pipe valve in Figure 8.7 can be understood by referring to the set position and the tripped position of the component parts and clappers. When a sprinkler head fuses, the air pressure is reduced above the air clapper. Once this pressure reaches the tripping point of the valve, the water pressure under the water clapper (11 in Figure 8.7) raises the clapper and the clapper link (9), thereby raising the air clapper (5) from its seat (6). Then, the flow of water through the valve carries the air and the water clapper assembly upward and backward out of the valve and riser waterway. The clapper assembly is automatically secured in this vertical position by the clapper latch (15).

In Figure 8.7, it should be noticed, the main drain valve outlet (0–1) is located below both the air and the water clappers. This arrangement of the main drain outlet *below* the clappers is standard design for dry pipe valves. However, on the alarm check valves, the main drain connection is always *above* the clapper of the alarm check valve.

A latching mechanism is always required on all of the dry pipe valves to prevent the clapper or clapper assembly from swinging back into the waterway, thus causing pressure surges or even more severe water hammers that could damage the valve housing or the sprinkler system piping. Thus, with the latching of the clapper assembly, the valve housing cover must be removed, the clapper assembly released from the latch, and the clapper manually reset on the valve seats. Therefore, a dry pipe valve cannot be reset without physically entering the valve housing and manually resetting the clapper assembly. The

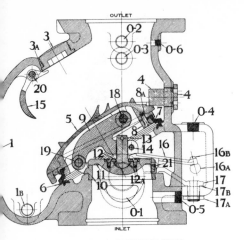

CLOSED (SET DRY) POSITION OF PARTS

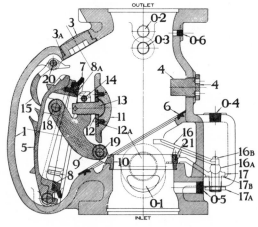

OPEN (TRIPPED) POSITION OF PARTS

Fig. 8.7. Differential dry pipe valve with two separate and connected clappers. (Star Sprinkler Corp.)

dry pipe valve is not self-resetting, as is the alarm check valve on the wet pipe automatic sprinkler system.

In Figure 8.7, note the provision for a positive set of the clapper assembly on the air and water seats by the provision of the stop plug (4). The stop plug protrudes from the side of the valve assembly. To be set properly, the air clapper must fit under the stop plug at the forward side of the clapper. The positive action of the stop plug helps to prevent minor water surges and intermittent flows from raising the water clapper from its seat.

Figure 8.8 illustrates the principle of operation for the differential dry pipe valve. In the set position, the priming water can be seen above the air clapper, with the sprinkler system air pressure above it. The priming water is important to dry pipe valves, since it provides a positive set and seal of the air and water clappers. If the clappers are not properly cleaned and reset when the priming water is applied to the system side of the air clapper, the water will flow around the air clapper into the intermediate chamber where it will be readily seen flowing from the velocity drip drain.

The dry pipe valve acts as an interface device between the water in the supply main and the air in the sprinkler system piping. Thus, the operating condition of the dry pipe valve is critical, and the priming water above the air

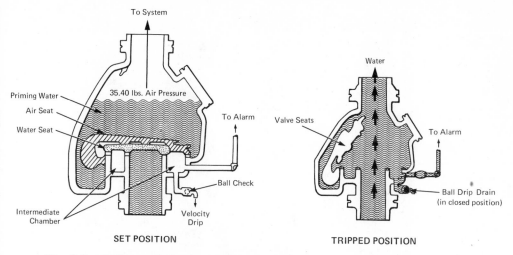

Fig. 8.8. *Differential dry pipe valve of single clapper design in both the set and tripped positions.* (Insurance Services Office)

clapper tends to help preserve the rubber or neoprene seals that are installed on the air clapper; these seals provide a positive seal against the metallic air seat. The air and water seats of all dry pipe valves should be cleaned only with soft cloths; any abrasive might damage the seats.

The inherent time delays involved with the operation of the dry pipe valves were discussed in Chapter 7, "The Wet Pipe Automatic Sprinkler System," relative to the comparison of the number of sprinkler heads opening on dry pipe and wet pipe automatic sprinkler systems. In an article titled "The Number of Sprinklers Opening and Fire Growth," Baldwin and North describe their study concerning the number of sprinklers opening and the sizes of fires in England.[1] Their study showed that more sprinklers opened on dry pipe systems than on wet pipe systems; on the average, twice as many sprinklers operated on dry pipe systems. Also, most of the dry pipe systems considered in the study took approximately 50 percent longer to operate. Obviously, dry pipe systems should not be utilized in heated areas or in climates not subjected to freezing temperatures.

Figure 8.9 shows the installation of a differential dry pipe valve with typical valve trimming, but without a quick-opening-device (Q.O.D.) installed on the dry pipe valve. This valve is similar in basic configuration to the alarm check valve previously shown in Figure 7.12 of Chapter 7, "The Wet Pipe Automatic Sprinkler System."

The components that comprise the trimming on the dry pipe valve in Figure 8.9 are described in the following paragraph. The priming cup (23) is used to pour water into the priming chamber (22), which then allows the water to flow into the top of the valve housing and down onto the top of the air clapper. The drain cup and line for the drain from the velocity drip valve of the intermediate chamber and the priming chamber and priming level is identified as (24). The

Fig. 8.9. Dry pipe differential type valve with trimming without a Q.O.D. (Quick-Opening-Device). (Star Sprinkler Corp.)

main drain valve (A) is below the clappers in the dry pipe valve. (B) and (C) are the valves on the priming chamber; they are provided so that water may be added to the dry pipe valve housing without tripping and resetting the valve. (D) is the level of the air inlet and is the priming water level indicator. Upon periodic inspection of this dry pipe valve, the priming water level should be opened to drain off any additional water (from condensation) that might have collected in the system. If not drained off, such additional water could cause water columning. Water columning is an accumulation of water above the air clapper to a depth that prevents the dry pipe valve from operating when a sprinkler head fuses, even though all the air is exhausted from the system. (E) is the air valve for the introduction of the air into the sprinkler system from an air compressor. Note the two air gages on the dry pipe valve; one gage, *below* the clappers, indicates *water pressure*, and the top gage, *above* the clappers, indicates *air pressure* when the valve is set and *water pressure* when the valve has been tripped. (F) is a valve used to indicate when priming water should be added to the valve. When no water flows from this valve, water should be added through the priming chamber assembly and valves until water flows from this valve, and not from the air inlet valve. (G) is the alarm test valve which allows water from below the dry pipe valve to flow into the line to the alarm devices. The valve for connection (H) allows the alarm pipe line to be drained following the testing or operation of the valve, so that the line will not freeze during cold weather.

Following is a review of some of the basic characteristics of conventional dry pipe valves and their trimmings relative to the type of valve illustrated in Figure 8.9. The water for the operation of the alarm devices flows through the intermediate chamber on dry pipe valves, and the alarm device lines originate from this portion of the dry pipe valve assembly. The trip point of the dry pipe valve is determined by the water supply pressure, considering the design of the differential in the clappers of the conventional dry pipe valve is approximately 1 to 5 or 1 to 6. Excessive air pressure should not be carried on the valves. Excessive air pressure tends to create maintenance problems; it also increases the operating time of the valve, thus resulting in additional time required for the water to reach the sprinkler heads in the fire area. Unless the water supply mains are provided with high pressure, it is unusual to find modern differential dry pipe valves which normally require air pressure exceeding 35 to 45 psi on the system. Table 8.1 shows the manufacturer's recommended air pressure to be maintained on the system with the dry pipe valve shown in Figure 8.9.

Table 8.1.   Chart of Recommended Air Pressure
Relative to Water Pressure for Dry Pipe Valve
Shown in Figure 8.8*

WATER PRESSURE	AIR PRESSURE	
MAXIMUM	NOT LESS THAN	NOT MORE THAN
50 Lbs	15 Lbs	25 Lbs
75 Lbs	20 Lbs	30 Lbs
100 Lbs	25 Lbs	35 Lbs
125 Lbs	30 Lbs	45 Lbs
150 Lbs	35 Lbs	50 Lbs

*Courtesy of Star Sprinkler Corp.

The time delay involved in the operation of the dry pipe valve primarily involves the time for the air to be exhausted out of the fused sprinkler head opening to drop the air pressure to the tripping point of the dry pipe valve with the resulting transport time for the water to flow through the system to the fused sprinkler head. Obviously, the longer the sprinkler piping system, the greater the transport time for the water to reach the topmost point in the system. Therefore, quick-opening-devices (Q.O.D.s) are installed on many dry pipe valves. The *NFPA Sprinkler System Standard* requires the installation of Q.O.D.s on the dry pipe valve when the capacity of the system exceeds 500 gallons.[2] This capacity requirement is the quantity of water needed to fill all the riser, feed, and branch lines in the system, and not the flow capacity of the system.

QUICK-OPENING-DEVICES (Q.O.D.s)

Quick-opening-devices are of two principal types: (1) the exhauster, and (2) the accelerator. Quick-opening-devices are installed on dry pipe valves when a system exceeds certain limits, generally a water capacity of 500 gallons, and on any dry pipe system when speed of operation may be a critical factor.

Exhausters

The exhauster is initiated by a very small pressure drop in the air pressure maintained on a system, usually several ounces. It operates to provide a 2-inch opening which exhausts the air from the riser immediately above the dry pipe valve. This action of quickly dropping the air pressure to the tripping point of the dry pipe valve increases the speed of operation of the dry pipe valve. Figure 8.10 shows the dry pipe valve with the two water and air clappers, with trimming (previously shown in Figure 8.9) with an exhauster (BE-20) installed.

Although the dry pipe valve shown in Figure 8.10 has retained the trimming shown in Figure 8.9, the priming chamber and cup have been moved to the rear of the dry pipe valve (directly behind the riser) to allow the installation of the exhauster. The dry pipe valve will operate without operation of the exhauster. Therefore, when resetting a system following tripping, it is usually the most efficient procedure to replace any fused heads, manually reset the dry pipe valve, and return the system to the set condition before resetting the exhauster. In Figure 8.10, valves (2) and (14) isolate the exhauster from the dry pipe system. With these valves closed, the system may be returned to service before the exhauster is reset. The resetting of the exhauster, like the dry pipe valve, is a manual procedure which requires opening the exhauster housing in two places, and draining and resetting the exhauster.

In order to provide a better understanding of the principles of operation of an exhauster as installed on a dry pipe valve, the following two paragraphs describe the operation of the exhauster shown in Figure 8.10. It is important to realize the majority of these quick-opening-devices operate on a pressure

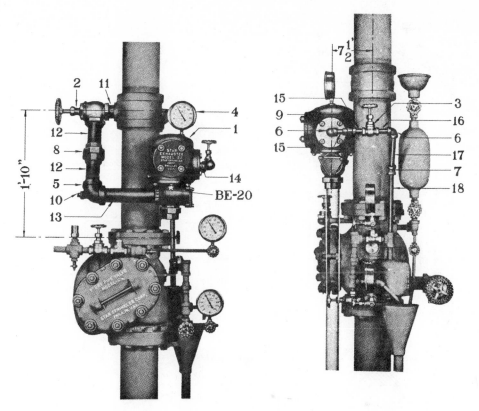

Fig. 8.10. Differential dry pipe valve with exhauster. (Star Sprinkler Corp.)

differential principle. Exhausters are, therefore, designed with at least two air chambers, and a diaphragm with air passage between the chambers through a restricted orifice or orifices. The location of the diaphragm and the division of the exhauster into two pressure chambers (areas A and B) is illustrated in Figure 8.11.

Operation of the exhauster may be initiated after the exhauster is in the set condition, with the air pressure indicated on the gage (4) equalized with the air pressure throughout the dry pipe sprinkler system above the valve. Valves (2) and (14) are open (as shown in Figure 8.10), and the exhauster is in the set condition (as shown in Figure 8.11) with the air outlet to valve (14) closed inside the exhauster. Upon operation of a sprinkler head and the initial drop of a few ounces in air pressure, the pressure within the exhauster becomes unbalanced. The pressure in the 2-inch inlet pipe and area B of the exhauster (as shown in Figure 8.11) drops below the pressure in area A and cannot be equalized rapidly enough through the restricted orifice. Thus, with the area A pressure being higher, the diaphragm is forced downward with the central shaft (BBE-7), then the protrusion (BE-14A) on the central shaft strikes the toggle valve (BE-14) opening the outlet (R) to valve (14) which is already open. Thus,

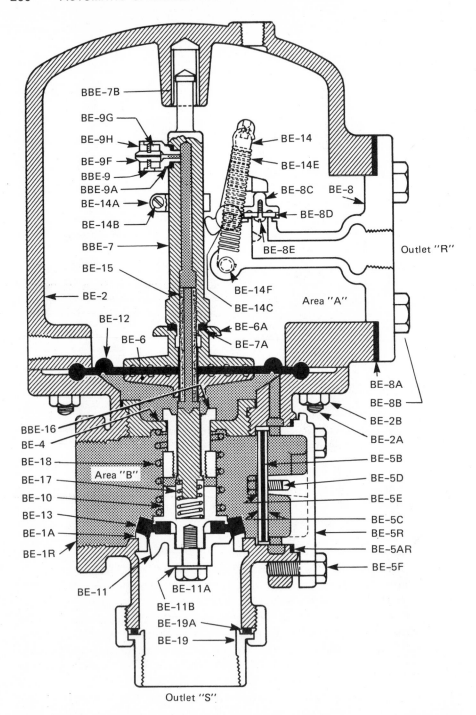

Fig. 8.11. Two-chamber design of exhuaster and component parts. (Star Sprinkler Corp.)

the pressure in area A is immediately dropped by the discharge through outlet (R), making the pressure in area B higher. The diaphragm is now moved upward, opening the outlet seat (BE-13) which allows the air from the 2-inch inlet pipe and the sprinkler system immediately above the dry pipe valve to pass out through the outlet (S). Once the dry pipe valve has tripped, water flows into the inlet and area B of the exhauster through the 2-inch pipe, while water also flows from the intermediate chamber up the pipe from outlet (R) into area A. The water pressure is now equalized in the exhauster. The springs on the center stem (BE-18) close the valve seat (BE-13) to outlet (S), preventing water flow through the 2-inch exhauster outlet.

The operation of the exhauster is automatic and extremely rapid, occurring within seconds. The opening of the 2-inch exhauster outlet from the area immediately above the dry pipe valve has the effect of lowering the air pressure in the system, which is equivalent to the operation of 15 sprinkler heads. As previously explained, the resetting of the exhauster is a manual process which, due to the possibility that sediment and particles of dirt might have been carried into the exhauster by the water flow, must always include thorough cleaning procedures in accordance with the manufacturer's recommendations. All Q.O.D.s utilized on dry pipe valves are subject to similar maintenance problems, due to the restricted orifices in these pneumatic-hydraulic devices.

Accelerators

The accelerator is similar to the exhauster, and most accelerators are also designed with at least two pressure chambers. Operation of the accelerator is initiated with the pressure differential condition created by the fusing of a single sprinkler head. However, unlike the exhauster which discharges the air from the dry pipe system into the atmosphere outside the sprinkler system, the accelerator takes in the air from immediately above the dry pipe valve and discharges it into the intermediate chamber, which is directly under the air clapper of the dry pipe valve.

Thus, the operational speed of the dry pipe valve is effectively increased due to: (1) a more rapid reduction in the air pressure above the valve to the valve trip point, and (2) the effective increase in the pressure on the water supply side of the air clapper of the dry pipe valve.

The piping used in the installation of an accelerator is considerably smaller, since there is no need for the 2-inch pipe to exhaust the air outside of the system. The piping for the accelerator will, of course, be initiated on the riser above the dry pipe valve in a location similar to the exhauster piping, and will terminate with discharge into the intermediate chamber. The piping should always be connected on the riser at a height which effectively prevents priming water or drainage from submerging the Q.O.D. Figure 8.12 shows a differential dry pipe valve and trimmings with an accelerator. Note the piping for the accelerator is of the $\frac{1}{2}$-inch size, and the inlet for the accelerator piping is at the air inlet immediately above the valve, while the outlet from the accelerator connects to the intermediate chamber.

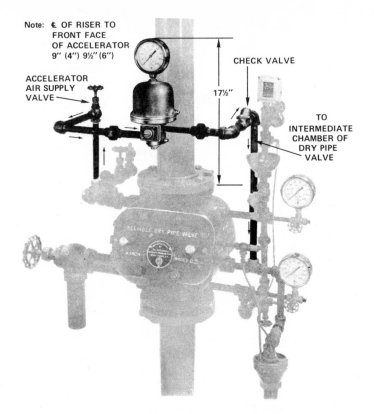

Fig. 8.12.   Differential dry pipe valve with trimmings including accelerator. (Reliable Automatic Sprinkler Company, Inc.)

As with the dry pipe valve, the accelerator must also be manually reset following tripping. The accelerator, like the exhauster, should always be isolated from the system while the dry pipe valve is being reset and the system is being put back into service. Thus, in Figure 8.12, the accelerator air supply valve should be closed to isolate the accelerator from the dry pipe system. If the valve is not closed once the dry pipe valve has been reset and the air turned into the system, the air passing through the accelerator into the intermediate chamber would probably cause the dry pipe valve to move off its seat. The operation of the accelerator on the dry pipe valve is illustrated in Figure 8.13 in the set (closed) position while being pressurized with air.

The operation of the accelerator requires both the dry pipe valve and the accelerator to be in the set condition with the pressure equalized between the sprinkler system piping air pressure above the dry pipe valve and both chambers of the accelerator. The pressure reading should be identical for the air gage on top of the accelerator and the dry pipe system air gage.

The accelerator illustrated in Figure 8.13 is designed with chambers at the top and in the middle. Thus, the air pressure must be equalized in both chambers for the accelerator to be in the set condition. The resetting of this partic-

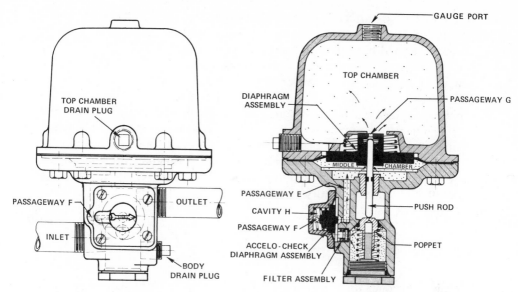

Fig. 8.13. Accelerator in the closed position. (Reliable Automatic Sprinkler Company, Inc.)

ular accelerator does not require manual entry into it, but does require the removal of drain plugs from both of its chambers. Therefore, manual operations are definitely required in order to place this accelerator back into the set condition.

The operation of the accelerator in Figure 8.13 is initiated with the fusing of a single sprinkler head. Thus, the pressure in the middle chamber of the accelerator is reduced with the drop in air pressure in the system due to the operation of the sprinkler head. The pressure in the top chamber is consequently higher due to the restricted orifice, and the higher pressure forces the diaphragm assembly and the push rod downward. This downward movement of the diaphragm assembly and push rod exerts pressure on the poppet, forcing the poppet open and allowing air to pass through the accelerator. The outlet air pressure then flows up and closes the accelo-check diaphragm assembly, preventing water from entering the accelerator. With the opening of the accelerator, the air immediately above the dry pipe valve is allowed to pass through the accelerator to the intermediate chamber of the dry pipe valve. Once the air reaches the intermediate chamber, it exerts pressure on the water supply side of the dry pipe valve clapper from the intermediate chamber. The air thus equalizes the pressure, negating the differential effect of the dry pipe valve clapper. The dry pipe valve then trips, allowing water to flow into the riser and to the fused head on the sprinkler piping system. The dry pipe valve will operate without the operation of the accelerator; however, the time of operation will be increased. Thus, Q.O.D.s are utilized on dry pipe systems to facilitate operation and decrease the time required for the operation of the differential dry pipe valve.

In the mid-1960s the U.S. Government expressed concern about the problem of maintaining fire protection with existing wet pipe sprinkler systems in structures that had been placed on an unoccupied standby basis. The problem existed because, in an unoccupied or standby condition, the buildings—all of which were equipped with standard wet pipe sprinkler systems with alarm check valves—would not be heated. A project was initiated to convert these wet pipe sprinkler systems to dry pipe systems by the addition of air compressor and air lines. As previously discussed, because the alarm check valve is basically a free-swinging clapper valve with essentially no appreciable differential effect for the pilot valve type and only a slight differential effect for the divided-seat ring type, a relatively high air pressure was required. Thus, the concept of a low-differential dry pipe valve was conceived and developed from the alarm check valve.

LOW-DIFFERENTIAL DRY PIPE VALVE

The air or nitrogen pressure maintained on the low-differential dry pipe valve is usually 15 to 20 pounds above the static water supply pressure. Thus, the sprinkler system piping must be maintained in good condition where relatively high static water pressure is prevalent. The low-differential dry pipe valve, as presently developed, is basically a modified alarm check valve. Wayne E. Ault, in an article titled "The Low-Differential Dry Pipe Sprinkler System," stated that as long as the alarm check valve clapper has a rubber or neoprene gasket on the clapper, it may function as a low-differential dry pipe valve.[3] Given the low differential of about 1.1 to 1, the tripping point of the valve will usually be approximately 10 percent less than the water pressure. Figure 8.14 illustrates a low differential dry pipe valve.

Ault reported because its design makes it almost impossible to water column a low-differential dry pipe valve, a necessary feature is signaling equipment that indicates any significant water level above the valve or an automatic drain. However, excessive water above the valve, once beyond the heated dry pipe valve enclosure, could be subjected to freezing temperatures and ice blockage of the riser. In some situations, a larger air compressor capacity is required due to the higher air pressure required to be maintained in the system as compared to the conventional dry pipe system.

Ault indicated the principal advantage of the low-differential dry pipe system is the reduction of the initial velocity of the water rushing into the system, as occurs with the conventional dry pipe valve with a differential of approximately 1 to 5 or 1 to 6.

Another advantage is the apparent faster operation of the valve with a reduction of the transport time to the fused sprinkler heads. Figure 8.15 illustrates the installation of a low-differential dry pipe valve with the necessary trimming for the valve.

Ault's article also described the advantage of the low-differential dry pipe valve relative to the reduction of the inrush velocity of the water with the

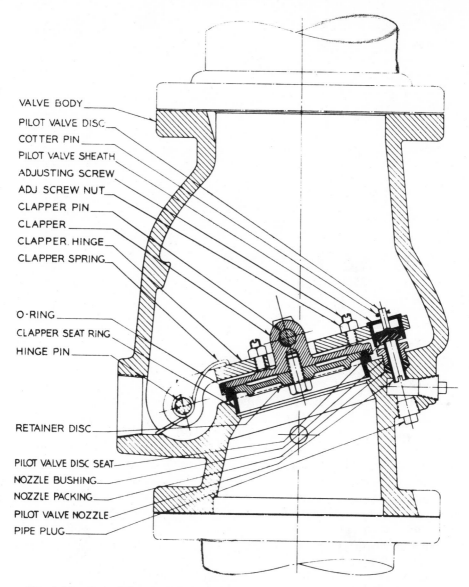

VALVE BODY

PILOT VALVE DISC

COTTER PIN

PILOT VALVE SHEATH

ADJUSTING SCREW

ADJ SCREW NUT

CLAPPER PIN

CLAPPER

CLAPPER HINGE

CLAPPER SPRING

O-RING

CLAPPER SEAT RING

HINGE PIN

RETAINER DISC

PILOT VALVE DISC SEAT

NOZZLE BUSHING

NOZZLE PACKING

PILOT VALVE NOZZLE

PIPE PLUG

Fig. 8.14. Low-differential dry pipe valve cross section. (From "The Low-Differential Dry Pipe Sprinkler System," by Wayne E. Ault)

example of a water supply to a conventional differential dry pipe valve with a 100 psi static pressure, and a 20 psi trip point on the valve. For a 6-inch pipe, these conditions produced an inrush velocity of approximately 7.15 feet per second and a flow of 645 gpm. The comparable low-differential dry pipe valve on a 6-inch pipe with the identical water supply system and a static pressure of 100 psi would have a trip point for the valve of 91 psi, with an inrush velocity of 2.2 feet per second and a flow of 198 gpm.

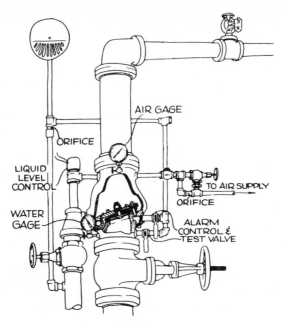

AIR GAGE

ORIFICE

LIQUID
LEVEL
CONTROL

TO AIR SUPPLY

ORIFICE

WATER
GAGE

ALARM
CONTROL &
TEST VALVE

Fig. 8.15. Low-differential dry pipe valve and trimming. (From "The Low-Differential Dry Pipe System," by Wayne E. Ault)

Ault further reports that with identical-size sprinkler systems consisting of a water capacity of 570 gallons, the low-differential dry pipe system would require only 15 seconds for the valve to respond to the trip point from a set condition with a 100 psi water static pressure. In addition, it would take approximately 26 additional seconds for the low-differential dry pipe system to completely pressurize the system with water, for a total response time of 41 seconds. Compare this with a conventional differential dry pipe valve with a trip point of 20 psi with 40 psi of air on the system, and a 100 psi water static pressure: the conventional dry pipe system would require approximately 100 seconds for the dry pipe valve to trip and 53 seconds to pressurize the system with water, for a total response time of 153 seconds, which is about $3\frac{1}{2}$ times as long as the low-differential dry pipe system. Thus, it should be noted, Ault did not consider quick-opening-devices on the conventional dry pipe system. However, an advantage of the low-differential dry pipe system is the elimination of the usual need for quick-opening-devices that are susceptible to maintenance problems. The continued exposure of the devices to the water from the initial inrush flow at a very high velocity (which tends to carry foreign matter in with the water) is one cause of maintenance problems. Because of this flooding, most of the quick-opening-devices approved by testing laboratories since 1970 have antiflooding features that prevent water from entering them.

Due to the need to reduce maintenance on existing quick-opening-devices and to prevent the flow of water into the accelerator, most sprinkler manufac-

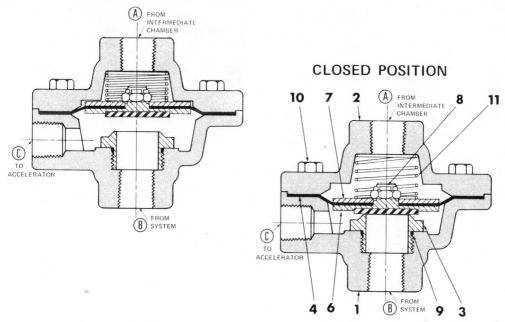

OPEN POSITION

Ⓐ FROM INTERMEDIATE CHAMBER

Ⓒ TO ACCELERATOR

Ⓑ FROM SYSTEM

CLOSED POSITION

10 7 2 Ⓐ FROM INTERMEDIATE CHAMBER 8 11

Ⓒ TO ACCELERATOR

4 6 1 Ⓑ FROM SYSTEM 9 3

Fig. 8.16. Antiflooding device to prevent water from entering accelerators. (Reliable Automatic Sprinkler Company, Inc.)

turers also provide antiflooding devices as separate valves which may be installed on existing dry pipe valves equipped with accelerators. Figure 8.16 illustrates this type of valve in both the open and closed positions.

The installation of valves similar to the one shown in Figure 8.16 tends to increase the operational reliability of the accelerator on older dry pipe automatic sprinkler systems.

ANTIFREEZE SOLUTION SPRINKLER SYSTEMS

When a limited number of automatic sprinkler heads are to be installed in an unheated area (such as along a loading dock or in a display window) a small dry pipe valve or an antifreeze solution sprinkler system may be used. The antifreeze solution sprinkler system is usually an extension of a wet pipe sprinkler system for a limited area which may be subjected to freezing temperatures, and use of the antifreeze solution must meet all of the local health department requirements if the sprinkler system is supplied or connected to the public water system. In the *NFPA Sprinkler System Standard,* the installation of an antifreeze solution sprinkler system is recommended only for 20 or fewer sprinkler heads.

Table 8.2 illustrates the types of antifreeze solutions that are recommended for use in antifreeze solution sprinkler systems. Note the difference in the formulations required when potable water supplies are involved.

Table 8.2.   Antifreeze Solutions for Sprinkler System Application*

TO BE USED IF PUBLIC WATER IS CONNECTED TO SPRINKLERS

MATERIAL	SOLUTION (BY VOLUME)	SPEC. GRAV. AT 60°F.	FREEZING POINT °F.
Glycerine	50% Water	1.133	−15
C.P. or U.S.P. Grade[1]	40% Water	1.151	−22
	30% Water	1.165	−40
	Hydrometer Scale 1.000 to 1.200		
Propylene Glycol	70% Water	1.027	+ 9
	60% Water	1.034	− 6
	50% Water	1.041	−26
	40% Water	1.045	−60
	Hydrometer Scale 1.000 to 1.120 (Subdivisions 0.002)		

[1]C.P. —Chemically Pure.
U.S.P.—United States Pharmacopoeia 96.5%.

SUITABLE FOR USE IF PUBLIC WATER IS NOT CONNECTED TO SPRINKLERS

MATERIAL	SOLUTION (BY VOLUME)	SPEC. GRAV. AT 60°F.	FREEZING POINT °F.
Glycerine	If glycerine is used, see Table 5-5.2.1.		
Diethylene Glycol	50% Water	1.078	−13
	45% Water	1.081	−27
	40% Water	1.086	−42
	Hydrometer Scale 1.000 to 1.120 (Subdivisions 0.002)		
Ethylene Glycol	61% Water	1.056	−10
	56% Water	1.063	−20
	51% Water	1.069	−30
	47% Water	1.073	−40
	Hydrometer Scale 1.000 to 1.120 (Subdivisions 0.002)		
Propylene Glycol	If propylene glycol is used, see Table 5-5.2.1.		
Calcium Chloride 80% "Flake"	Lb CaCl$_2$ per Gal of Water		
Fire Protection Grade[1]	2.83	1.183	0
Add corrosion inhibitor	3.38	1.212	−10
of sodium bichromate	3.89	1.237	−20
$\frac{1}{4}$ oz per gal water	4.37	1.258	−30
	4.73	1.274	−40
	4.93	1.283	−50

[1]Free from magnesium chloride and other impurities.

*From NFPA 13, *Standard for the Installation of Sprinkler Systems.*

Obviously, the antifreeze solution should be prepared to provide protection below the minimum expected temperature for the climate involved. The preferred piping arrangement at the interface between the antifreeze solution sprinkler system and the water sprinkler system is to have all of the antifreeze piping below the water-connected feed main, since all the antifreeze solutions are heavier than water.

Thus, it is not unusual to see antifreeze solution sprinkler system piping for loading-dock areas supplied from the second floor. However, in single-story buildings or other situations where it is not possible to have the antifreeze solution sprinkler system's sprinkler heads below the existing system, a 5-foot loop connection should be provided, as illustrated in Figure 8.17.

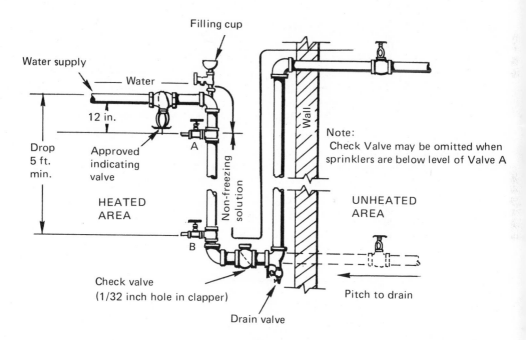

NOTE: The 1/32 inch hole in the check valve clapper is needed to allow for expansion of the solution during a temperature rise and thus prevent damage to sprinkler heads.

Fig. 8.17. Arrangement of supply piping and valves for antifreeze solution sprinkler system. (From **NFPA 13**, *Standard for The Installation of Sprinkler Systems*)

The antifreeze solution should be checked yearly to assure proper maintenance of the solution's capability to protect against the expected freezing temperatures. Note in Figure 8.17 that valve B may be used to obtain samples of the antifreeze solution needed to test the solution's ability to perform as originally mixed and installed within the piping system.

INSPECTION AND TEST OF THE DRY PIPE SPRINKLER SYSTEM

The initial test of a dry pipe automatic sprinkler system should be the hydro-static pressure test as conducted on the wet pipe sprinkler system of 200 psi for 2 hours, or 50 psi above the maximum static pressure, when the static pressure exceeds 150 psi. As with wet pipe systems, the supply mains to the dry pipe system should also be flushed to remove all foreign materials before the sprinkler system is connected to the supply main. Should hydrostatic testing be required at a time of the year when water freezes, an interim test with a minimum air pressure of 40 psi for 24 hours should be conducted. The required hydrostatic test should then be performed when permitted by the weather. In addition, the dry pipe sprinkler system should be subjected to an air pressure test. The air pressure on the system should be raised to a pressure of 40 psi for 24 hours, and repairs should be made to any leaks which cause the pressure to drop more than $1\frac{1}{2}$ pounds in 24 hours. An operating test of the main drain facility and a trip test of the dry pipe valve, as illustrated in Figure 8.19, should be conducted prior to acceptance of any new dry pipe sprinkler system.

Due to the complex design of the dry pipe valve, especially when equipped with quick-opening-devices, the inspection and testing of the dry pipe automatic sprinkler system is a complicated process. One of the advantages cited by Ault ("The Low-Differential Dry Pipe Sprinkler System"[3]) for the low-differential dry pipe valve was the relatively simple operating test procedure, since manual resetting of the valve was not required (as is the case with the conventional differential dry pipe valve). In 1965, it was estimated that 20,000 dry pipe valve trip tests were being conducted annually by the Insurance Services Offices.

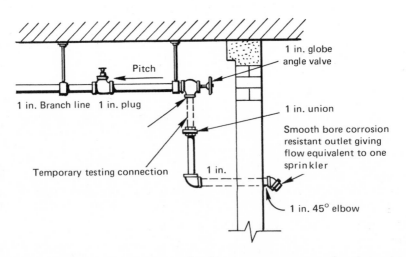

Fig. 8.18. Inspector's test pipe and valve arrangement on a dry pipe automatic sprinkler system. (From "The Inspector's Test-Pipe and Valve," by James M. Hammack[4])

James M. Hammack, in his article "The Inspector's Test Pipe and Valve," indicates that one of the principal purposes of the inspector's test valve on the dry pipe system is to enable the simulation of the operation of a single sprinkler head on the topmost branch sprinkler line.[4] Figure 8.18 illustrates the recommended installation procedure for the inspector's test valve on a dry pipe automatic sprinkler system. This illustration should be compared with the arrangement for the wet pipe automatic sprinkler system previously shown in Figure 7.15 of Chapter 7, "The Wet Pipe Automatic Sprinkler System."

Dry Pipe Valve Tripping Procedure

Due to the inspection procedures required and the need to completely restore the system to an operational condition with the dry pipe valve in a set condition, the dry pipe valve trip test procedure, without any operational or maintenance problems, often involves several hours. As previously explained, the trip testing of the dry pipe valve is a procedure for determining the actual operation of the system when a fire occurs, and for determining the pressure trip point of the differential dry pipe valve. The operation of the accessory equipment, including quick-opening-devices and alarm devices, is also observed.

The recommended dry pipe valve tripping procedure, as provided by some insurance authorities, is illustrated in Figure 8.19.

Most insurance authorities and engineers are carefully instructed to request that the actual manual operation of the valves on the sprinkler system always be performed by a representative of the building's occupant or owner. The representative of the owner (usually a maintenance person) will conduct the actual trip test of the system. Such a procedure reduces the possibility of any legal liability problems that might arise should a portion of the system fail under test conditions, and cause damage or injury.

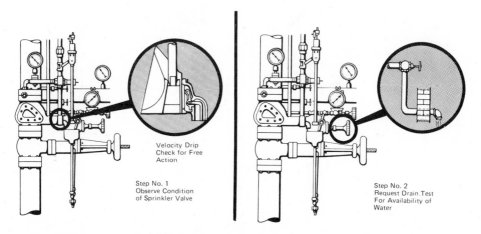

Velocity Drip
Check for Free
Action

Step No. 1
Observe Condition
of Sprinkler Valve

Step No. 2
Request Drain Test
For Availability of
Water

Fig. 8.19. Dry pipe valve tripping procedure. (Insurance Services Office)

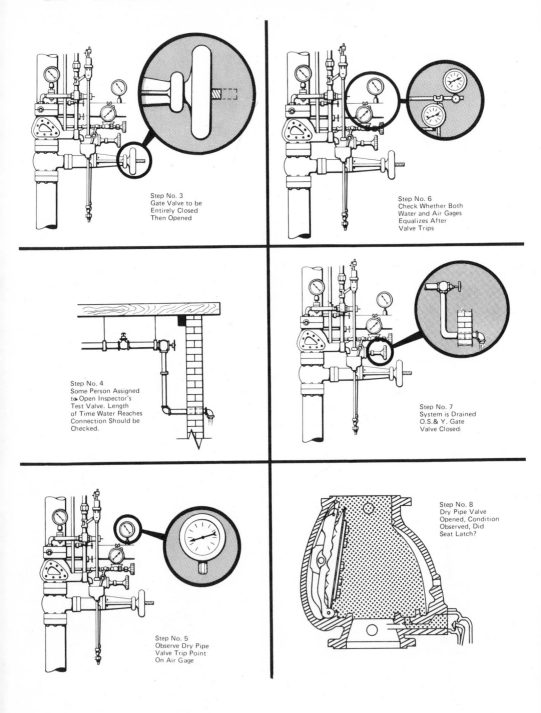

Step No. 3
Gate Valve to be
Entirely Closed
Then Opened

Step No. 4
Some Person Assigned
to Open Inspector's
Test Valve. Length
of Time Water Reaches
Connection Should be
Checked.

Step No. 5
Observe Dry Pipe
Valve Trip Point
On Air Gage

Step No. 6
Check Whether Both
Water and Air Gages
Equalizes After
Valve Trips

Step No. 7
System is Drained
O.S.& Y. Gate
Valve Closed

Step No. 8
Dry Pipe Valve
Opened, Condition
Observed, Did
Seat Latch?

Fig. 8.19.   (Continued)

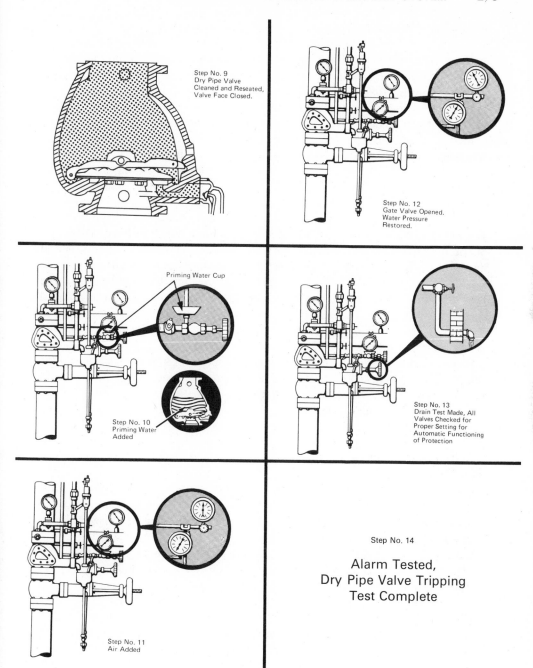

Step No. 9
Dry Pipe Valve
Cleaned and Reseated,
Valve Face Closed.

Step No. 12
Gate Valve Opened.
Water Pressure
Restored.

Priming Water Cup

Step No. 10
Priming Water
Added

Step No. 13
Drain Test Made, All
Valves Checked for
Proper Setting for
Automatic Functioning
of Protection

Step No. 11
Air Added

Step No. 14

Alarm Tested,
Dry Pipe Valve Tripping
Test Complete

Fig. 8.19. (Continued)

Restoration Procedures for the Dry Pipe Sprinkler System

Once the dry pipe sprinkler system has operated and it is certain that the fire has been completely extinguished, the water control valve on the main riser may be closed. However, personnel must always be stationed at the valve, preferrably with radio equipment (as previously indicated in Chapter 2, "Fire Department Procedures for Automatic Sprinkler Systems"), to be available to immediately reopen the valve should the fire rekindle. The resetting of a dry pipe valve is a complex procedure involving a considerable period of time due to the need to open and blow out all the low-point drains on the system (such as areas under stairs). Thus fire department personnel are not usually involved in the resetting of the dry pipe valve and placing the system back into service. However, fire department personnel should never leave the premises of a property without having first notified a responsible representative of the owner of the property that the dry pipe sprinkler system has been in operation and needs to be reset. Sprinkler manufacturers and their representatives provide service or maintenance contracts that include resetting dry pipe valves. If the weather is mild, the system can be retained in a wet condition after the fused sprinkler heads are replaced. This procedure will provide protection similar to that provided by a wet pipe sprinkler system without the alarm devices, until the appropriate personnel arrive to reset the system.

The restoration procedure for the dry pipe sprinkler system is similar to the restoration procedure for the wet pipe sprinkler system. After the system's main riser water control valve has been closed (with personnel standing by), the main drain and all auxiliary drains should be opened. The fused sprinkler heads should be replaced with identical heads relative to type and temperature rating. The conventional differential dry pipe system valve cover must then be removed, the clapper unlatched, the seat and clapper seal cleaned with a soft cloth, and the clapper set on the seat. The valve cover should be replaced, and the priming water should be added to the valve until it flows from the priming water level valve. Some air should then be admitted to the system in order to blow out the low point drains, and all the drains should then be closed. The air pressure should be increased on the system to approximately 20 pounds above the trip point of the dry pipe valve. A general approximation of the system's air pressure can be made if the manufacturers' instruction chart (as shown earlier in this chapter in Table 8.1) is not available, by dividing the water pressure by 5 and adding 20.

When the required air pressure is on the system and no water is leaking out of the velocity drip valve from the intermediate chamber, the dry pipe valve clapper is properly seated. At this point the main riser water control valve should be reopened. However, in order to avoid creating a pressure surge which might trip the dry pipe valve and thus require the draining of the system and the manual resetting of the valve, it is a good procedure to open the 2-inch main drain valve. Tripping can usually be avoided if the main drain valve is opened several turns until a full stream is flowing, and then slowly closed while observing the water pressure gauge so that pressure can be steadily applied on the dry pipe valve clapper. Once the main drain valve is closed, the main riser water control valve can be fully opened and sealed, or locked in the open position. The alarm devices should be tested by operating the alarm test valve which allows water to pass from below the dry pipe valve and into the alarm line to the water motor gong and the electrical pressure switch in exactly the same way as the alarm test valve on the wet pipe sprinkler system. However, the inspector's test valve should *not* be operated: since the inspector's test valve simulates the operation of a sprinkler head, the dry pipe valve will trip and the entire restoration procedure will have to be repeated.

SUMMARY

The dry pipe automatic sprinkler system should be installed only in areas or occupancies that are unheated and are subjected to freezing temperatures. Due to the design of the dry pipe system—without water in the sprinkler pipe distribution system—the time it takes for the system to operate is increased because of the extra time it takes for the valve to operate, and for the additional time it takes for the water to fill the system and reach the head which has fused from the heat of the fire. Thus, more sprinkler heads will always be fused on dry pipe systems than on wet pipe sprinkler systems. Because of these variables, the time of operation for the dry pipe automatic sprinkler system can be expected to be increased by at least 50 percent.

The operating time of the dry pipe valve may be shortened, and the time for the water pressurization of the system reduced with quick-opening-devices such as accelerators or exhausters installed on the dry pipe valve. More accelerators than exhausters are usually found on dry pipe systems. However, depending primarily on the design preferences of the manufacturer, either device can be used. Because their dual chamber pressure differential design requires restricted orifices, maintenance requirements for all quick-opening-devices are extensive.

The low-differential dry pipe valve, being primarily a modified alarm check valve, offers some advantages relative to reduced water velocity and flow into the system with the faster operation of the valve, and reduced maintenance requirements since quick-opening-devices are not usually required.

Frequent inspections of dry pipe valves are a necessity, and automatic maintenance of the air supply to the system is usually recommended. The tripping of dry pipe valves is an annual procedure usually required by insurance authorities. Dry pipe valves presently require extensive maintenance and testing to insure operational reliability. However, the performance record of the dry pipe sprinkler system is commendable, although admittedly not equivalent to the performance record of the wet pipe sprinkler system.

SI Units

The following conversion factors are given as a convenience in converting to SI units the English units used in this chapter:

1 inch	= 25.400 mm
1 pound (force)	= 4.448 N
1 pound (mass)	= 0.454 kg
1 psi	= 6.895 kPa
1 gallon	= 3.785 litres
1 gpm	= 3.785 litres/minute

ACTIVITIES

1. What type of automatic sprinkler system do you think would be best for installation on a department store loading platform in Halifax, Nova Scotia? Explain your reasoning, and describe how the system you have chosen would be properly maintained.
2. Outline the steps necessary for restoring to operation a dry pipe automatic sprinkler system following its use during a fire. Choose one of the steps and describe in detail how each procedure is carried out, including the necessary precautions to be observed.
3. Prepare a written statement explaining the principles of operation for both an accelerator and an exhauster. In your statement, explain the importance of Q.O.D.s on sprinkler piping systems.
4. Explain the different locations on the main sprinkler riser for the fire department connection on the dry pipe sprinkler system as compared with the wet pipe sprinkler system.
5. In outline form, list the advantages and the disadvantages of the low-differential dry pipe sprinkler system as compared to the conventional dry pipe sprinkler system.
6. Determine the locations in your area that have dry pipe automatic sprinkler systems. Then choose at least four, and compile a listing of the manufacturer's name, model number, and year of manufacture for each dry pipe valve.
7. Explain the importance of the use of priming water in dry pipe valves.
8. Discuss some of the major differences between wet pipe automatic sprinkler systems and dry pipe automatic sprinkler systems. Prepare a written comparison of both types of systems.

SUGGESTED READINGS

Ault, Wayne E., "The Low-Differential Dry Pipe Sprinkler System," *Fire Journal,* Vol. 59, No. 4, July 1965, pp. 12–16.

Automatic Sprinkler Corporation of America, "Low Differential Dry Pipe Systems," Bulletin 119-7-65, 1965, Automatic Sprinkler Corporation of America, Cleveland.

Baldwin, R. and North, M. A., "The Number of Sprinklers Opening and Fire Growth," *Fire Technology,* Vol. 9, No. 4, Nov. 1973, pp. 245–253.

Dry Pipe, Deluge and Pre-Action Valves for Fire Protection Service, 1972, 3rd edition, Underwriters Laboratories, Northbrook, Ill.

Reliable Automatic Sprinkler Company, Inc., "Reliable Accelo-Check," Bulletin 321, 1970, Reliable Automatic Sprinkler Company, Inc., Mount Vernon.

——, "Reliable Accelerator, Model B," Bulletin 322, 1972, Reliable Automatic Sprinkler Company, Inc., Mount Vernon.

Star Sprinkler Corporation, "Star Model C Dry Pipe Valve," Bulletin 6-C, 1965, Star Sprinkler Corporation, Philadelphia.

——, "Star Model BB Exhauster," Bulletin EBB, 1965, Star Sprinkler Corporation, Philadelphia.

BIBLIOGRAPHY

[1]Baldwin, R. and North, M. A., "The Number of Sprinklers Opening and Fire Growth," *Fire Technology*, Vol. 9, No. 4, Nov. 1973, pp. 245–253.

[2]NFPA 13, *Standard for the Installation of Sprinkler Systems*, 1975, NFPA, Boston.

[3]Ault, Wayne E., "The Low-Differential Dry Pipe Sprinkler System," *Fire Journal*, Vol. 59, No. 4, July 1965, pp. 12–16.

[4]Hammack, James M., "The Inspector's Test Pipe and Valve," *Fire Journal*, Vol. 14, No. 2, March 1970, pp. 28–30.

Chapter 9

Deluge and Preaction
Automatic Sprinkler Systems

CHARACTERISTICS OF THE DELUGE AUTOMATIC SPRINKLER SYSTEM

The basic design of the deluge automatic sprinkler system is a modification of the design of the wet and the dry pipe automatic sprinkler systems previously described in Chapters 7 and 8 of this text. The deluge sprinkler system has two important and distinct modifications. First, the system is equipped with open sprinkler heads or nozzles. Thus, since the sprinkler heads and nozzles do not have heat-responsive elements, and the system is also designed to operate automatically, a separate detection system is provided. This detection system is designed to activate the deluge valve which, when activated, discharges water from every open head that is connected to the deluge system piping controlled by the specific deluge valve. It is the characteristic multihead large area water discharge which identifies the deluge automatic sprinkler system as a "deluge" system.

The deluge automatic sprinkler system is designed for and installed in areas of severe fuel hazard where a fast-spreading fire can be expected to occur. Buildings or occupancies with flammable liquid hazards, woodworking occupancies, cooling towers, explosive or ordnance plants, and aircraft hangars are the major areas in which deluge automatic sprinkler systems are employed. In aircraft hangars, the deluge automatic sprinkler system can often be found combined with a mechanical foam agent capability. If a fast-spreading fire occurred in one of these types of occupancies, a wet or dry pipe automatic sprinkler system with the individually fusing standard sprinkler heads could not control the rapid fire spread, while the deluge automatic sprinkler system could.

Because of the resultant large quantities of water discharged from all the sprinkler heads on the deluge piping system when the deluge valve has been tripped, deluge automatic sprinkler systems, when installed to protect large areas such as aircraft hangars, require extremely large water supply capacity and pumping equipment. Figure 9.1. illustrates three deluge valves on a water supply manifold, as installed in an aircraft hangar. Figure 9.2 shows a series of manifolded fire pumps to supply deluge systems that have a mechanical protein foam capability in another aircraft repair and overhaul hangar.

Fig. 9.1. Deluge valves installed on a water-supply manifold for an aircraft hangar. (Duane McSmith)

Deluge systems that have nozzles on the system branch lines in addition to, or in place of, the open sprinkler heads, are advantageous in special hazard situations. Nozzles on the piping system are typically utilized with the deluge valve-operated water-spray systems, and will be reviewed in detail in Chapter 11, "Exposure Sprinkler and Water Spray Systems." In some occupancies, environmental conditions may cause the open sprinkler heads of deluge automatic sprinkler systems to become loaded with accumulations of foreign matter. Such loading of the sprinkler heads may cause them to fail to operate. Therefore, blow-off caps have been designed to protect the heads. Figure 9.3 illustrates a blow-off cap provided for open sprinkler heads on deluge systems.

Fig. 9.2.   Fire pumps manifolded to supply a deluge system with foam capability.
(Michael W. Magee)

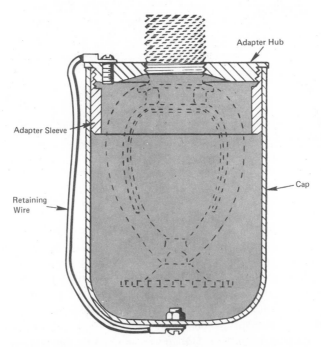

Fig. 9.3.   Blow-off cap for open sprinkler heads.
(Grinnell Fire Protection Systems Company, Inc.)

The deluge automatic sprinkler system consists of the following essential components:

1. A reliable water supply, usually provided with stationary fire pumps because of the high pressure and flow requirements from the simultaneous flow of numerous heads.

2. An indicating water control valve of the O.S. & Y., P.I.V., or P.I.V.A. type on the main riser.

3. An approved deluge valve. Deluge valves are all manufactured with a manual release, and provision for remotely located manual releases. The valves may be used as provided for operation by various detection devices that are electrically, hydraulically, or pneumatically operated.

4. An approved detection system which is listed for service as a releasing device. The detection system and devices may be selected from devices that operate electrically, hydraulically, or pneumatically. It is desirable for the detection system to be supervised, and the *NFPA Sprinkler System Standard* requires supervision if the system contains more than 20 sprinkler heads.[1]

5. A system of riser, feed mains, and branch lines, with air in the system above the deluge valve at atmospheric pressure, since the piping system is open to the atmosphere.

6. Open sprinkler heads and nozzles installed on the branch lines of the sprinkler piping system.

There is a modification of the standard deluge sprinkler system, usually adopted when faster system operation is required, as in ordnance or explosive plants. This modified system is called a preprimed deluge system, since the piping system is filled with water prior to activation of the deluge valve.

Preprimed Deluge System

The preprimed deluge system adapts the principle utilized in many fire departments of carrying an attack line fully charged with water, thereby reducing the time required to discharge water on the fire. In the preprimed deluge system, the piping system above the deluge valve is filled with water after the valve is set. To prevent the water in the piping system from flowing out of the open sprinkler heads or nozzles, rubber or neoprene plugs are placed in the orifice of the head or nozzle. The piping system is filled with water in a static pressure condition and, as the valve trips, the system is pressurized and water flows from the sprinkler heads and nozzles.

The preprimed deluge system has effectively reduced the time for the water to travel from the valve to the sprinkler heads or nozzles. Preprimed deluge systems on relatively small systems, utilizing pneumatic detection devices provided with heat-collecting canopies over the detectors, have been installed for many years in propellent plants, with a total water-discharge response time of 0.5 second. Therefore, preprimed deluge systems are not suitable for installation in unheated structures or buildings. In summary, note the preprimed deluge sprinkler system involves no new or additional equipment other than orifice plugs for the heads and nozzles. The preprimed system utilizes all the standard deluge system equipment, and essentially involves modification of the equipment and its installation.

Ultra High-Speed Automatic Deluge Sprinkler System

The ultra high-speed deluge sprinkler system, a modification of the pre-primed deluge sprinkler system, is a result of the technological advances of the past several decades. G. J. Grabowski, in "Explosion Protection Operating Experience," stated that in 1958 the explosion suppression system was introduced in the United States with the application of very high-speed detection devices and the utilization of explosively activated suppression devices.[2] These improvements were adapted to the preprimed deluge sprinkler system in the early 1960s.

The ultra high-speed deluge system has the basic components of the pre-primed deluge system. However, the ultra high-speed deluge system employs a detection system of the electronic type, with flame- or pressure-sensitive detectors of the type developed for explosion suppression systems. These systems have been installed for the protection of mixing vessels or reactors, and cutting, drilling or machining equipment involved with propellents or explosives. The ultra high-speed systems are usually installed for the protection of the equipment, while preprimed deluge systems are utilized to protect the structure or building.

The ultra high-speed deluge valve is primarily a small explosive squib-operated deluge valve that is electronically activated by the detection system. To eliminate the water transport time, the piping system usually contains water above the valve, as in the preprimed deluge system. The total water discharge response time for the ultra high-speed deluge system is approximately 0.030 to 0.040 seconds. Thus, sprinkler systems are available that operate and actually prevent explosive pressure generation. Figure 9.4 shows an ultra high-speed deluge valve.

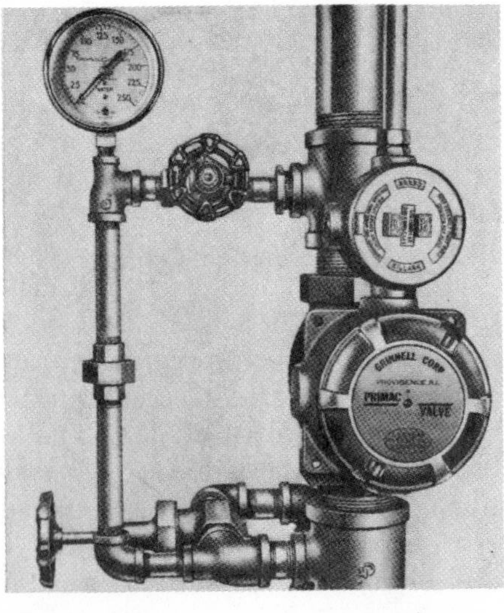

Fig. 9.4. Ultra high-speed deluge valve. (Grinnell Fire Protection Systems Company, Inc.)

MODES OF DELUGE VALVE OPERATION

Deluge valves have been designed, approved, manufactured, and installed on systems with various modes of operation. Thus, deluge valves are in service and are available for new systems that operate on an electrical, hydraulic, or pneumatic principle. Also available are deluge valves that are hydraulic in design and in principle of operation, which may be connected to any of the various types of detection systems. These valves are compatible to multimode detection and, since they are basically hydraulically operated valves, will be discussed later in this chapter in the section on hydraulically operated valves.

Electrically Operated Deluge Valves

The electrically operated deluge valve is primarily designed to be used as a portion of a total electric system, with both the detection system and the detectors of the electrical type. The major advantage of the electrically operated deluge valve is the speed of operation for the total response system, including the operation of the detectors, the transmission of the signal through the detection system, and the operation of the deluge valve. Figure 9.5 illustrates the installation of an electrically operated deluge valve.

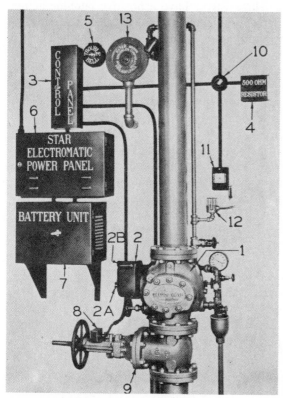

Fig. 9.5. Electrically operated deluge valve. (Star Sprinkler Corp.)

In Figure 9.5, the standard operating components of the electrically operated deluge valve can be identified as follows:

Deluge valve—(1)
The electrical release for the deluge valve clapper—(2)
The manual activation of the deluge valve release—(2B)
The manual reset for the deluge valve release—(2A)
The release control panel—(3)
The end line resistor for the detection system, which provides supervisory electrical current for the detection system—(4)
Trouble bell for the supervisory circuits faults—(5)
Power supply panel, providing primary electrical power—(6)
Storage batteries, supply secondary electrical power—(7)
Tamper switch, supervision of O.S. & Y. water control valve—(8)
O.S. & Y. water control valve—(9)
Detector unit of rate-of-rise and fixed-temperature or fixed-temperature only operation—(10)
Remotely located manual station—(11)
Electric pressure switch for water flow signal—(12)
Water motor gong—(13)

The main drain valve is located directly behind the riser on this valve and is not visible in Figure 9.4. Most deluge valves have a positive latching action which involves a mechanical latch arrangement. However, one of the hydraulically operated deluge valves does not have this mechanical latched design feature. Thus, the majority of the deluge valves have the mechanical latch design which requires the removal of the valve cover, and the manual resetting of the valve. Most of the deluge valves involve a single clapper provided with latches or stops to prevent it from swinging shut and causing a water surge or hammer. The main drain valve is located below the clapper; therefore, the latches allow the system to drain.

In some hazardous areas electrically operated deluge valves may require the installation of explosion proof electrical equipment and detectors. The provision and maintenance of standby batteries or alternate power sources is an important design factor developed to ensure the operation of the electrically operated deluge valve.

The provision of manual activation stations is an important factor on deluge systems: remember, the detection system provides the basic and primary activation of the deluge valve and, should it fail or become deactivated, the manual activation stations become the only means of tripping the deluge valve. There are many situations in hazardous occupancies when it may be desirable to activate the deluge system as a preventative measure, making manual activation a necessity. Typical preventative situations may involve equipment malfunctions and flammable liquid or chemical spills.

It is usually desirable to provide remotely located manual activation stations for the deluge system at exits from the protected area so that personnel may activate the deluge system as they evacuate the area. Figure 9.6 illustrates a

Fig. 9.6. Remotely located manual release station for an electrically operated deluge valve. (Star Sprinkler Corp.)

remotely located manual activation station for an electrically operated deluge valve. Note the appearance of the releasing device is similar to a manual fire alarm box station.

The electrically operated deluge system illustrated in Figure 9.5 operates the detection system on a 28 volt dc operation, unless higher voltages are permitted by the insurance authority having jurisdiction. However, the power in the power supply panel usually is line voltage of 120 volts.

It should be noted that since most deluge valves mechanically latch and are manually reset, priming water is not required above the valves and water is usually not provided above the clapper unless the deluge valve is being used on a preprimed deluge system. On most deluge valves, the water line to the alarm devices leaves the deluge valve immediately above the clapper. Therefore, when the clapper is released and water flows into the sprinkler piping system, water is also flowing to the alarm devices.

Figure 9.7 shows the electrically operated deluge valve in both the set and tripped positions.

Pneumatically Operated Deluge Valves

The pneumatically operated deluge valve is usually installed as a complete system with pneumatically operated rate-of-rise detectors. The deluge valve is essentially a pneumatically released mechanically operated valve.

Generally, the pneumatic detector tubing consists of hollow copper tubing approximately $\frac{1}{8}$ inch in diameter, arranged in circuits with the rate-of-rise heat

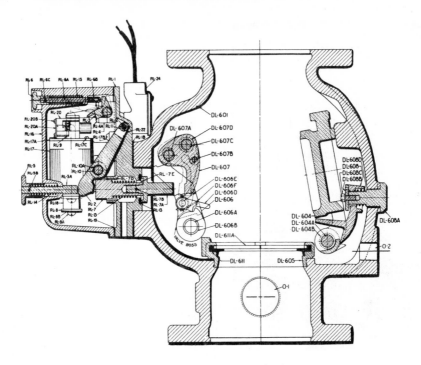

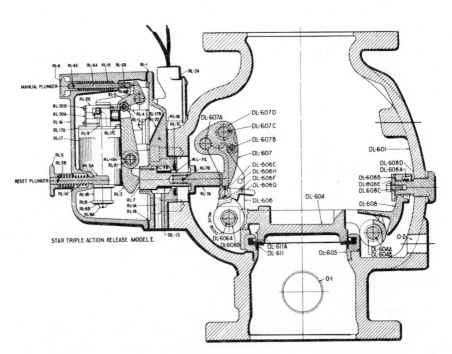

Fig. 9.7. Electrically operated deluge valve in set and tripped positions. (Star Sprinkler Corp.)

actuating devices limited to five or six detectors per circuit. The *NFPA Sprinkler System Standard* requires the supervision of the detection circuits for any deluge sprinkler system that has 20 or more sprinkler heads or nozzles.[1] The pneumatic detectors and the copper tubing circuits are supervised by maintaining a small amount of air or nitrogen (usually about $1\frac{1}{2}$ psi) in the tubing circuits. Should the detector be damaged, develop a leak, or should the tubing leak, the loss of the supervisory air or gas causes a drop in the supervisory pressure, thus activating a trouble signal.

When a fire occurs, the heat of the fire causes the air in the rate-of-rise detector (identified as a Heat Actuating Device—HAD) to increase in pressure, due to the restricted volume of the HAD and the copper tubing. This impulse of increased pressure is transmitted through the tubing back to a manifold or header containing mercury checks and a diaphragm-operated release. When this release is operated, a weight is dropped which releases the latch on the deluge valve clapper. The resulting water supply pressure causes water to flow into the riser. Figure 9.8 illustrates the operation of the pneumatic tubing system from the heat actuated device to the diaphragm valve release.

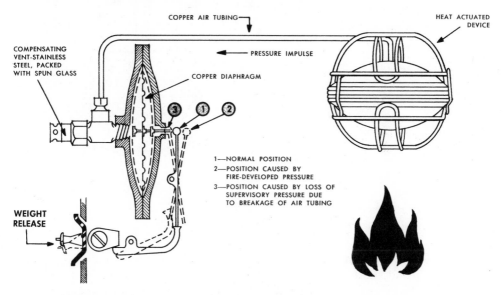

Fig. 9.8.   *HAD and tubing system for diaphragm-operated valve release on pneumatically operated deluge valve.* (Automatic Sprinkler Corp.)

The diaphragm release mechanism will be equipped with a compensating vent for systems with a single circuit of pneumatic detectors. However, deluge systems with multiple circuits of pneumatic detectors have the compensating vent for each circuit located with the mercury checks.

Essentially, the mercury checks are pneumatic check valves that prevent the pressure increase generated in the circuit exposed to the fire from being dissipated through the other detection circuits instead of operating the diaphragm valve release.

The compensating vents are installed on the detection circuits to allow for the relief of air expansion which may occur in the system due to ambient temperature changes. This prevents the activation of the deluge valve from atmospheric conditions.

Figure 9.9 illustrates the arrangement of the mercury checks and the compensating vents on a pneumatically operated deluge valve with four detector circuits. In Figure 9.9, note the arrangement of the latching mechanism for the deluge valve clapper. Also note the weight which is allowed to drop down the metal guide rod to release the clapper by the increase of pressure in a detection circuit.

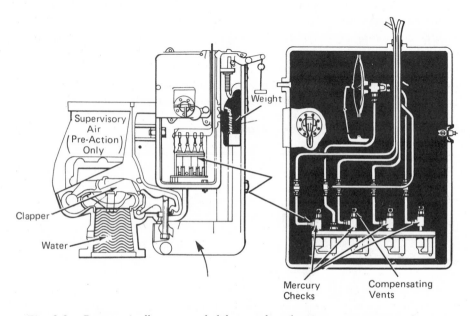

Fig. 9.9. Pneumatically operated deluge valve showing arrangement of mercury checks and compensating vents. (Insurance Services Office)

The required manual activation for the pneumatically operated deluge valve consists of the short chain and handle adjacent to the weight in Figure 9.9. It should be noted the pneumatically operated deluge valve may have the manual activation stations remotely located with a copper tubing circuit to the manual station. Activation of the handle at the manual station creates a pressure increase through a small diaphragm arrangement which is transmitted back to the diaphragm valve release. However, some facilities find it more efficient or economical to utilize a remotely operated mechanical activation by running cables through conduit to the manual activation station. Thus, activation of the handle attached to the cable trips the weight, which releases the clapper, allowing water to flow into the system.

The proper placement of the HADs may be critical to the operation of the deluge system. Thus, it is essential to place these detection devices in locations where they will immediately operate from the expected fire occurrence. It has been demonstrated that canopies, acting as heat collectors, installed immediately above the heat activating devices, can reduce the response time. (See Chapter 6, "The Automatic Sprinkler Head," section titled "Automatic Sprinkler Head Installation and Operating Variables," and article in that section by Richard E. Rice titled "Effects of Protective Materials on the Operation of Sprinkler Heads"[18]). However, since these detectors are rate-of-rise devices, care should be exercised when installing them in occupancies where cold air is allowed into the occupancy; for example, in occupancies having large doors, such as warehouses, aircraft hangars, and storage buildings. The cold exterior air cools the detectors when the doors are open; then, after the doors are closed, the interior heaters provide a temperature rise which exceeds the tolerance level of the detector, thus tripping the deluge valve.

Hydraulically Operated Deluge Valves

Hydraulically operated deluge valves have several unique, standard features that can be considered as advantages that pneumatically or electrically operated valves may not have. There is available a hydraulically operated deluge valve which does not require a positive latch arrangement. Instead, this valve operates on a differential principle. This means that the valve does not require the resetting of the clapper with physical and manual entry into the deluge valve housing. Figure 9.10 shows a diagram of the differential hydraulically operated deluge valve in both the set and the tripped positions.

Figure 9.10 shows the hydraulic deluge valve in the set position with the pressure in chambers A and B equal. However, since the surface area of the clapper exposed to the pressure in chamber B is greater, this pressure keeps the clapper closed and the water out of the deluge system riser. The water pressure is released from chamber B through an actuation valve connected to the detection system. Since the pressure cannot be equalized through the restricted inlet orifice, the clapper moves upward, allowing the water to flow through the valve into the deluge system riser, the sprinkler piping, and the open heads or nozzles throughout the deluge system area. It is important to note that a reduced pressure is maintained within chamber B by the restricted orifice on the water bypass line from below the valve into chamber B.

The only actions required in resetting the deluge valve illustrated in Figure 9.10 are: (1) turning off the water at the riser water control valve, (2) resetting the actuating control from the detection system, (3) draining the system above the deluge valve, and (4) turning the main water control valve into the operating position after pressurizing chamber B with the bypass valve. The water can be drained out of chamber B of the hydraulically operated differential deluge valve by operating the actuation valve by means of a pilot line of air-or water-operated closed sprinkler heads that are installed on $\frac{1}{2}$-inch pipe or electrical detectors connected to an electrically operated solenoid valve. Thus, the hydraulically operated deluge valve is the only type of deluge valve which may be routinely activated by pneumatically, hydraulically, or electrically operated

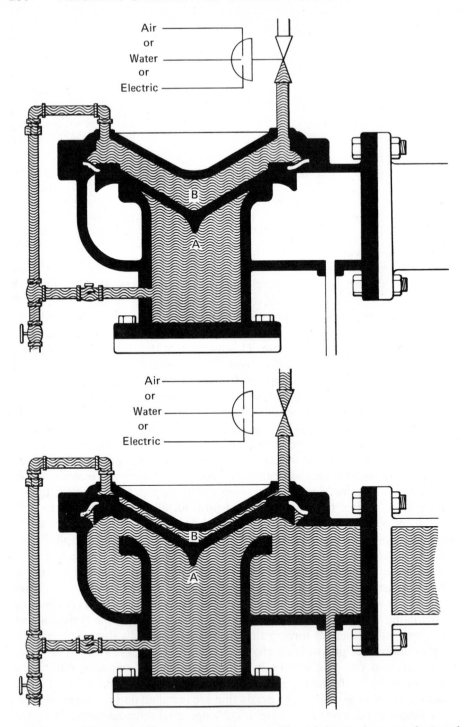

Fig. 9.10. Hydraulically operated differential deluge valve in the set and tripped positions. (The Viking Corp.)

detection devices. The previously discussed electrically and pneumatically operated deluge valves, especially the latter, can be modified to provide more than one mode of detection although they are usually used in the single mode.

The application of the pneumatic operation on a hydraulic differential deluge valve with the rate-of-rise and fixed-temperature pneumatic release with a closed sprinkler head on the ½-inch air pilot line is shown in Figure 9.11. An accelerator can be installed on the pneumatic detection system line to speed up the operation of the hydraulic deluge valve. Since this differential-type hydraulic deluge valve operates on a drop in air pressure similar to the differential dry pipe valves previously discussed in Chapter 8, "Dry Pipe Automatic Sprinkler Systems," an accelerator will also reduce its response time. An air pilot line with pneumatic releases for the hydraulically operated differential deluge valve is shown in Figure 9.11.

In Figure 9.11, the manual release for the deluge valve is located on the pneumatic pilot line between the deluge valve and air reservoir adjacent to the top air gauge. When hydraulic operation of the detection system is desired, a

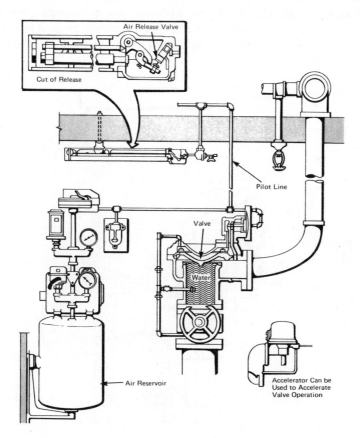

Fig. 9.11. Differential hydraulically operated deluge valve with pneumatic detection and activation of the valve. (Insurance Services Office)

water-filled pilot line is utilized with closed sprinkler heads on the line. Thus, a drop in the water pressure rather than a drop in air pressure actuates the valve; the operation of the valve relieves the water pressure in the topmost chamber eliminating the valve differential, and the valve operates with water flow into the sprinkler system.

The unique features of the hydraulically operated differential deluge valve shown in Figures 9.10 and 9.11 are the elimination of the need to manually remove the cover of the valve housing, and physically release and manually reset the clapper on the valve seat.

There are a number of hydraulically operated deluge valves which have a positive latching feature. Resetting of these valves requires manual entry into the valve housing, release of the clapper from the latch, and manually placing of the clapper back into the set position with a positive latching of the assembly. Figure 9.12 is a diagram of a hydraulically operated deluge valve of the positive latching design.

The hydraulically operated deluge valve illustrated in Figure 9.12 may be provided with pneumatically operated detectors consisting of closed sprinkler heads on an air or nitrogen pilot line, hydraulic operation with closed sprinkler heads on a water pilot line, or electric detector and manual station operation with a solenoid valve to operate a drain to relieve the hydraulic pressure against the diaphragm.

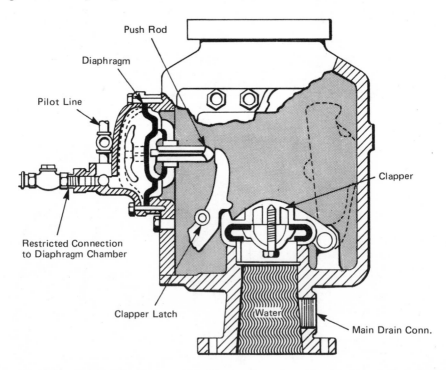

Fig. 9.12. Hydraulically operated deluge valve of positive latching design. (Insurance Services Office)

In the set condition, the water pressure on the left side of the valve (see Figure 9.12) is equal to the water pressure under the deluge valve clapper. The push rod is connected to the diaphragm and applies pressure to the clapper latch, thus providing positive latching action on the clapper. The water pressure is supplied through a restricted orifice to the diaphragm chamber by a $\frac{1}{2}$-inch pipe bypass from below the main water control valve on the riser.

When the detection system or a remotely located manual activation station is operated, the water pressure in the diaphragm chamber drops because the water is released from the chamber faster than it can be replaced through the restricted orifice. Therefore, the diaphragm moves, causing the push rod to release the clapper latch. The valve trips when the water pressure in the diaphragm chamber is approximately 40 percent of the water pressure in the water supply main on the underside of the deluge valve clapper. Note the diaphragm chamber is activated with a release of water from the chamber through the operation of the detection system. Therefore, even with air or nitrogen in the pilot line for pneumatic operation, the system will operate on a decrease and loss of pressure. However, the pneumatically operated deluge valve previously described in this chapter operates on an increase in the pneumatic pressure. Figure 9.13 illustrates the component parts of the hydraulically operated deluge valve with the positive latch feature, and arrangement of solenoid valve on the water piping to the diaphragm chamber to enable activation by electrical detection devices and manual stations.

The hydraulically operated deluge valves illustrated in Figures 9.12 and 9.13 require manual entry into the valve with the removal of the valve cover plates,

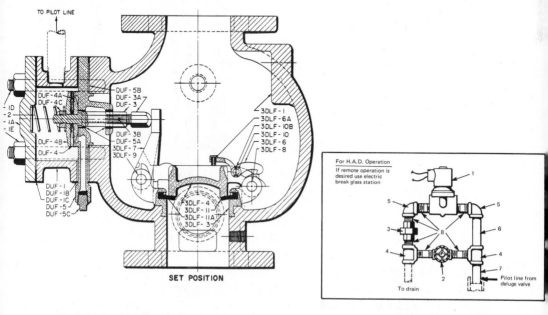

Fig. 9.13. Hydraulically operated deluge valve with positive latch feature, and arrangement of solenoid valve for electrical detector operation. (Star Sprinkler Corp.)

the release of the clapper, and the manual resetting of the clapper on the valve seat.

One of the established uses of the deluge sprinkler system is the distribution of mechanical foam through the system. The application of the protein or the fluoroprotein type of foam through the deluge sprinkler system requires the use of the foam-water sprinkler heads and a proportioning or pumping system to properly mix the water and the foam liquid into solution.

FOAM APPLICATION THROUGH DELUGE SPRINKLER SYSTEMS

For approximately thirty years, mechanical foam has been successfully applied through deluge sprinkler systems equipped with foam-water sprinkler heads. The foam-water sprinkler head is required to induct the air into the flow of the foam liquid and water solution that produces the mechanical foam. Figure 9.14 shows the design of typical foam-water sprinkler heads of the standard sprinkler design for both the upright and the pendent types.

Mechanical foam liquid is proportioned into the water in the sprinkler system with balanced proportioning equipment or foam liquid pumps. A typical arrangement for a foam pump system is shown in Figure 9.15.

Fig. 9.14. Standard foam-water sprinkler heads of upright and pendent types. (Automatic Sprinkler Corp.)

"Super 'Garages' for Jumbo Jets Require Super Fire Protection Systems," an article written by Charles Averill, reports on the mechanical protein foam system that discharges foam through the foam-water sprinkler heads installed on the deluge sprinkler systems in the American Airlines Hangar at Los Angeles.[3] The system uses low-expansion protein mechanical foam distributed through 16 deluge sprinkler systems with pneumatic rate-of-rise heat-actuated devices, and 2,480 foam-water sprinkler heads. These systems are designed to

Fig. 9.15. Foam liquid pump and mechanical foam liquid concentrate tank for protein foam application through deluge sprinkler system. (Grinnell Fire Protection Systems Company, Inc.)

discharge foam at a solution rate of 0.2 gpm/sq ft, with each of the sprinkler heads in the hangar area designed to cover from 85 to 90 square feet of floor area.

The water supply for the hangar system described by Averill consisted of a maximum estimated design demand of 22,750 gpm provided from suction tanks supplying seven 4,000 gpm pumps. The foam liquid supply consisted of mechanical foam liquid concentrate of the three percent type stored in two 6,000 gallon tanks with two 750 gpm centrifugal pumps. The foam liquid pumps supplied the foam liquid concentrate to the four water deluge valve locations; the foam concentrate entered the deluge system risers above the deluge water valves. A calibrated metering orifice in the connection to the deluge system risers controlled the flow of the concentrate liquid into the riser to maintain a three percent solution. Figure 9.16 shows the arrangement of the foam concentrate liquid control valves and the deluge valves on the water system at one of the four water deluge valve locations in the aircraft hangar system described by Averill.

As indicated, the deluge system described by Averill was activated by manual releases or the pneumatic rate-of-rise type of detection device on pneumatic tubing. Extensive tests of the system with flammable liquid fires of various sizes indicated the detection system would operate to trip the system after approximately 48 seconds with a fire area of 100 square feet, involving both 30 and 42 gallons of fuel.

David E. Breen, in an article titled "Hangar Fire Protection with Automatic AFFF Systems," reported that Aqueous Film Forming Foam (AFFF), applied through deluge systems equipped with the standard sprinkler heads rather

Fig. 9.16. Deluge valve and foam concentrate valve arrangements, with foam valves on the right in hangar system. (From "Super 'Garages' for Jumbo Jets Require Super Fire Protection Systems," by Charles F. Averill)[3]

than foam-water sprinkler heads, appeared to extinguish the fire 1.3 to 1.6 times faster with the identical application rates of 0.16 gpm/sq ft of floor area.[4] Breen also indicated an additional advantage of applying the aqueous film forming foam through the standard sprinkler head is a lower operating discharge pressure at the sprinkler head of approximately 10 psi as opposed to 25–50 psi for the foam-water sprinkler head. Breen explained the use of foam monitor nozzles to automatically sweep the lower fuselage areas that are shielded from the overhead discharge of the sprinklers appeared to be a highly effective technique. He stated that application rates for the foam monitor nozzles of between 0.1 and 0.2 gpm/sq ft seemed adequate. Figure 9.16 illustrates one of the automatically activated and controlled foam monitor nozzles mounted on the wall of a hangar as utilized in addition to an aqueous film forming foam discharge from the deluge sprinkler system through standard sprinkler heads, or the protein-type foam as discharged through foam-water sprinkler heads.

Averill further reported the foam liquid concentrate system which supplied the liquid to the deluge systems in the American Airlines Hangar also supplied 16 monitor nozzles, each of which discharged 400 gpm of foam-water solution.[3]

Breen explained the superior results achieved with the aqueous film-forming foam when discharged through the standard sprinkler heads seemed to result

Fig. 9.17.   Foam monitor nozzle utilized in connection with foam discharge from
deluge sprinkler system to protect large aircraft in hangars. (Michael W. Magee)

Fig. 9.18.   Deluge valves arranged for foam-water system protecting large aircraft
hangar. Note outside arrangement of valves, a standard procedure in California.
(Michael W. Magee)

from the greater density of the water droplets. The greater density is achieved without the aerating effect of the foam-water sprinkler head, which allowed the droplets to achieve a lighter density and thus be affected to a greater extent by the fire plume and thermal column. The improved "fall" mechanism of the droplets from the standard sprinkler head, when used with the aqueous film-forming foam, should be considered when there is a need to improve fire protection in hangars with existing water deluge systems and the standard sprinkler heads.

APPLICATIONS OF DELUGE SPRINKLER SYSTEMS

Deluge sprinkler systems are primarily installed in special hazard situations where it is necessary to apply water over a sizable area to suppress a fast-developing fire. They are also used to apply mechanical foam of both the protein and aqueous film forming types. Both the water and foam-water systems may be used to prevent the ignition of a fire following the spilling of a flammable liquid or other hazardous material.

David E. Breen, in the report *Suppression of Class A Fuel Fires in Simulated Ship Cargo Holds Using High-Expansion Foam and Water Deluge Systems*, reported on the utilization of deluge sprinkler systems with high-expansion foam systems for the suppression of class A cargo fires, as tested on a fire situation simulated to be similar to the holds of merchant ships.[5] The fire test situation involved 8-foot high stacks of oak pallets. The most effective extinguishment procedure involved the application of high-expansion foam (expansion ratio of 700 to 1, specific gravity 1.4×10^{-3} gm cm^{-3}) to suppress the diffusion flame combustion, followed by a three-minute deluge system application with a water discharge density rate of 0.2 gpm/sq ft. Breen found the high-expansion foam and the water deluge discharge were more effective when used together than either system when used alone.

Deluge systems are often utilized for the protection of cooling towers of combustible construction. The cooling towers with serious fire records are those with combustible fill sections and those of complete combustible construction. Since cooling towers are inaccessible for manual fire fighting activities, deluge systems are often utilized to protect them, especially in climates where temperature conditions prevent the use of wet pipe sprinkler systems. The deluge system is often preferred over the dry pipe system because it overcomes the discharge delay aspects of the standard dry pipe system. Figure 9.19 shows a nozzle specially designed for use on wet pipe, dry pipe, or deluge systems in cooling towers, with and without diffusion decks. When sprinkler heads with activating elements are used on deluge systems, the activating elements are removed following the hydrostatic test of the system piping.

David Shpilberg, in an article titled "Mathematical Model for Analyzing the Trade-Offs in Aircraft Hangar Deluge Sprinkler Systems Design," presented a study of the application of mathematical models to optimize the design of deluge foam-water sprinkler systems in aircraft hangars in relation to the cost

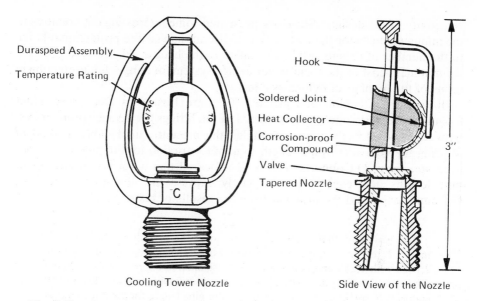

Fig. 9.19. *Sprinkler head for the protection of cooling towers.* (Courtesy of Grinnell Company, Inc.)

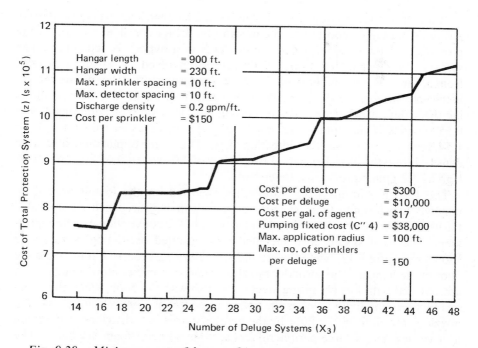

Fig. 9.20. *Minimum cost of hangar fire protection system as a function of the number of deluge systems.* (From "Mathematical Model for Analyzing the Trade-Offs in Aircraft Hangar Deluge Sprinkler System Design," by David Shpilberg)

sensitivity of the design.[6] Shpilberg presented some interesting data relative to the minimum costs for the installation of aircraft hangar fire protection relative to the number of deluge systems installed in the hangar. Figure 9.20 presents the minimum cost of the foam-water deluge system for hangar fire protection in terms of the number of deluge systems.

In Francis J. O'Connor's article, "Pressure Losses in Deluge Sprinkler Fittings," the hydraulic pressure losses in the fittings for deluge sprinkler systems are studied.[7] O'Connor concluded the presently acceptable method of utilizing the equivalent pipe length to express friction loss in pipe tees is not accurate. From O'Connor's studies, it appeared that a variety of equivalent pipe lengths are required to adequately express the flow from pipe tees. O'Connor explained the problem from the results of his research as follows:*

> The test show that a 150 gpm flow out of the branch of a 6-inch × 6-inch × 2-inch tee will cause a pressure drop in the side outlet varying from 2.7 psi with 2.250 gpm flowing in the cross main to 0.6 psi with 1,000 gpm flowing in the cross main. In a 2-inch pipe, the pressure loss at a flow of 150 gpm is 0.20 psi/ft; therefore, the equivalent pipe length for the 2-inch outlet will vary from 13.5 to 3 feet. Various tables used by the sprinkler industry prescribe an equivalent pipe length of approximately 9 feet of 2-inch pipe for this fitting regardless of the flow in the main.

"Fire Protection System for a Hyperbaric Chamber," an article written by Arthur Widawsky, reported on the design of a deluge sprinkler system for the protection of the occupants and the contents of a hyperbaric chamber.[8] The system, activated by IR (infrared detectors), involved 14 sprinkler nozzles from a 3-inch water supply riser. There is an approximate delay of one second in the activation of the IR optical flame detector. To prevent false signals following this delay, water is discharged from the nozzles within an additional 0.3 to 0.55 seconds. Thus, total water discharge delivery times varied from 1.32 to 1.54 seconds, including the detection time. The water application discharge density inside the chamber varied from a low of 2.0 gpm/sq ft on the table, to as high as 3.2 gpm/sq ft on the upper bunk.

The British Aircraft Corporation has reported on the installation of an aqueous film-forming foam system in an aircraft hangar which incorporated the unique feature of ground sprinklers in place of the conventional ceiling sprinklers. The total system also included wall-mounted oscillating monitors as shown in Figure 9.17, while the ground sprinklers are recessed in the concrete floor of the hangar. The sprinkler nozzles are housed in cast-iron cylinders and are covered with loosely fitting steel caps flush with the floor. When the aqueous film forming foam reaches the sprinkler nozzle at a pressure of 50 psi, the nozzle is raised from the floor by the hydraulic pressure, distributing the foam in a vertical jet as three peripheral jets provide a lateral foam discharge. The

---

*From O'Connor, T. Francis, "Pressure Losses in Deluge Sprinkler Fittings," *Fire Technology,* Vol. 2, No. 3, Aug. 1966, p. 210.

total area of discharge per nozzle is approximately 16.5 feet vertically, and a circle approximately 26 feet in diameter.

One of the frequent problems with testing deluge automatic sprinkler systems is water damage from the water discharged through the open heads over large areas. J. C. Abbott, in his article "Testing Deluge Systems," suggested the use of polyvinyl chloride tubes placed over 1,540 sprinkler heads for testing of the systems.[9] The polyvinyl tubes, acting as sleeves, enabled the selective measurement of the discharge from the individual heads, and also allowed water to be diverted away from areas where expensive electrical equipment was being installed in the hangar at the time of the tests. Using the polyvinyl sleeves, Abbott successfully tested nine deluge systems designed to provide a discharge density of 0.20 gpm/sq ft of floor area.

J. R. DeMonbrun and J. A. Parsons, in their report *Cooling Tower Sprinkler Systems*, reported on the investigation of the most suitable type of sprinkler system for installation on cooling towers. They recommended the deluge type of sprinkler system. The deluge system they recommend involved the use of a pneumatic detection system with the detectors located around the periphery of the fan ring on the cooling tower.[10]

There are fire protection situations involving a limited number of sprinkler heads where the deluge system type of application is most desirable. Such situations often involve electrical transformers, switch gear, or a single piece of equipment. Small deluge valves have been developed which incorporate the thermal detection device on the valve. These valves are often identified as thermal-control valves or limited-control valves. Figure 9.21 illustrates one type of small thermal-control valve of the 1- or 1½-inch size. (The 1-inch size and the 1½-inch size look alike, the only difference being the pipe size.) The 1-inch size is limited to a system containing three or less heads, while the 1½-inch size is limited to a system consisting of from three to eight heads.

Fig. 9.21. Small, thermal-control valve for use on limited size deluge systems. N.B.: The 1-inch size and the 1½-inch size look alike, the only difference being the pipe size. (Grinnell Fire Protection Systems Company, Inc.)

PREACTION AUTOMATIC SPRINKLER SYSTEMS

The preaction automatic sprinkler system uses a standard deluge valve with the customary detection system and detectors required for the operation of the deluge valve. However, the preaction system has standard sprinkler heads with fusible elements installed on the piping system supplied by the deluge valve. Since it is a deluge valve, it is activated by the detection system, thus allowing water into the piping system. For any preaction sprinkler system with more than 20 sprinkler heads, the detection system must be supervised. In addition to the standard sprinkler heads installed on the piping system, the preaction system has a rubber-faced check valve installed above the deluge valve. This check valve is primed with water to provide a positive air check, and supervisory air or nitrogen is provided within the piping system, usually at a relatively low pressure of approximately $1\frac{1}{2}$ psi. Thus, with the preaction system, in addition to the supervision of the detection system, the piping system and the standard sprinkler heads on the piping system are also supervised.

The preaction system, since it utilizes the standard deluge valve, can be provided with a pneumatic, electrical, or hydraulically operated valve. The mode of the detection system can also be electrical, pneumatic, or hydraulic, and may involve thermally activated detectors of the rate-of-rise, fixed-temperature, or a combination rate-of-rise and fixed-temperature types. Smoke detectors may also be utilized on the detection system for preaction systems. All detection devices should, of course, be listed as releasing devices for application on preaction systems. The preaction system, with a low air or nitrogen pressure to supervise the sprinkler piping system, has a trouble alarm to indicate loss of this supervisory pressure. Thus, any disruption of the continuity of the sprinkler system piping or the sprinkler heads will be indicated with a trouble signal.

The loss of the supervisory air in the detection system, with the pneumatically operated deluge valves or the hydraulically operated valves with the pneumatic pilot valve and pressure pilot line operation, will trip the deluge valve and allow water to enter the piping system, thus transmitting a waterflow fire alarm. Therefore, with preaction systems, physical damage to the detection system or the detector elements consisting of pilot-line sprinkler heads or HADs, will trip the valve and result in a waterflow fire alarm signal. Thus, waterflow signals from preaction systems may also be trouble signals for the detection system which activates the valve. Fire department personnel who respond to properties protected by preaction sprinkler systems should realize the waterflow signal on a preaction system may also be a trouble signal for the detection system.

Loss of supervisory pressure in the piping system will result in a trouble signal which is distinct from the waterflow fire alarm signal. This unique feature of the preaction system—the waterflow fire alarm activation when the valve has tripped, allowing water into the sprinkler system and effectively converting the preaction system into a wet pipe system—is the basic reason for identifying it as a preaction sprinkler system. The activation of the deluge valve sounds the waterflow fire alarm as the water fills the piping system. However, since the preaction system has standard fusible element sprinkler heads, the water is not discharged into the area. The time between the detection system activation of the valve allowing water into the system, and the fusing of the sprinkler head to discharge the water, allows personnel in the vicinity to take some positive action toward the fire occurrence with hand extinguishers or other types of fire fighting equipment.

Basic Characteristics of the Preaction Sprinkler System

The preaction sprinkler system consists of the following essential components:

1. A reliable water supply, usually connected to private or public water mains.

2. An indicating water control valve of the O.S. & Y.; P.I.V.; or P.I.V.A. type on the main riser.

3. An approved deluge valve. The valves may be utilized as designed for operation on an electrical, pneumatic, or hydraulic principle.

4. An approved detection system which is listed for service as a releasing device. The detection system and devices may be selected from devices that operate electrically, pneumatically, or hydraulically. Those having more than 20 sprinkler heads must be supervised.

5. An approved and listed check valve with a rubber-faced clapper and provisions for priming water to be admitted above the check valve to provide a positive air seal to the piping system. The check valve is installed above the deluge valve on the system riser.

6. Air or nitrogen under low pressure, usually approximately $1\frac{1}{2}$ psi, is provided in the piping system to supervise it and to activate a suitable and distinctive trouble signal upon loss of the supervisory pressure. Thus, the complete piping system of riser, feed mains, and branch lines above the air check valve are under low-supervisory pressure.

7. Standard sprinkler heads with fusible elements are installed on the branch lines of the sprinkler piping system.

Some of the preaction sprinkler systems utilize a hydraulically operated air pump to provide the supervisory air for the sprinkler piping system.

Applications of the Preaction Sprinkler System

The preaction sprinkler system is usually installed in locations where accidental damage to the sprinkler piping or heads on a wet or dry pipe system would result in water discharge that could cause severe damage to facilities or equipment. Dan W. Jacobsen, in his article "Automatic Sprinkler Protection for Essential Electrical and Electronic Equipment," recommended the installation of preaction sprinkler systems for U.S. Air Force electronic data-processing facilities.[11] As explained with the preaction system, any substantial damage to the sprinkler piping or the heads results in the loss of supervisory pressure and a trouble signal. Damage to the detection system sounds a trouble signal. With some deluge valves, damage results in a valve activation and water flow *into* the piping, but not *outside* the piping, because of the standard sprinkler heads with fusible elements installed on the piping system. Thus, it takes a severe thermal exposure (sufficient enough to activate the standard sprinkler heads) to discharge water from a preaction system, even if the system utilizes a detection system designed to operate on smoke conditions.

INSPECTION AND TESTING OF DELUGE AND PREACTION SPRINKLER SYSTEMS

The standard flow tests for the water supply mains, as indicated in the previous chapters for both wet and dry sprinkler systems, should be conducted on the supply water mains before they are initially connected to the sprinkler system riser. The sprinkler system piping should be hydrostatically tested to 200 psi or 50 psi above the static water pressure if the static water pressure in the supply mains to the system exceeds 150 psi. With the deluge sprinkler system, standard fusible sprinkler heads are usually installed and used during the hydrostatic test; the fusible elements are removed following the test. However, should the deluge system include nozzles, suitable caps or plugs are installed in place of the nozzles.

The operational acceptance tests of deluge and preaction systems are essentially response tests of the detection system that determine if the detection system will respond and activate the deluge valve. If the sprinkler system utilizes a thermal detection system with the system located in a flammable liquid or vapor area, explosion-proof test equipment, or warm water in a closed container, should be utilized to test the detection system. Some of the HAD rate-of-rise devices, especially those with electrical detectors, are sensitive enough to be activated by the heat transmitted by human hands.

Operational tests of the deluge or preaction system to determine if the expected fire conditions will activate the detectors in their installed locations may have to involve fire or smoke tests within the premises where the systems have been installed. A review of the criteria utilized by Underwriters Laboratories in their approval of detection devices for releasing service, as summarized in UL's *Fire Protection Equipment List*, will provide some indication of the types and sizes of fire conditions that should be adequate for test purposes without creating a hazardous or damaging condition.[12] As previously discussed, Averill's ("Super 'Garages' for Jumbo Jets Require Super Fire Protection Systems"[3]) testing of the deluge systems in the American Airlines Hangar in Los Angeles included fires involving 35 and 42 gallons of fuel, 100 square feet in area, due to the height and size of the structure.

With both deluge and preaction sprinkler systems, the supervisory provisions for the detection system should be tested. With the preaction system, the supervisory aspects of the sprinkler piping system may be tested by opening a low-point drain or by removing a sprinkler head.

Procedures should be utilized to prevent damage to facilities or equipment when the deluge system is operationally tested in an acceptance test procedure where it is necessary to discharge water or mechanical foam over an extensive area to check the discharge density and foam consistency or stability. An example of this is Abbott's use of polyvinyl sleeves from deluge heads in an airplane hangar to prevent water damage to some susceptible equipment which could not be moved from the area ("Testing Deluge Systems"[9]). Despite the polyvinyl sleeves, Abbott was still able to determine operational efficiency and discharge density from the sprinkler heads.

However, when testing the detection systems to ensure the operational capability of the detection circuits, the riser water control valve for the circuits connected to the deluge valve may be closed if an individual is stationed at the valve. Once the water control valve is closed, the main drain valve should be opened. The operation of the deluge valve can be effectively tested without water discharge through the open sprinkler heads.

Restoration of Deluge and Preaction Systems Following Activation

Most of the deluge sprinkler systems utilize deluge valves which have a positive latch feature. To reset the deluge valve, the water control valve must be turned off and an individual stationed at the valve. After draining the system piping through the main drain, the cover of the deluge valve must be removed. Next, the deluge valve clapper must be physically and manually reset with the latching mechanism in place. The valve cover is then replaced while the detection system is reactivated, and the water control valve is then turned back on. Due to the involved arrangement of the detection devices, including the supervisory procedures, it is not recommended that fire department personnel attempt to restore deluge valves; such restoration should be the responsibility of the occupying agency or the owner of the structure.

The preaction sprinkler system, since it involves a deluge valve and a detection system with supervisory features as well as a supervised system of sprinkler piping, should not be reset by fire department personnel. However, in some situations, a wet pipe sprinkler system may be permissible in locations with preaction systems. Thus, replacement of the fused heads and the retaining of the system with the valve tripped as a wet pipe system does retain automatic sprinkler system protection, although without waterflow alarm service. Therefore, in such situations, personnel should be provided as watchmen in premises affected by the utilization of the system in a wet pipe condition. Most fire departments do not attempt to restore deluge or preaction systems to an operational condition.

SUMMARY

Deluge and preaction automatic sprinkler systems are highly effective sprinkler systems that are utilized for special hazard situations. The deluge sprinkler system is utilized where a severe, fast-spreading fire is expected which could propagate at a velocity greater than the response capabilities of the wet or dry pipe sprinkler systems with the standard fusible sprinkler heads. The modification of the deluge system to a preprimed deluge system can reduce the operational response and water-discharge time to as low as 0.5 second.

Deluge systems may be utilized to discharge water with additive agents in a solution form, and foam-water sprinkler heads are provided on deluge systems to discharge mechanical foam of the protein type. However, the open standard sprinkler head has been found to be more effective with the discharge of the aqueous film forming foam from deluge systems.

The preaction automatic sprinkler system is a system with a deluge valve and standard fusible-element sprinkler heads installed on the piping system. The sprinkler piping system and the fire detection system are both supervised. Thus, damage or leaks in the sprinkler heads or piping provide a trouble signal, while damage or leaks in the detection system may provide still another trouble signal. However, with some deluge valves, damage or leaks to the detection system will trip the valve and thus activate the waterflow fire alarm signal. Preaction systems are primarily utilized in areas with high concentrations of property which are susceptible to water damage.

Both deluge and preaction sprinkler systems are very versatile systems that can be utilized to protect unusual and often critical facilities that are not suitable for protection by other types of sprinkler or agent suppression systems.

SI Units

The following conversion factors are given as a convenience in converting to SI units the English units used in this chapter:

1 square foot =	0.0929	m²
1 foot	= 0.305	m
1 inch	= 25.400	mm
1 psi	= 6.895	kPa
1 gallon	= 3.785	litres
1 gpm	= 3.785	litres/min

ACTIVITIES

1. In outline form, list the essential differences between the deluge and the preaction automatic sprinkler systems. Then compile your outline into an explanatory statement that describes these differences. Include in your statement recommended uses for each.
2. List the locations of some special hazard situations in your area where deluge or preaction sprinkler systems are employed. Determine the types of detection systems utilized on these systems.
3. Explain the reasoning for the identification and terms "deluge sprinkler system," and "preaction sprinkler system."
4. Explain the difference between a deluge sprinkler system and a preprimed deluge sprinkler system.
5. Having responded to a property protected by a preaction automatic sprinkler system, you discover there is no fire but the valve has been activated. What could this have been caused by? Why?
6. List the essential components of the deluge automatic sprinkler system.

Then list the essential components of the preaction automatic sprinkler system.

7. For what use was the electrically operated deluge valve primarily designed? What is the major advantage for using it?

8. What are some of the problems encountered when testing deluge automatic systems? Explain some possible solutions for such problems.

SUGGESTED READINGS

Averill, Charles F., "Super 'Garages' for Jumbo Jets Require Super Fire Protection Systems," *Fire Journal*, Vol. 66, No. 2, Mar. 1972, pp. 24–29.

Breen, David E., "Hangar Fire Protection with Automatic AFFF Systems," *Fire Technology*, Vol. 9, No. 2, May 1973, pp. 119–131.

NFPA 16, *Standard for the Installation of Foam-Water Sprinkler Systems and Foam-Water Spray Systems,* 1975, NFPA, Boston.

Shpilberg, David, "Mathematical Model for Analyzing the Trade-Offs in Aircraft Hangar Deluge Sprinkler Systems Design," *Fire Technology*, Vol. 10, No. 4, Nov. 1974, pp. 304–313.

BIBLIOGRAPHY

[1]NFPA 13, *Standard for the Installation fo Sprinkler Systems,* 1975, NFPA, Boston.

[2]Grabowski, G. J., "Explosion Protection Operating Experience," *NFPA Quarterly*, Vol. 52, No. 2, Oct. 1958, pp. 109–119.

[3]Averill, Charles F., "Super 'Garages' for Jumbo Jets Require Super Fire Protection Systems," *Fire Journal*, Vol. 66, No. 2, Mar. 1972, pp. 24–29.

[4]Breen, David E., "Hangar Fire Protection with Automatic AFFF Systems," *Fire Technology*, Vol. 9, No. 2, May 1973, pp. 119–131.

[5]Breen, David E., *Suppression of Class A Fuel Fires in Simulated Ship Cargo Holds Using High-Expansion Foam and Water Deluge Systems*, Factory Mutual System, Norwood, Mass., May 1974.

[6]Shpilberg, David, "Mathematical Model for Analyzing the Trade-Offs in Aircraft Hangar Deluge Sprinkler Systems Design," *Fire Technology*, Vol. 10, No. 4, Nov. 1974, pp. 304–313.

[7]O'Connor, T. Francis, "Pressure Losses in Deluge Sprinkler Fittings," *Fire Technology*, Vol. 2, No. 3, Aug. 1966, pp. 204–210.

[8]Widawsky, Arthur, "Fire Protection System for a Hyperbaric Chamber," *Fire Technology*, Vol. 9, No. 2, May 1973, pp. 85–90.

[9]Abbott, J. C., "Testing Deluge Systems," *Fire Journal*, Vol. 65, No. 1, Jan. 1971, pp. 24–25, 27.

[10]DeMonbrun, J. R. and Parsons, J. A., *Cooling Tower Sprinkler Systems,* Union Carbide Corporation, Y-12 Plant, Oak Ridge, Tenn., 1972.

[11]Jacobson, Dan W., "Automatic Sprinkler Protection for Essential Electrical and Electronic Equipment," *Fire Journal*, Vol. 61, No. 1, Jan. 1967, pp. 48–53.

[12]*Fire Protection Equipment List,* Northbrook, Ill., Underwriters Laboratories, Jan. 1975.

Chapter 10

Specialized
Automatic Sprinkler Systems

COMBINED DRY PIPE AND PREACTION AUTOMATIC SPRINKLER SYSTEM

There are several variations and adaptations of the basic types of automatic sprinkler systems which have been examined in the previous chapters of this text. The sprinkler systems to be considered in this chapter have generally been designed for use in unique occupancy or hazard situations, or to fulfill specific economic needs.

The combined dry pipe and preaction sprinkler system was originally designed to provide fire protection in occupancies with relatively heavy fuel loads and restricted access to water supply mains, such as piers and other waterfront structures. It is used in unheated structures that would normally be protected by conventional dry pipe sprinkler systems. The combined dry pipe preaction system has an economic advantage in the elimination of numerous dry pipe valves, and the required valve enclosures throughout the entire length of the structure. Operationally, the combined dry pipe preaction sprinkler system has the advantage of allowing water into all of the sprinkler piping with the initial fire detector activation. Thus, thermal damage to an exposed dry sprinkler piping system by the high heat environment, heavy fuel load fire is avoided. In addition, with the entire sprinkler system piping being filled with water upon the initial activation of the detection system, the system is then effectively converted to a wet pipe sprinkler system and water is immediately discharged by the activation of the fusible element in the sprinkler head.

The combined dry pipe and preaction sprinkler system involves the conventional features of a preaction sprinkler system, as outlined previously in Chapter 9, "Deluge and Preaction Automatic Sprinkler Systems." The sprinkler piping system is equipped with the standard sprinkler heads containing fusible elements. An automatic detection system, installed to provide the initial response, allows water to enter the piping system to the sprinkler heads prior to the activation of the fusible elements. However, due to the inherent configurations of piers and other waterfront structures, the supply and feed mains have to travel the full length of the structure from the shore to the seaward end. The combined dry pipe and preaction system, therefore, provides for the location of the dry pipe valves in a tandem arrangement at the shore end of the structure.

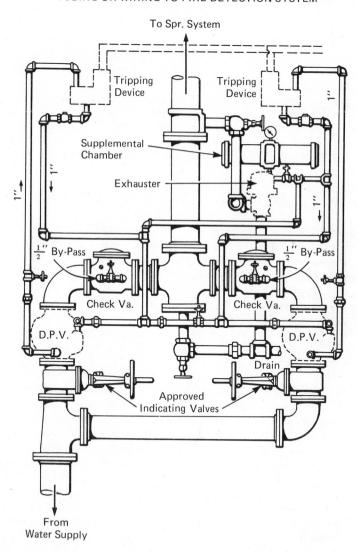

Fig. 10.1.   Tandem dry pipe valve arrangement for combined dry pipe and preaction system. (From **NFPA No. 13,** *Standard for The Installation of Sprinkler Systems*)

The activation of the tandem dry pipe valves allows the water to flow from the shore end to the seaward end of the structure, converting the supply main and all the sprinkler system piping into a wet pipe sprinkler system. Figure 10.1 illustrates the arrangement of the tandem dry pipe valves. This procedure prevents a fire situation on the pier from traveling at a velocity beyond the response capability of individual dry pipe systems, with a resulting discharge delay and water transport time.

A classic illustration of a fire occurrence progressively overpowering successive dry pipe sprinkler systems is the Luckenbach Steamship Company Pier Fire in Brooklyn, New York, on December 3, 1956,[1] as described by W. G. Hayne:*

On Monday, December 3, 1956, at 3:15 pm. the Fire Department received a telephone call of fire on the pier from the Luckenbach Steamship Company's telephone operator. The Brooklyn fire alarm dispatcher notified an engine and truck to respond, but at 3:17 p.m. a box alarm was received from a box in front of the pier which sent a full assignment to the pier. In the meantime, the A.D.T. Company had received a low air signal at 3:15 p.m. and a water flow signal from dry pipe valve G at 3:16 p.m. which was transmitted at once to the Fire Department. It appears the street box and A.D.T. signal were received by the Fire Department almost simultaneously.

When the Fire Department arrived at the pier a few minutes later, the fire had gained such headway and was of such proportions that a second alarm was requested immediately after the first alarm, allowing only enough time for the companies to move from their quarters to the pier and estimate the severity of the fire. There seems to be no doubt that the fire was already of major proportions when the first alarm companies arrived at the pier. One minute before the second alarm was transmitted or at 3:20 p.m., the A.D.T. received a water flow signal from dry valve D and at 3:25 p.m. a similar signal was received from dry pipe valve B. Dry pipe valve G controlled sprinklers on the north side of the pier in the area where the fire started. Dry valve D is located on the south side of the pier and controlled sprinklers on that side of the pier in the second section from valve G toward the shore end of the pier.

No signal was received from either dry pipe valve E or F that were the valves next in the line of travel of the fire after valve G. The records of the A.D.T. Company indicate that one minute after the first waterflow signal was received from valve G or at 3:17 p.m., the circuit was grounded. During the time that was required to transmit the original alarm to the Fire Department and to restore the circuit by switching over to the McCullough circuit, it is probable that both E and F valves operated, but because of the grounding, no signal would have been received. The signals from dry pipe valves D and B were received after the switchover was made to the McCullough circuit. Sometime after 3:25 p.m., the entire circuit was completely out of service and could not be restored so that no additional signals were received.

*Hayne, W. G., Brooklyn, N.Y. Waterfront Fire and Explosion, December 3, 1956, Luckenbach Steamship Co., Inc., Pier, Foot of 35th St., Brooklyn, N.Y., The New York Board of Fire Underwriters Bureau of Fire Prevention and Water Supply, New York, 1957, pp. 8–9.

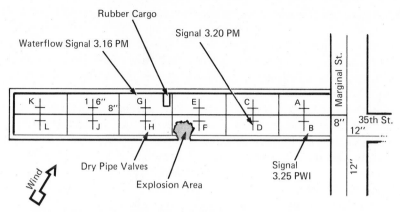

Fig. 10.2. Arrangement of dry pipe systems with operational times from Lucken-bach Steamship Co. pier fire, Brooklyn, New York, December 3, 1956. (From Brook-lyn, N. Y. Waterfront Fire and Explosion, December 3, 1956, Luckenbach Steamship Co., Inc., Pier, Foot of 35th St., Brooklyn, N.Y., by William G. Hayne)

From the preceding description of the activation of the dry pipe valves on the successive sprinkler systems, the progress of the fire can be traced for almost two-thirds the length of the pier through four dry pipe sprinkler systems in the interval from 3:16 to 3:25 p.m., or approximately nine minutes. The illustration of the pier in Figure 10.2 shows the location of the dry pipe sprinkler systems and the operational times of the waterflow signals.

The combined dry pipe and preaction sprinkler system, with the activation of the detection system, trips accelerators that are located with the dry pipe valves at the shore end of the pier, with exhausters on the end of the main water supply line at the seaward end of the pier. The operation of the accelerators causes the activation of the dry pipe valves, thus allowing a resultant high rate of water to travel into the supply main and along the full length of the pier. This prevents any damage to the supply main by thermal exposure.

The combined dry pipe and preaction sprinkler systems have rather limited applications and are primarily utilized where sprinkler system protection is desired for unheated piers. These systems have been most effective and popular since their initial installation in the New York City area some 30 years ago.

The primary economic advantage of the combined dry pipe and preaction sprinkler system appears to be the elimination of numerous dry pipe valves that require servicing and maintenance. As shown in Figure 10.2, the Luckenbach Steamship Company pier was protected by a total of twelve dry pipe systems, six on each side of the pier, with a total of from 196 to 280 sprinkler heads on each dry pipe valve.

THE CYCLING AUTOMATIC SPRINKLER SYSTEM

An automatic sprinkler system that will turn water off as well as on has been a concept of automatic sprinkler system design for many years. As previously indicated in Chapter 6, "The Automatic Sprinkler Head," the on-off sprinkler head has provided this feature since its introduction and approval in 1972. However, prior to the on-off sprinkler head, an automatic sprinkler system was developed and approved by both Underwriters Laboratories and Factory Mutual. This system was trademarked as the "Firecycle System" by the manufacturer, the Viking Sprinkler Corporation, due to the valve's ability to turn back on if the fire regained intensity.[2]

The cycling sprinkler system is basically a modified preaction sprinkler system. It utilizes a hydraulically operated deluge valve which is activated by the operation of 135°F rate compensated thermal sensors that are equipped with high-temperature armored cable, and are especially designed to withstand exposure to high temperatures. Figure 10.3 illustrates the external appearance of the cycling sprinkler system. Note in Figure 10.4, the system has a check valve on the riser above the deluge valve. This check valve retains the supervisory air within the sprinkler piping system at approximately 10 psi. The main drain valve located above the check valve for the supervisory air allows for the draining of the water from the sprinkler system riser and piping following the cycling of the deluge valve back to the closed position, since water in the piping system is then trapped above the check valve.

The set position of the cycling sprinkler system is illustrated in Figure 10.4. Note the detector on the insulated cable; the detector serves to activate the solenoid valve located in the valve trim box. The firecycle valve is a modification of the hydraulic differential deluge valve discussed in Chapter 9, "Del-

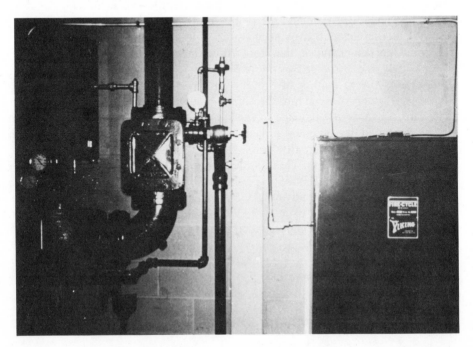

Fig. 10.3. The Firecycle sprinkler system deluge and check valves. (The Viking Corp.)

uge and Preaction Automatic Sprinkler Systems," and illustrated in Figure 9.10. The firecycle hydraulically operated valve has the water pressure equalized on both sides of the clapper through the bypass water line which has a ⅛-inch restricted orifice.

The area of the clapper facing the upper chamber is approximately twice the area of the clapper facing the lower chamber; thus, the clapper will remain in the closed position as long as the water pressure is balanced with equal pressure in both the upper and lower chambers.

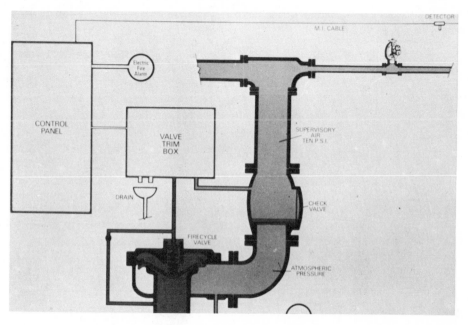

Fig. 10.4. Set position of the firecycle sprinkler system with component elements. (The Viking Corp.)

In Figure 10.4, note the supervisory air pressure above the check valve, the valve trim box which contains the electrically operated solenoid valves, the electrical control panel for the thermal detection system, and the drain from the valve trim box (solenoid valves) with no water flowing from the drain. The piping system, as is standard with preaction systems, has standard fusible element sprinkler heads.

The occurrence of a fire should activate the thermal rate compensated type detectors at approximately 140°F. The detectors transmit an electrical signal to the control panel that activates the electrical fire alarm signal. It also transmits an electrical signal to the electrical solenoid valves located in the valve trim box. The electrical solenoid valves open to drain the water from the upper chamber of the hydraulically operated firecycle deluge valve. With the relieving of the water pressure above the clapper, the restricted orifice in the bypass line prevents the bypass line from restoring the pressure fast enough and the differential advantage of the valve clapper is destroyed, raising the valve clapper upward and allowing the supply main to flow water into the sprinkler system riser.

Figure 10.5 illustrates the fire occurrence that has been detected by the thermal detection system. Note the discharge of water from the valve trim box, which indicates the solenoid valves have operated to relieve the water pressure from above the clapper of the firecycle deluge valve.

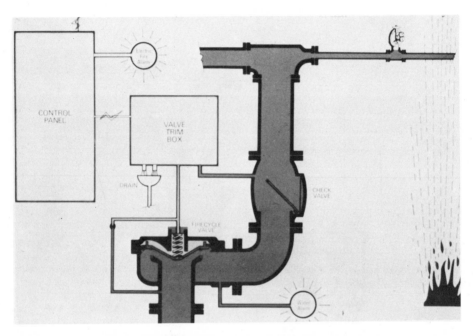

Fig. 10.5. Firecycle sprinkler system with valve activated by thermal detection system due to fire. (The Viking Corp.)

Also note in Figure 10.5 that the waterflow alarm is activated once the water flows from the supply main into the sprinkler system riser, and the sprinkler heads on the system have not fused. The system is responding in the designed operational mode of a preaction sprinkler system, converting the system to a wet pipe sprinkler system prior to thermal energy accumulation sufficient to activate the fusible element in the standard sprinkler heads. Figure 10.6 illustrates the operation of the sprinkler head and the resulting water discharge in the fire area.

With the operation of the fusible element in the standard sprinkler head and the discharge of water into the fire area, the heat output to the thermal detector is effectively reduced, if the fire propagation is controlled. When each thermal detector again reaches approximately 130 to 140°F, the contacts will close and a signal is transmitted to the control panel. Once the last detector has again reached approximately 130 to 140°F, a timer is activated in the electrical control panel, thus allowing the water to continue to flow from the solenoid control valves, and through the sprinkler system for a period varying from 0 to 5 minutes, depending on the design requirements of the system.

The operation of the timer following the signals from the detection system is necessary to assure adequate control of the fire occurrence, and to prevent inefficient rapid cycling of the system if the detectors have been cooled by

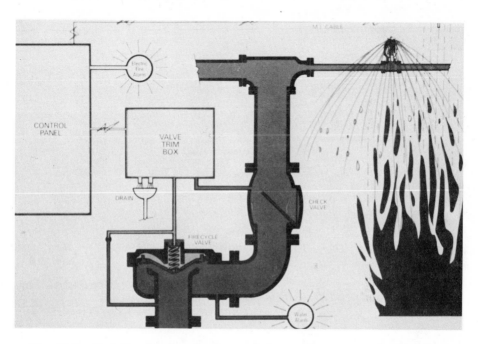

Fig. 10.6. *Sprinkler head activation and the resulting water discharge in the fire area.* (The Viking Corp.)

the water discharge from the sprinkler heads. Therefore, once the timer is in operation with waterflow continuing, if a detector should again signal for a fire occurrence, the timer is aborted until all the detectors again indicate that they are at a temperature in the 130 to 140°F range, and all the thermal detector circuits are closed. Then, the timer operation is again initiated and, once the timer cycle is completed, the electrical current to the solenoid valves is stopped, resulting in the closing of both valves. The closing of the solenoid valves allows the water bypass to again increase the water pressure above the deluge valve clapper, and within one minute the firecycle deluge valve will again close.

Figure 10.8 on the following page shows the firecycle system with the timing cycle completed, the solenoid valves closed (note that no water is flowing from the valve trim box drain), and the deluge valve clapper on the firecycle system also closed.

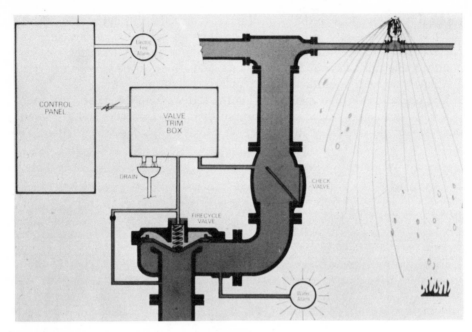

Fig. 10.7.   Firecycle system with fire occurrence under control and water discharge continuing under timer operation. (The Viking Corp.)

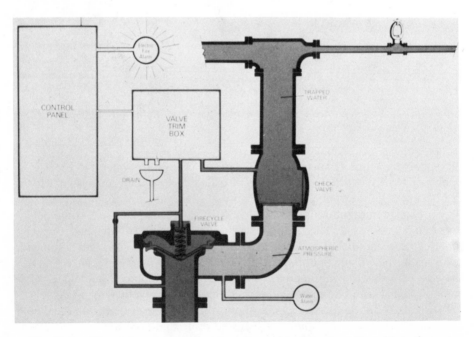

Fig. 10.8.   Firecycle system with valve returned to the set position. Note the water above the check valve. (The Viking Corp.)

Note in Figure 10.8 that although the deluge valve on the firecycle system has returned to the set position with the closing of the solenoid valves (closed by the thermal detection system following the timing cycle), the electrical alarm bell on the detection system is still operating. This bell must be manually reset, just as the water above the check valve must be manually removed by opening the drain valve, before the supervisory air can be restored to the sprinkler piping after the fused sprinkler heads are replaced.

With the firecycle system in the set condition illustrated in Figure 10.8, should the fire rekindle, the thermal detectors would again send a signal to the control panel. This signal would then cause the opening of the solenoid valves, activating the firecycle deluge valve, and again allowing water to flow into the sprinkler system piping and out of the sprinkler heads which have previously fused. Under these conditions, it is unlikely that additional sprinkler heads would open, since the thermal detectors are more sensitive than the fusible elements utilized in the standard sprinkler heads.

An important advantage and operating feature of the firecycle sprinkler system involves the complete restoration of the system, including the re-placement of the sprinkler heads, without ever closing a water control valve on the system. Thus, the firecycle system provides a means of controlling the continual problem of closed water control valves on automatic sprinkler systems. In addition, two solenoid valves are installed in the valve trim box to increase the reliability of the system.

The Firecycle Detection System

The firecycle system uses a thermal system with detectors of the rate-compensated type with a fixed temperature setting of approximately 140°F. The detection system is electrically operated on low voltage current of 24 volts. These detection circuits are provided with standby battery power; if the normal line power sources fail, the batteries operate the system for a period of 96 hours.

The detector utilized on the firecycle system must be in an operable condition after the initial fire exposure to transmit a signal to the control panel to initiate the timing cycle prior to the closing of the solenoid drain valves and the closing of the firecycle deluge valve. Therefore, the thermal detectors have been especially designed to withstand exposures to 1500°F temperatures for five, two-minute time periods. Figure 10.9 shows the thermal rate-compensated detector used on the firecycle system.

In Figure 10.9, note the copper-clad, mineral-insulated cable that provides the electrical circuit to the detector. The cable must withstand the same intense fire exposure as the detector, and must remain in operational condition to transmit the signals for the closing of the valve. The thermal setting of the firecycle detector for ambient installations and conditions is provided with a 140°F operating temperature. However, where design conditions require the installation of detectors with higher operating temperatures, detectors with 160°F, 190°F, and 225°F operating temperatures are available for the firecycle system.

Fig. 10.9. Rate-compensated firecycle system detector. (The Viking Corp.)

The horizontal piece of metal attached to the wire guards of the firecycle detector, as shown in Figure 10.9, is constructed of eutectic metal which will melt at 850°F. Following the operation of the firecycle system, if detectors are observed with the eutectic metal bars missing, they should be removed from the system. A replacement detector should be placed in the circuit and the exposed detector should be thoroughly tested to assure that there has been no thermal exposure damage. Figure 10.10 shows a firecycle detector installed immediately under a roof adjacent to a sprinkler head in an industrial location. The sprinkler head appears to be a High temperature rated head with a blue color, with an operating temperature in the range of 250°F to 300°F.

Note again in Figure 10.10, the copper-clad, mineral-insulated cable which is used to protect and maintain the continuity of the thermal detection system. This cable contains a single 14-gage conductor covered by magnesium oxide insulation encased in a copper tube. When manufactured, the cable is equipped with plug-in connections at both ends; these connections are tested for a positive hermatic seal, and are x-rayed to ensure circuit continuity. The cable will not only withstand high temperatures and mechanical or physical damage, but will also continue to transmit signals when pounded flat.

The detector in Figure 10.10 was intentionally installed adjacent to the sprinkler head (approximately 12 inches away) so that it will be protected by the sprinkler discharge once the head fuses, since this firecycle system is located in a paper storage warehouse where the detector may be subjected to a very rapid increase in temperature. Figure 10.11 shows a firecycle detector and sprinkler head installed on the ceiling of an office area.

Fig. 10.10.   Firecycle thermal rate-compensated detector, installed approximately 12 to 18 inches from sprinkler head. (The Viking Corp.)

Fig. 10.11.   Firecycle detector and sprinkler head mounted on finished ceiling. (The Viking Corp.)

The firecycle system is basically a modified preaction sprinkler system; thus, loss of the supervisory air pressure from the sprinkler system piping will result in a low air pressure trouble alarm. Also, the detection system is supervised; damage to it will activate a fire signal by transmitting an electrical signal to the solenoid valves. This results in the operation of the firecycle deluge valve and a waterflow alarm. However, since the fusible elements in the sprinkler heads would not have fused, no water would be discharged from the firecycle system. Thus, as previously indicated with the preaction sprinkler systems considered in Chapter 9, "Deluge and Preaction Sprinkler Systems," a fire or waterflow signal on a preaction or cycling sprinkler system may also be a trouble signal for the thermal detection system.

The firecycle sprinkler system may be converted to a low air-pressure dry pipe system if the thermal electrically operated detection system becomes inoperative for a period of time. With the conversion to a low air-pressure operation, the fusing of the sprinkler head causes the system air pressure to drop below 7 psi, and the differential firecycle deluge valve allows water to enter the system.

The primary advantage of the cycling sprinkler system is the elimination of the need to close any water control valves on the sprinkler system, either to restore the system and replace heads following an activation, or to make repairs and modifications to the system.

DETECTOR ACTUATED SPRINKLER SYSTEMS

By utilizing the sprinkler head actuator previously discussed in Chapter 6, "The Automatic Sprinkler Head," and shown in Figure 6.9, research has been initiated on the activation of standard wet pipe sprinkler heads by various types of smoke detectors. In his article titled "Detector-Actuated Automatic Sprinkler Systems—A Preliminary Evaluation,"[3] Richard L. P. Custer indicated that smoke detector actuated sprinkler systems were primarily designed for occupancies of the health care type, with installation planned for patient rooms. Custer describes the purpose of this evaluation at the National Bureau of Standards, as follows:*

> The purpose of this test series was to evaluate the effectiveness of a combination detection-suppression system in providing both early warning and suppression capabilities when exposed to fires initiated in bedding by both smoldering and flaming ignition sources. In addition to system effectiveness, the problem of false alarms presently associated with existing smoke detection hardware was also considered. Finally, an attempt has been made to provide tentative design requirements based on engineering judgement and the data available from these tests for the use of detector-actuated, automatic sprinkler systems in nursing homes and care-type facilities.

---

*Custer, Richard L. P., *Detector-Actuated Automatic Sprinkler Systems—A Preliminary Evaluation*, Technical Note 836, July 1974, National Bureau of Standards, U.S. Department of Commerce, Washington, D.C., p. 2.

The studies conducted by Custer involved two types of smoke detectors: a products-of-combustion detector of the ionized type, and a smoke detector of the photoelectric type, operating on the light scattering principle. Both were spot type detectors, and in preliminary evaluations they were found to have acceptable response characteristics when exposed to smoke with low movement velocities. The detectors were exposed to smoke at test movement velocities of 15, 30, 50, and 100 ft/min. The development of the smoke at these velocities, in terms of light obscuration, was approximately one percent/ft/min. These smoke detectors were connected to the sprinkler head actuator devices, which resulted in times of operation for a standard 135°F sprinkler head of approximately 3 to 5 seconds. Figure 10.12 shows the arrangement of the smoke detector, the pendent type standard sprinkler head, the electrical-chemical sprinkler head actuator, the photoelectric light cell, and the light source for the measurement of the smoke density adjacent to the detector, as utilized in Custer's evaluation.

The arrangement of the sprinkler system test procedure in a simulated hospital or nursing home patient's room, with the bed in relation to the location of the sprinkler head and the detector, is illustrated in Figure 10.12. All of Custer's fire situations were initiated on the bed. Of the total of nine fire situations, four situations involved open flame ignition from a dropped cigarette lighter and five situations involved smoldering ignition from a lit cigarette. In the latter situation, the cigarette was placed on the bottom sheet and covered with the top

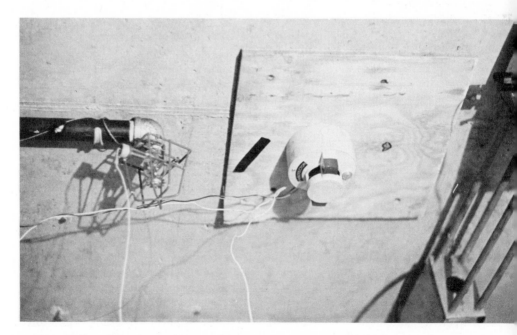

Fig. 10.12. Arrangement of smoke detector actuated sprinkler head in experimental study. (From "Detector-Actuated Automatic Sprinkler Systems—A Preliminary Evaluation," by Richard L. P. Custer)

sheet. The sheets were 50 percent cotton and 50 percent polyester fiber. All fire situations took place on single-sized beds and mattresses. Each mattress had cotton ticking and padding interiors. The mattresses were in the flat horizontal position for the smoldering ignition tests, while they were arranged at an angle 45 degrees from the horizontal for the flaming-ignition test series.

Custer has recommended that the smoke detector for use with a detector-actuated sprinkler system in a health care patient's room be preferably of the

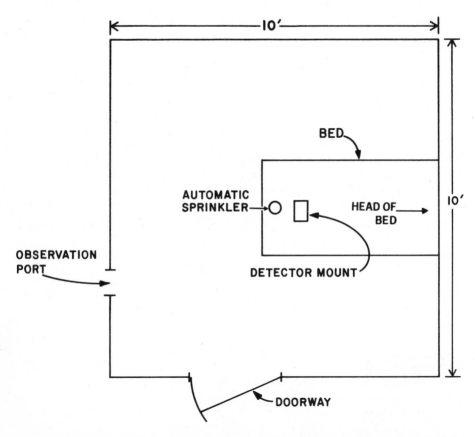

Fig. 10.13. Arrangement of test room to simulate health care facility for the smoke detector actuated sprinkler system tests. (From "Detector-Actuated Automatic Sprinkler Systems—A Preliminary Evaluation," by Richard L. P. Custer)

photoelectric type. These detectors should be capable of responding to a minimum smoke obscuration level not exceeding two percent per foot at a velocity of 15 ft/min. Smoke detector placement and location is extremely important when the detectors are used for the actuation of sprinkler heads. Therefore, Custer recommended the detector be placed on the ceiling above or between the beds, approximately three feet from the head of the bed. Recommended spacing between the detectors is not to exceed 10 feet.

In his evaluation of the smoke detector actuated sprinkler system, Custer also evaluated the operational response times of five standard sprinkler heads and one old style sprinkler head relative to the open flaming ignition test. The open flaming ignition test involved the dropping of a flaming cigarette lighter on the sheets and mattresses at the crevice formed by the sheets and mattress, with the bed in the 45-degree position, as often occurs with the patient sitting in bed.) The tests used Ordinary temperature rated sprinkler heads; however, the operating temperatures of the sprinkler heads varied between 135° and 165°F. Table 10.1 presents the operational response times, ranging from 2 minutes and 30 seconds to 3 minutes and 45 seconds. The two flaming ignition tests with the smoke detector actuated sprinkler system resulted in operational response times of one minute, 45 seconds, and 36 seconds.

Table 10.1. Results of Sprinkler Operation Test With Open Flame Ignition Source*

HEAD[a] NUMBER	RATING °F (°C)	OPERATION TIME (MIN:SEC)	TEMPERATURE AT OPERATION °F (°C)		SMOKE LEVEL AT OPERATION OB % FT⁻¹ (ODm⁻¹)		TEMPERATURE AT HEAD OF BED °F (°C)
			CEILING	5 FEET	CEILING	5 FEET	
1	135 (56.7)	2:30	161 (72)	79 (37)	4.8 (0.07)	3.2 (0.05)	1089 (587)
2	135 (56.7)	2:52	252 (122)	136 (58)	5.8 (0.08)	5.2 (0.07)	1206 (652)
3	165 (73.2)	3:01	259 (126)	147 (64)	6.1 (0.09)	4.5 (0.06)	1263 (684)
4	135 (56.7)	3:18	261 (127)	169 (76)	6.7 (0.10)	4.0 (0.057)	824 (440)
5	165 (73.2)	3:14	259 (126)	163 (73)	6.1 (0.09)	3.9 (0.055)	900 (482)
6	135 (56.7)	3:45	239 (115)	174 (79)	8.3 (0.12)	4.5 (0.06)	606 (319)

[a] 1 = Prototype low thermal lag sprinkler
2 = Center strut with heat collectors
3 = On-off sprinkler
4 = Center strut
5 = Old-style 165°F (73.2°C) center strut
6 = Glass bulb

LIMITED WATER SUPPLY AUTOMATIC SPRINKLER SYSTEMS

Previous to the 1975 edition the *NFPA Sprinkler System Standard* defined the limited water supply sprinkler system as a system which has automatic sprinklers and conforms to the NFPA standard with the exception of the water requirement section of the standard.[4] The water supply requirements of the *NFPA Sprinkler System Standard* were discussed in Chapter 4, "Water Specifications for Automatic Sprinkler Systems," and presented in Figure 4.5 for the hydraulically designed automatic sprinkler systems.

*From "Detector Actuated Automatic Sprinkler Systems—A Preliminary Evaluation," by Richard L. P. Custer.

Limited water supply sprinkler systems have been installed in Light Hazard Occupancies which are located in areas not provided with an adequate public or private water main supply system. Such occupancies include restaurants, health care facilities, country clubs, and motels. Edward J. Reilly, in his article titled "Fire Protection Where There Was None," reported on the installation of a private water supply system for the automatic sprinkler system and private hydrants utilized to protect a motel and convention center complex.[5] The system was designed with the primary water supply from a 9,000-gallon pressure tank. The secondary water supply came from the swimming pool, with a 500 gpm fire pump taking suction from the swimming pool with a capacity of 113,500 gallons. As previously indicated in Chapter 4, the utilization of large sources of water provided for functional or operational purposes such as swimming pools, air-conditioning cooling ponds, and other reservoirs as secondary water supplies for automatic sprinkler systems is efficient and effective fire protection design.

The basic operational and design characteristics of most limited water supply systems required the total water supply for the sprinkler system be limited to the water available in a tank connected to the sprinkler system. In addition to the standard pressure tank with air pressure on the tank, systems have been designed and installed with nitrogen gas pressurization of a water tank, and the chemical tank systems, which were utilized on systems during the 30-year period from 1910 to 1940.

The Chemical Tank Systems

The oldest type of limited water supply system to a sprinkler system was provided from what was essentially an enlarged soda and acid fire extinguisher. This chemical tank system is illustrated with the principal component parts in Figure 10.14.

The chemical tank system involved a solution of water with bicarbonate of soda contained in the storage tank with a sulphuric acid container mounted on the top of the tank. The system was basically a wet pipe sprinkler system installed with the fusible element sprinkler heads on the branch lines. Upon activation of a sprinkler head, the drop in water pressure in the system actuated the syphoning of sulphuric acid into the soda-water solution tank. The resulting generation of carbon dioxide pressure forced the soda-water solution through the riser and piping to the fused sprinkler head.

As indicated in Figure 10.14, the chemical tank systems were operated on the principle of the old chemical tank system utilized on fire apparatus and the soda-acid fire extinguisher. The system tank had a capacity of 200 gallons and was listed as a chemical extinguisher by Underwriters Laboratories in 1916.

Pressure Tank Systems

The standard pressure tank system utilizes a tank filled to 70 percent of the capacity of the tank with water. The remaining 30 percent capacity of the tank

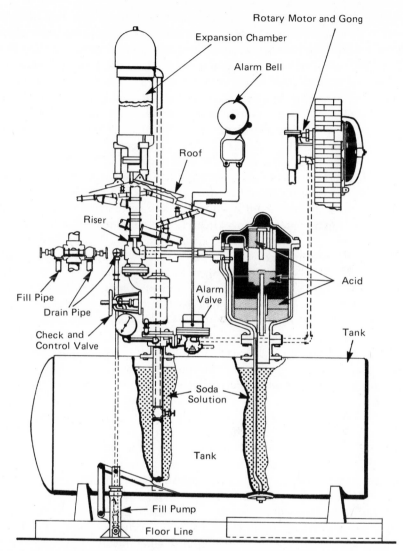

Fig. 10.14. General view of Sypho Chemical system. (From *Automatic Sprinkler Protection,* by Gorham Dana)

is filled with air to a minimum pressure of 75 psi, plus the head pressure required for the difference in elevation between the discharge outlet at the bottom of the tank and the topmost sprinkler head. In Chapter 4, "Water Specifications for Automatic Sprinkler Systems," it was explained that 0.434 psi equals one foot of head or elevation difference, and 2.34 feet of height equals one pound of head pressure.

Where it is desirable to utilize a deluge system or preaction system, many industrial and some other limited water supply systems utilize nitrogen gas stored in cylinders to provide the discharge pressure for the water flow to the sprinkler heads. The nitrogen cylinders are actuated by a thermal detection system which releases the nitrogen gas to pressurize the water storage tank. This converts the system to a standard pressure tank operation, as the nitrogen gas provides the pressure to force the water from the storage tank into the sprinkler system piping. Figure 10.15 shows a drawing relative to the operational design of a nitrogen-water tank system. One major manufacturer of this type of limited water supply system also provides the system with water spray nozzles for the protection of hazards involving flammable liquids or energized electrical equipment. The nitrogen cylinders are generally located adjacent to the water storage tank. The thermal detection system usually releases the gas from the nitrogen cylinders by the activation of a discharge head similar to the operational procedure for high-pressure carbon dioxide extinguishing systems. Relative to the design of the nitrogen-water tank systems, the ratio of nitrogen gas to each gallon of water is approximately one cubic foot of gas per gallon of water. As shown in Figure 10.15, the nitrogen gas is usually provided in Interstate Commerce Commission (I.C.C.) cylinders, with a pressure of 1800 psi and a capacity of 175 cubic feet of nitrogen per cylinder.

There are critical engineering variables of major importance. These variables relate to the size of sprinkler piping, the number of nozzles, the length of piping, and the size of the water supply tank to provide an adequate discharge pressure at the nozzle or sprinkler head.

Another important design variable is sufficient water discharge time to achieve complete extinguishment. Thus, limited water supply systems are usually designed on a pre-engineered basis by the manufacturer.

J. E. King, in his report titled *Packaged Automatic Fire Protection Systems for Remote Buildings*, indicated the Naval Facilities Engineering Command had studied and recommended the utilization of limited water supply sprinkler systems of a modified nitrogen-water tank type for the protection of isolated facilities.[6] The basic system design allowed the use of 500-, 750-, or 1,000-gallon water supply tanks. This design was basically a dry pipe sprinkler system with standard fusible element sprinkler heads. The piping system was pressurized to approximately 60 psi with nitrogen gas, while the identical gas pressure was maintained in the space above the water level in the water supply pressure tank. Water was prevented from flowing into the piping system by a rupture disc at the supply tank discharge outlet. When a sprinkler head fused, the nitrogen pressure in the sprinkler piping was lowered. Due to the 60 psi of nitrogen gas in the supply tank, the resulting pressure differential ruptured the disc in the supply line, allowing water to flow into the sprinkler system piping. With the discharge of water into the sprinkler system piping, the nitrogen cylinders maintained a 60 psi pressure on the water supply pressure tank through the nitrogen cylinder regulators.

The system described by King is basically a pressure tank system with nitrogen used in place of air in the pressure tank and in the dry pipe sprinkler system. King indicated the absorption of the nitrogen gas by the water in the supply tank is inconsequential, since 1,000 gallons of water will only absorb approximately $2\frac{1}{2}$ cubic feet of nitrogen.

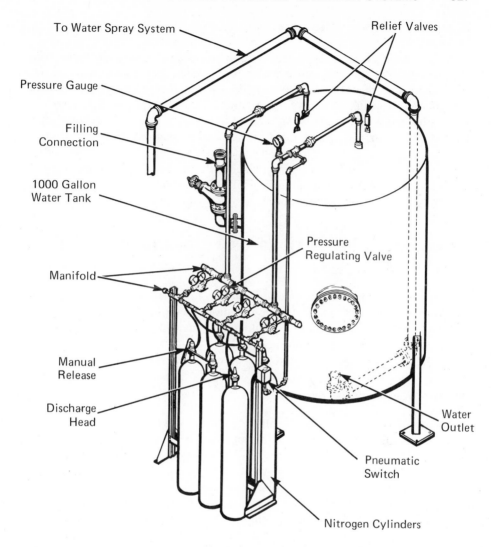

To Water Spray System

Relief Valves

Pressure Gauge

Filling
Connection

1000 Gallon
Water Tank

Pressure
Regulating Valve

Manifold

Manual
Release

Discharge
Head

Water
Outlet

Pneumatic
Switch

Nitrogen Cylinders

Fig. 10.15. Operational and design arrangement of nitrogen-water tank limited supply system. (Grinnell Fire Protection Systems Company, Inc.)

The use of the standard pressure tank with air as the pressurization agent on limited water supply automatic sprinkler systems will not be considered in this chapter, since the air-water pressure tank as a type of water supply was thoroughly discussed in Chapter 4, "Water Specifications for Automatic Sprinkler Systems." However, as explained in Chapter 4, the limited water supply sprinkler system differs from the standard types of systems only in the arrangement of the water supply since this system is usually supplied from a water tank, as the property being protected is located in an area without reliable private or public water supply mains. The limited water supply systems have been most effective in the light hazard type of occupancies with a high degree of compartmentation in the structure, including health care and residential occupancies.

The concept of the limited water supply sprinkler system with the water supply tank and the nitrogen cylinders as a preengineered system was once applied by one manufacturer to provide a preengineered packaged sprinkler system of 100 sprinkler heads for temporary hazard situations. The temporary automatic sprinkler system was designed to be easily installed by using schedule five aluminum piping with most of the pipe connections of the grooved and snap type, with a spring wire clasp and gaskets. The sprinkler riser and the water supply pipes were of $2\frac{1}{2}$-inch size, with the 2-inch size utilized for the cross mains and the branch lines. The sprinkler system was preengineered, but presently is not being manufactured.

SPECIALIZED APPLICATIONS OF AUTOMATIC SPRINKLER SYSTEMS

There are various unique and diverse occupancies that require sprinkler systems with features that vary from the standard types of automatic sprinkler systems previously considered. Sprinkler systems will be considered that have been especially developed and designed for the discharge of ablative water, the protection of combustible doors, race track properties, and dwellings.

Ablative Water Sprinkler Systems

Factory Mutual has been involved with the development of an ablative water automatic sprinkler system for approximately seven years. The ablative water is added to the water flow in the sprinkler system riser by a proportioner that injects the ablative agent into the water stream. With the sprinkler head's discharge of the ablative water, the water droplets tend to be larger. Therefore, they survive and penetrate the extreme gas and air flow velocity of the thermal updraft from high-intensity fires, as often found in the high-rack storage fire situation. In "The Fireball Tests," Factory Mutual has reported that once thermal flow velocities in the convection column exceed 40 feet per second, the water droplets from the standard sprinkler head are primarily diverted upward to the ceiling areas, and do not penetrate the fire plume.[7] However, due to the larger droplet sizes, the ablative water is deflected less often, the fall velocity of the droplet is maintained, and the droplets are able to penetrate the fire plume, thus coating and protecting the fuel surface areas. The high-speed photograph in Figure 10.16 illustrates the size and configuration of ablative water discharge droplets.

Factory Mutual reported the ablative water sprinkler system is most effective when applied at high flow rates. The first ablative water nozzle to operate is usually approximately 20 to 30 percent of the total flow demand. The discharge from the nozzle is directed downward, with a maximum discharge area of ablative water of approximately 130 degrees. This discharge pattern

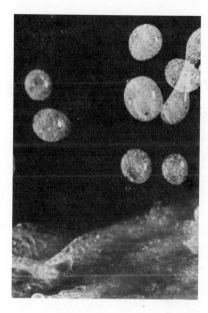

Fig. 10.16. Ablative water droplets. (Factory Mutual System)

results in intensive coverage of an area approximately 300 square feet at 10 feet below the ablative water nozzle. In the experimental ablative water systems utilized by Factory Mutual, the nozzles have been spaced 15 feet apart on the branch lines which have been spaced 15 feet apart, which results in an area coverage of 225 square feet per head.

In addition to the reported advantage of the ablative water droplets to penetrate the fire plume, the ablative water adheres to any fuel surface; thus, the ablative water tends to remain on the fuel surface, or to flow very slowly. This quality provides a lasting protective covering against radiant and convective heat exposure to the fuel surface. The ablative water has been tested on identical fuel arrangements and configurations involving the rack storage of cardboard cartons, polyurethane foam, and wood pallets. Figure 10.17 shows an ablative water discharge on a rack storage arrangement consisting of cardboard cartons. Note the adhesion of the ablative solution to both the steel supports and components of the rack system as well as to the cartons.

In "Ablative Fluids in the Fire Environment," John J. Stratta and William L. Livingston explained the principles of operation of the ablative water solution in providing protection to the fuel surface area involve the mechanisms of thermal energy absorption, thermal energy accumulation, temperature degradation, and thermal energy expulsion.[8] Figure 10.18 is a diagram illustrating these heat transfer mechanisms relative to the layer of ablative water solution which is developed on top of the fuel surface.

Stratta and Livingston indicated the ablative water solution provided improved fire protection and suppression when compared to water discharged from standard sprinklers. Since more of the ablative water is retained in the fire area, the improved performance resulted from: (1) the improved penetra-

Fig. 10.17.   Ablative water system discharging on rack storage arrangement with cardboard cartons. (Factory Mutual System)

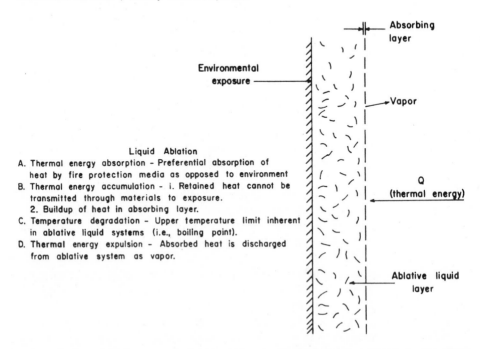

Liquid Ablation

A. Thermal energy absorption - Preferential absorption of heat by fire protection media as opposed to environment
B. Thermal energy accumulation - 1. Retained heat cannot be transmitted through materials to exposure.
   2. Buildup of heat in absorbing layer.
C. Temperature degradation - Upper temperature limit inherent in ablative liquid systems (i.e., boiling point).
D. Thermal energy expulsion - Absorbed heat is discharged from ablative system as vapor.

Environmental exposure —

Absorbing layer

Vapor

Q (thermal energy)

Ablative liquid layer

Fig. 10.18.   Mechanisms of liquid ablation. (From "Ablative Fluids in the Fire Environment," by John J. Stratta and William L. Livingston)

tion of the fire plume, and (2) the thermal resistance of the solution to the evaporation of the agent. Thus, much of the radiant and conductive heat capacity of the fire relative to total energy output is largely negated by the ablative water layers.

In an article titled "Ablative Water," Factory Mutual reported comparative fire tests for automatic sprinklers discharging plain water, and the ablative water system with 20-foot high rack storage of polyethylene-wrapped, multi-walled cartons containing sheet-metal liners.[9] With the standard automatic sprinklers, a total of 102 sprinkler heads operated with a total water supply demand of 3,100 gpm. However, with the ablative water sprinkler system, only three nozzles operated with a total water flow demand of approximately 500 gpm. With the automatic sprinkler water discharge, the temperatures in the rack steel four feet from the point of ignition reached 1900°F. With the ablative water discharge, the steel temperatures at the same point reached only 400°F. With the automatic sprinkler water discharge, the temperature of the air at ceiling level was above 1,000°F for 10 minutes and 50 seconds, while the ablative water discharge allowed the air temperature at the ceiling to exceed 1,000°F for only 48 seconds. With the automatic sprinkler water discharge after 61 minutes of sprinkler operation and fire extinguishment, 75 percent of the cartons had been consumed in the fire test. However, after 41 minutes of operation with the ablative water system and fire extinguishment, only 15 percent of the cartons had been consumed in the fire test situation.

Thus, it would appear the ablative water system utilizes the water more efficiently than the presently designed automatic sprinkler system with standard sprinkler heads in the rack storage fire situation. Therefore, the ablative water system would appear to offer some design and operational advantages in areas with high-intensity fire situations and relatively limited water supplies.

Sprinkler Systems for Door Protection

"Use of Simplified Sprinkler Systems to Protect Wood Doors," an article written by T. E. Waterman, reported on the design and evaluation of a simplified sprinkler system to protect hollow-core wood doors.[10] The three experimental fire conditions that were utilized in Waterman's study involved the ignition and burning of typical furniture in a single room. The experimental tests involved one test situation with no water spray on the door, and two tests with the water spray application to the door. The fires were developed with a contents loading of 5.1 pounds per square foot, and a total fuel load of 5.8 pounds per square foot.

All three of the experimental fire situations were ignited with the ignition of a wastebasket adjacent to the bed, producing rapid ignition of the bed and total involvement of the room contents. Room flashover occurred in about 7 minutes after ignition in all three of the fire situations. However, the fire penetrated the unprotected door in approximately 11.5 minutes, causing the corridor ceiling temperature to reach 1,400°F almost immediately. The radiation levels on the corridor wall across from the door reached 1.1 calories per square centimeter

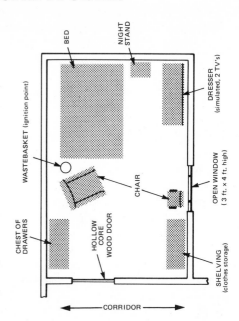

Fig. 10.19.   Arrangement of room contents for protection of door with sprinkler spray. (From "Use of Simplified Sprinkler Systems to Protect Wood Doors," by T. E. Waterman)

per second for over 1 minute, and .8 calories per square centimeter per second for 10 minutes. More importantly, from the consideration of life safety, the hot gases and smoke that exhausted through the door opening brought the entire corridor to 600°F or higher, producing total visual obscuration in the corridor. Figure 10.19 illustrates the arrangement of the furniture in these test fires, the open window in the experimental room, and the hollow-core wood door.

The second experiment utilized the installation of a small spray nozzle directed to discharge approximately 0.75 gpm. Since the spray discharge from the nozzle did not fully cover the door, some of the inner face of the door was burned away. However, damage on the corridor (unexposed) side of the door only consisted of minor discoloration. Negligible radiation was measured in the corridor, and fire gases and smoke leaking around the door (due to a warpage of approximately ¾ inch) produced some smoke accumulation in the corridor. However, ceiling temperatures in the corridor only reached a maximum temperature of 120°F immediately adjacent to the door. The spray nozzle was mounted adjacent to a 165°F rated sprinkler head to determine the time of actuation. When the sprinkler head fused after 9 minutes from ignition, the water spray nozzle was turned on manually. Figure 10.20 shows the installation arrangement of the spray nozzle relative to the sprinkler head (not connected to a water supply) and the door.

The final room fire experiment involved a spray nozzle with a flow rate of approximately 1.66 gpm. The nozzle was initiated manually after the actuation of the sprinkler head, which occurred approximately 7.5 minutes following ignition of the wastebasket fire. As in the previous fire experiment with the use of a spray nozzle, the corridor ceiling remained cool; less than 100°F was the maximum temperature. There was no measurable increase in radiation, and the surface of the door showed only a slight increase in warming. In both

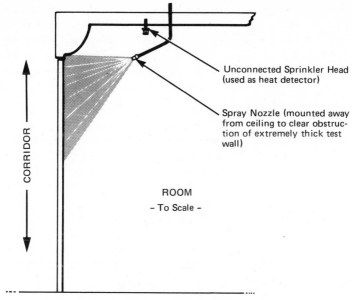

Fig. 10.20. Location of spray nozzle for protection of door. (From "Use of Simplified Sprinkler Systems to Protect Wood Doors," by T. E. Waterman)

experiments with the spray nozzle, it was noticed the chest in the fire room located adjacent to the door did not exhibit active flaming. Thus, a secondary benefit of the door sprinkler system appeared to be the inhibition of the burning adjacent to the door, produced by the steam generation from the water spray.

Waterman's conclusions indicated the water flow rate of approximately 1.66 gpm., consisting of less than 4 gal/sq ft/hr of door surface, appeared to be capable of protecting for an unlimited time a hollow-core wood door from a room fire typical of a heavily loaded residential occupancy. A fusible link provides sufficient activation time, and the water supplies required can be provided with small lines connected to the domestic public water system. Thus, the water pressure and flow requirements provide a system suitable for installation in existing structures consisting of a water line off the domestic system with a sprinkler head protecting the door.

Sprinkler Systems for Stables

In an article titled "Fire Protection for Race Horses," James M. Hammack reported in 1971 there were 103 race tracks in the United States; these tracks were in operation for a total of 4,980 days, and it was not unusual to find approximately 1,000 horses stabled at a major track during the racing season.[11] In addition, Hammack stated the National Association of State Racing Commissioners indicated that between 1960 and 1970 there were 138 fires in stables at race tracks which resulted in fatalities to 12 persons and 1,400 horses; the value of the horses were estimated at approximately $30,000,000. Fire tests were conducted in typical stall configurations of 12 × 12 feet with ceiling heights of 15 feet. The fast-fire situation involved oat hay; the air temperatures reached 375°F at the 15-foot ceiling level, with an activation time of one minute for a 165°F-rated automatic sprinkler head. Utilizing the same stall configuration, slow-burning barley straw developed an air temperature of 250°F at the 15-foot ceiling level, resulting in the operation of the 165°F-rated automatic sprinkler head in 3.5 minutes.

However, to avoid critical injury to horses, the fire must not exceed one foot in diameter, and the temperature at the ceiling should not rise above 150°F. Thus, it was apparent while the standard automatic sprinkler system with the heads located at the ceiling could protect the stable structure, it was not adequate for the protection of the race horses. In addition to the lack of protection for the immediately exposed horses in the stall of fire involvement, the water discharge during the fire produces heated air, steam, and smoke which may drift to other horses in the stable, causing lung damage and pneumonia.

Hammack also indicated a preprimed deluge sprinkler system with specially designed nozzles has been designed to provide protection for horses in stables. The system utilizes thermal rate detectors which operate when the temperature rises 25°F above the ambient temperature in the stable. In actual tests, the detector operated in 4.6 seconds with a rate of temperature rise of 320°F/minute and in 2.5 minutes when the rate of temperature rise was 10°F/minute. The detection system will also respond with an alarm signal at a fixed temperature of 135°F.

The entire sprinkler system for the protection of the stables is hydraulically calculated for uniform water distribution through the open nozzles. The open nozzles are not listed for sprinkler system use by a testing laboratory; however, they are of an industrial type, providing a square pattern of large water droplets. This special nozzle—a nozzle with the water level in the piping to the nozzle—used on these unique preprimed electrically operated deluge systems for race horse stable protection is shown in Figure 10.21.

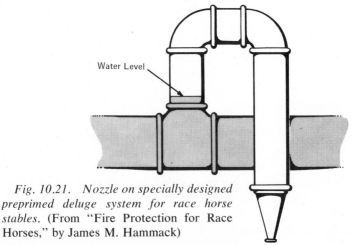

Water Level

Fig. 10.21. Nozzle on specially designed preprimed deluge system for race horse stables. (From "Fire Protection for Race Horses," by James M. Hammack)

Sprinkler Systems for Dwellings

In "Protection for Dwellings," Rolf Jensen has reported that over 82 percent of dwelling fires occur in the normal use areas of the residence: the living room, kitchen, bedroom, and basement.[12] He further reported that 88 percent of the causes of dwelling fires involve heating equipment, electrical failures, careless use of smoking materials or matches, and the misuse of cooking equipment or flammable liquids. In addition, Jensen explained that sleeping persons and unattended children are involved in 80 percent of the dwelling fires, and that 86 percent of these dwelling fire fatalities occur between the hours of 9:00 p.m. and 9:00 a.m.—the hours when people are most likely to be asleep. Thus, in 1973 the NFPA Committee on Automatic Sprinklers appointed a subcommittee to develop a standard for the installation of sprinkler systems in one- and two-family dwellings. The basic design objectives for the dwelling sprinkler system were: (1) to provide a system which would allow the occupants sufficient time to assure their survival, and (2) a system that would be inexpensive.

The NFPA Committee on Automatic Sprinklers has drafted a standard for the installation of sprinkler systems in one- and two-family dwellings and mobile homes. The draft recommends a sprinkler system to provide a discharge density rate from the sprinkler heads of 0.1 gpm/sq ft of floor area for the single sprinkler head with the largest area of coverage. The standard also recommends a stored water supply of 250 gallons where an adequate public water supply is not available. The sprinkler standard for dwelling encourages the use of the extended-coverage and the quick-response sprinkler heads previously discussed in Chapter 6, "The Automatic Sprinkler Head."

John M. Foehl, in his article titled "In Quest of an Economical Automatic Fire Suppression System for Single-Family Residences," has reported on the design and cost aspects of providing a complete automatic sprinkler protection in a single-family dwelling utilizing copper tubing in the distribution system.[13] He reported that copper tubing of the $\frac{3}{4}$-inch size was utilized in the pre-engineered or packaged "junior" sprinkler system provided by Grinnell in 1932 for the protection of basements in dwellings and other Light Hazard Occupancies. The copper tubing was provided in 12-foot lengths, with a tee fitting provided at one end of the tubing to accommodate the $\frac{1}{2}$-inch pendent 135°F-rated sprinkler head. The system was supplied from the domestic water supply, and an alarm valve provided the shutoff and alarm capabilities.

Foehl adopted an upper limit of economic feasibility of $1,000 for the cost of an automatic sprinkler system to protect a one-story, three-bedroom home with full basement and attached two-car garage. The first floor of Foehl's design dwelling had approximately 1,050 square feet of space used for living room, dining room, kitchen, three bedrooms and two baths. The garage consisted of 480 square feet, and the undivided basement consisted of 1,200 square feet.

Foehl indicated with a conventional 20-year mortgage having an interest rate of $8\frac{1}{4}$ percent, the $1,000 cost for an automatic sprinkler system adds $8.68 per month, or a total of $104.16 per year, to the mortgage payments.

The design limitations imposed by Foehl in his study in order to maintain the cost at $1,000 assumed a municipal water supply with a residual pressure at the service entry to the home between 30 and 50 psi, with a maximum of 60 gpm flow. The sprinkler system was designed with copper tubing to provide a Hazen-Williams friction loss formula C factor of 150. The sprinkler piping system was independent of the domestic water supply and beyond the water meter or the service entrance to the residence. Foehl used type M copper tube throughout his design, and the design purpose of the system was to increase the life safety of the residents from fire.

Foehl's final design specified the installation of eight horizontal sidewall sprinkler heads and six pendent sprinklers protecting all areas of the dwelling except the garage, closets, and the stairwell, thus providing a system costing a

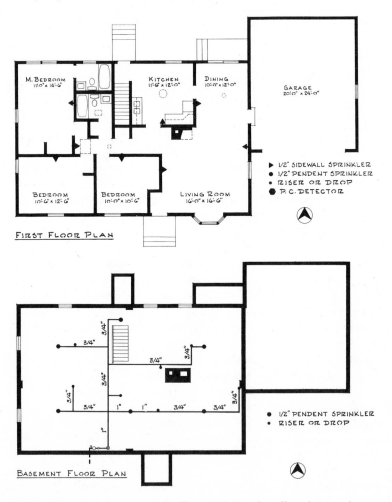

Fig. 10.22. Sprinkler system in dwelling with 8 sidewall and 6 pendent sprinkler heads. (From "In Quest of an Economical Automatic Fire Suppression System for Single-Family Residences," by John M. Foehl)

total of $894 (including 210 feet of ¾-inch type M copper tubing). This system resulted in a total cost of $63.86 per sprinkler head, or $0.40 per square foot of protected area.

Foehl indicated one sprinkler head protecting the bedroom end of the hallway might be eliminated in favor of a particle of combustion detector located in the hall. The alarm function of the sprinkler system could be delegated to the detector. The total cost of the sprinkler system with this smoke detector modification becomes $904. It should be noted that Foehl's economic computations were based on 1974 contractor prices for fittings and tubing, and labor prices as of May 1, 1974.

Foehl indicated regardless of the type of sprinklers utilized, either pendent, sidewall, or a combination of both types, complete automatic sprinkler protection could not be provided for the entire dwelling (including the garage) within the economic constraint of $1,000. Figure 10.22 is a diagram showing the location of the eight horizontal sidewall and the six pendent sprinklers for the final design, with no sprinklers in the garage, closets, or stairwell, for the estimated total cost of $894.00.

SUMMARY

The combined dry pipe and preaction automatic sprinkler systems were initially recognized in the NFPA standards in 1956 and have been primarily utilized for the protection of piers, wharfs, and other waterfront structures. The limited water supply sprinkler systems have been available since the early 1900s with the chemical tank type of system patterned after the operation of the soda and acid fire extinguisher, and the chemical tank on fire apparatus of that era. Limited water supply systems are found in the protection of health care and residential facilities located beyond public water supply mains.

The cycling type of automatic sprinkler system, developed in 1966, provides a sprinkler system with automatic operation of the sprinkler system control valve. This system provides on-off operation of the automatic sprinkler system where a basic preaction sprinkler system is required or desired.

Aluminum piping has been used in temporary sprinkler systems for construction projects and temporary storage or warehousing operations; copper tubing has been studied for application in the design of automatic sprinkler systems for single-family dwellings. It is apparent that plastic pipe, as previously examined in Chapter 5, "Basic Design of Automatic Sprinkler Systems," will be applied to the design of sprinkler systems for dwellings in the future.

SI Units

The following conversion factors are given as a convenience in converting to SI units the English units used in this chapter:

$$
\begin{array}{lll}
\text{1 square foot} & = & 0.0929 \text{ m}^2 \\
\text{1 foot} & = & 0.305 \text{ m}
\end{array}
$$

1 inch	= 25.400 mm
1 pound (force)	= 4.448 N
1 psi	= 6.895 kPa
$\frac{5}{9}$ (°F − 32)	= °C
1 cubic foot	= 0.0283 m³
1 gallon	= 3.785 litres
1 gpm	= 3.785 litres/min

ACTIVITIES

1. Discuss the cycling type of automatic sprinkler system, and compare it with the on-off sprinkler head previously discussed in Chapter 6, "The Automatic Sprinkler Head." Then write a brief statement of explanation describing the operation of the cycling type of automatic sprinkler system, including in your statement the most appropriate uses for such systems.
2. What type of detectors are used on firecycle detection systems? Why is it necessary for the detectors used on firecycle systems to be in operable condition after the initial fire exposure?
3. What are the possible benefits or advantages of the smoke detector actuated sprinkler systems? The disadvantages?
4. In a rack-storage fire situation, which sprinkler system would be more effective in controlling the fire: an automatic system with standard sprinkler heads, or an ablative water system? Defend your reasoning.
5. What are the advantages of installing a combined dry pipe and preaction automatic sprinkler system? Where are they predominantly used? Why?
6. Write your own definition of a limited water supply automatic sprinkler system. Include the types of occupancies best suited for such systems. Then list any limited water supply sprinkler systems in your community or area of responsibility.
7. List occupancies and specific buildings in your area where the sprinkler system for the protection of wood doors might be of practical value and economically beneficial.
8. What type of a specialized automatic sprinkler system installation do you think would be most effective for each of the following facilities? Why?

   (a) A pier.
   (b) A nursing home patient's room.
   (c) A country club.
   (d) A restaurant.
   (e) A paper storage warehouse.
   (f) A polyurethane foam warehouse.
   (g) A stable at a horse track.
   (h) A cardboard box storage warehouse.
   (i) A studio apartment.
   (j) A kitchen in a home.
   (k) A townhouse.
   (l) A motel.

SUGGESTED READINGS

Custer, Richard L. P., *Detector-Actuated Automatic Sprinkler Systems— A Preliminary Evaluation*, Technical Note 836, July 1974, National Bureau of

Standards, U.S. Department of Commerce, Washington, D.C.

"Fire Protection Breakthrough, A Sprinkler System That Makes Decisions," *Factory Mutual Record,* Vol. 44, No. 2, Factory Mutual System, Mar.–Apr. 1967, pp. 4–9.

Jensen, Rolf, "Protection for Dwellings," *Fire Journal,* Vol. 69, No. 1, Jan. 1975, pp. 33–35, 46.

Stratta, John L. and Livingston, William L., "Ablative Fluids in the Fire Environment," *Fire Technology,* Vol. 5, No. 3, Aug. 1969, pp. 181–192.

BIBLIOGRAPHY

[1]Hayne, W. G., *Brooklyn, N.Y. Waterfront Fire and Explosion, December 3, 1956, Luckenbach Steamship Co., Inc., Pier, Foot of 35th St., Brooklyn, N.Y.,* New York Board of Fire Underwriters, Bureau of Fire Prevention and Water Supply, New York, 1957.

[2]*Introducing the Fire Cycle,* The Viking Corporation, Hastings, 1968.

[3]Custer, Richard L. P., *Detector-Actuated Automatic Sprinkler Systems— A Preliminary Evaluation,* Technical Note 836, July 1974, National Bureau of Standards, U.S. Department of Commerce, Washington, D.C., p. 2.

[4]NFPA 13, *Standard for the Installation of Sprinkler Systems,* 1975, NFPA, Boston.

[5]Reilly, Edward J., "Fire Protection Where There was None," *Fire Journal,* Vol. 59, No. 1, Jan. 1965, pp. 32–33.

[6]King, J. C., *Packaged Automatic Fire Protection Systems for Remote Buildings,* Port Hueneue: U.S. Naval Civil Engineering Laboratory, Technical Report R520, April 1967.

[7]"The Fireball Tests," Factory Mutual Record, Vol. 49, No. 4, Factory Mutual Systems, July–Aug. 1972, pp. 5–9.

[8]Stratta, John J. and Livingston, William L., "Ablative Fluids in the Fire Environment," *Fire Technology,* Vol. 5, No. 3, Aug. 1969, pp. 181–192.

[9]"Ablative Water," Factory Mutual Record, Vol. 49, No. 1, *Factory Mutual Systems,* Jan.–Feb. 1972, pp. 5–9.

[10]Waterman, T. E., "Use of Simplified Sprinkler Systems to Protect Wood Doors," *Fire Journal,* Vol. 67, No. 1, Jan. 1973, pp. 42–44, 83–84.

[11]Hammack, James M., "Fire Protection for Race Horses," *Fire Journal,* Vol. 65, No. 6, Nov. 1971, pp. 35–37.

[12]Jensen, Rolf, "Protection for Dwellings," *Fire Journal,* Vol. 69, No. 1, Jan. 1975, pp. 33–35, 46.

[13]Foehl, John M., "In Quest of an Economical Automatic Fire Suppression System for Single-Family Residences," *Fire Journal,* Vol. 68, No. 5, Sept. 1974, pp. 42–48.

Chapter **11**

Exposure Sprinkler and Water Spray Systems

EXPOSURE SPRINKLER SYSTEMS

Exposure sprinkler systems are primarily utilized to protect the outside areas and exterior openings of buildings from fire involvement when fire occurs in nearby buildings, storage areas, or other exposure hazard situations. Although completely automatic types are available, exposure sprinkler systems may be arranged for manual operation with an indicating control valve or the dry standpipe type with a fire department connection as the only means of water supply. Automatic operation will usually be of the standard dry pipe sprinkler system type with fusible elements in the sprinkler heads and air or nitrogen in the sprinkler system piping, or the typical deluge sprinkler system with open heads, a deluge valve, and a separate detection system. The deluge system is usually preferred for exposure type automatic sprinkler systems.

Figure 11.1 shows an exposure sprinkler system installed to protect the window openings of an apartment building from fire exposure should fire occur in an adjacent shopping center which includes vehicle parking. In Figure 11.1, note the metal heat collector canopies installed over the pilot line sprinkler heads for the nitrogen gas pilot line operated deluge valve. The fusing of any one of the pilot line sprinklers will result in the operation of the deluge valve, and water being discharged from the open exposure window type heads on the three lines of open sprinklers.

Figure 11.2 shows the nitrogen gas pilot line operated deluge valve for the exposure sprinkler system previously shown in Figure 11.1. Note the required manual release located on the deluge valve, as previously discussed in Chapter 9, "Deluge and Preaction Automatic Sprinkler Systems."

340

Fig. 11.1.   Exposure sprinkler system installation for protection of window openings in an apartment building.

Fig. 11.2.   Deluge valve with nitrogen gas pilot line operation for exposure sprinkler system.

The exposure type sprinkler system will usually not be of the wet pipe design for automatic operation, except in mild climates where the system will not be exposed to freezing temperatures. Manually operated systems are not preferred unless the building is occupied 24 hours a day, or unless fire department personnel are thoroughly trained and familiar with the system design, operation, and, most importantly, the location and operation of the manual water control valves.

All sprinkler heads installed on exposure sprinkler systems, whether of the open or fusible element type, shall be approved for intended usage by a recognized approval laboratory. Sprinkler heads are normally approved and listed for use as window, cornice, sidewall, or ridgepole installation. However, other types of sprinkler heads and spray nozzles on deluge valves may be utilized where the coverage is suitable and adequate for the intended application. Sprinkler heads with orifice sizes varying from $\frac{1}{4}$, $\frac{5}{16}$, $\frac{3}{8}$, $\frac{1}{2}$, $\frac{5}{8}$, and $\frac{3}{4}$ inch are found on exposure sprinkler systems.

Design of Exposure Sprinkler Systems

Exposure sprinkler systems may be designed as hydraulically calculated sprinkler systems with a minimum of 7 psi at any sprinkler head, with all the sprinkler heads facing the exposure hazard flowing.

However, exposure sprinkler systems may also be designed as pipe schedule systems utilizing the *NFPA Sprinkler System Standard*.[1] The pipe schedule chart for the maximum number of sprinkler heads provided on each branch line, as determined by the size of the sprinkler head orifice and the size of the branch line, is shown in Table 11.1.

The *NFPA Sprinkler System Standard's* pipe schedule for the risers and central feed mains to exposure sprinkler systems is shown in Table 11.2.

Exposure sprinkler systems for many buildings will often require large quantities of water. The *NFPA Sprinkler System Standard* recommends the water supply demand be calculated, considering water flow from all the sprinkler heads on the portion of the building facing the exposure hazard for a minimum period of operation of at least one hour.

The primary objective of the exposure sprinkler system is always to prevent the ignition of combustible materials on the building, or the ignition of the interior of the building, including the building's contents. Therefore, the location of the sprinkler heads on the exposure sprinkler system is most critical. The objectives for locating the heads are: (1) to provide for thorough wetting of all exposed combustible surfaces, and (2) to provide a film of flowing water over all exposed glass surfaces. The specific location and the installation of the outside sprinkler heads on exposure sprinkler systems should follow the principles of the installation rules provided in Appendix B of the *NFPA Sprinkler System Standard*. Figure 11.3 illustrates an acceptable procedure for the installation of window and cornice types of sprinkler heads on an exposure sprinkler system.

In their article titled "Measurement of the Transmission of Radiation through Water Sprays," A. J. M. Heselden and P. L. Hinkley reported as a result of

Table 11.1. Maximum Number of Sprinkler Heads Supplied per Branch Line for Exposure Systems*

SIZE OF PIPE INCHES	ORIFICE SIZE—INCHES						
	$\frac{1}{4}$	$\frac{5}{16}$	$\frac{3}{8}$	$\frac{7}{16}$	$\frac{1}{2}$	$\frac{5}{8}$	$\frac{3}{4}$
1	4	3	2	2	1	1	1
$1\frac{1}{4}$	8	6	4	3	2	2	1
$1\frac{1}{2}$		9	6	4	3	3	2
2				5	4	4	3

Table 11.2. Maximum Number of Sprinkler Heads Supplied per Riser for Exposure Systems**

PIPE SIZE	NUMBER OF SPRINKLERS		
	$\frac{3}{8}$-in. OR SMALLER ORIFICE	$\frac{1}{2}$-in. ORIFICE	$\frac{3}{4}$-in. ORIFICE
$1\frac{1}{2}$	6	3	2
2	10	5	4
$2\frac{1}{2}$	18	9	7
3	32	16	12
$3\frac{1}{2}$	48	24	17
4	65	33	24
5	120	60	43
6		100	70

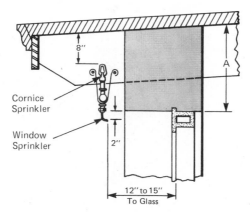

Fig. 11.3. Location of window and cornice sprinkler heads. (From NFPA 13, *Standard for the Installation of Sprinkler Systems*)

*From NFPA 13, *Standard for the Installation of Sprinkler Systems*.
**From *Ibid*.

their studies, exposure systems with relatively high nozzle pressure and a water flow of 4 to 5 gal/ft/min would absorb approximately 60 to 70 percent of the fire incident radiation.[2] Their studies also indicated that radiation transmission was inversely related to the nozzle pressure of the water flow from the orifice.

The manual control valves and the location of the automatic valves on all exposure sprinkler systems should be marked with suitable signs to ensure that fire department personnel will see them without difficulty at the time of a fire. To provide the fire department with suitable means to augment the connected water supply system, fire department connections to exposure sprinkler systems should always be installed. As previously explained, in some locations preaction and deluge sprinkler systems may have outside exposure heads installed off the interior system.

The outside sprinklers for installation on the exposure type of sprinkler systems are normally of the small orifice $\frac{3}{8}$-inch size for the protection of in-

· Cornice sprinkler

· Window sprinkler

Fig. 11.4. Window and cornice sprinkler heads for exposure system. (Grinnell Fire Protection Systems Company, Inc.)

dividual windows, and light hazard situations. Figure 11.4 illustrates one type of window and cornice sprinkler heads of the open design without fusible elements. The open sprinkler heads are usually installed on systems with a manual water control, a system of the dry standpipe type supplied by a fire department connection, or automatic systems with a deluge valve operated by an installed separate detection system.

Exposure type sprinkler systems—either automatically or manually operated—are often installed on the sides of buildings in conjunction with other exposure protection devices such as shutters or fire doors. The exposure sprinkler system should be subjected to the required hydrostatic tests to determine the continuity of the sprinkler piping system as previously indicated for the wet pipe, dry pipe, or deluge sprinkler system, as may be appropriate. In addition, the water supply mains to the system should always be flow tested and flushed, as previously described for the particular type of system, prior to the completion of the final connections to the exposure sprinkler system.

WATER SPRAY SYSTEMS

Water spray systems are automatically or manually activated systems primarily designed to provide protection from fire situations for equipment, tanks, vessels, or containers, and to achieve control or suppression of fire situations with high volume discharges of water spray. Water spray systems are utilized in many situations. However, they are primarily used in industrial situations involving the processing and storage of flammable liquids or gases, and in the protection of electrical equipment of the transformer or switchgear types that involve petroleum fire situations. NFPA 15, *Standard for Water Spray Fixed Systems for Fire Protection,*[3] hereinafter referred to in this text as the *NFPA Water Spray Standard,* defines a water spray system as follows:*

> A water spray system is a special fixed pipe system connected to a reliable source of fire protection water supply, and equipped with water spray nozzles for specific water discharge and distribution over the surface or area to be protected. The piping is connected to the water supply through an automatically or manually actuated valve which initiates the flow of water. An automatic valve is actuated by operation of automatic detection equipment installed in the same areas as the water spray nozzles.

One of the principal differences between the automatic sprinkler systems previously considered and a water spray system is the design of the spray system for the local protection of equipment or storage tanks. Water spray systems are not generally utilized for the protection of the total area of a building or structure, but to protect specific equipment and apparatus, and uncloseable openings in fire walls. Characteristically, a water spray system utilizes

---

*NFPA 15, *Standard for Water Spray Fixed Systems for Fire Protection,* 1973, NFPA, Boston, p. 8.

nozzles instead of sprinkler heads, thereby providing a directional flow of water often in the horizontal as well as in the vertical mode of discharge. In many situations, the water spray system will utilize nozzles of the open type; it is, therefore, activated with the deluge type of automatic sprinkler valve, or a special service type of total flooding valve. There is some overlapping of design features on deluge type sprinkler systems and water spray systems since deluge valves are often utilized to activate the water spray systems, and spray nozzles are often installed on deluge and preprimed deluge systems to protect apparatus or equipment. Figure 11.5 illustrates a water spray system discharging water to protect an electrical transformer.

In Figure 11.5, note the heat-actuating-devices (HAD) of the pneumatic type are installed with heat-collecting metal canopies above the detector. These canopies are similar to the heat-collecting canopies over the pilot line sprinkler heads for the exposure sprinkler system shown in Figure 11.1. The operational time of both detectors and sprinkler heads may be decreased on these types of exterior installations if the thermally activated devices are installed under heat-collector canopies as shown in Figures 11.1 and 11.5.

Fig. 11.5. Water spray system discharging water to protect an electrical transformer. (Factory Insurance Association)

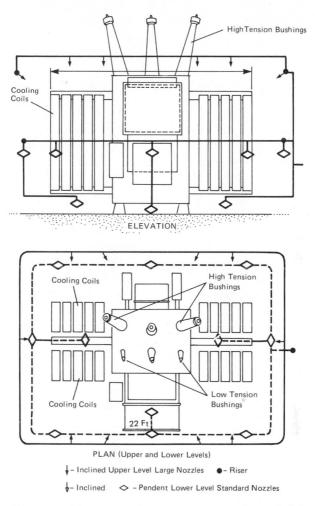

Fig. 11.6. *Diagram of large capacity water spray nozzles and piping for electrical transformer.* (Factory Mutual System)

Figure 11.6 illustrates for a typical design, the location of the water spray system piping, and the large capacity nozzles for the protection of an electrical transformer similar to the water spray system shown in operation in Figure 11.5.

Design of Water Spray Systems

The water spray systems, as utilized in industrial situations, are designed to accomplish one or more of the following objectives:
1. Extinguishment of the fire occurrence.
2. Control of the fire situation with a resultant decrease in the thermal exposure created by the fire situation.

3. Exposure protection for equipment, containers, vessels or tanks, usually containing flammable liquids or gases.

4. The prevention of fire occurrences as may be accomplished by the dispersion and diffusion of flammable liquid or gas leaks.

The extinguishment of the fire is accomplished by a water spray system with the cooling of the thermal output of the diffusion flame combustion. Also of importance is the smothering of the flame shield from the steam produced by the emulsification of some of the flammable liquid fuels having high flash points, and the cooling of these liquids below their ignition temperatures. Obviously, the mechanisms of the extinguishment of fire as explained in detail in Chapter 4, "Water Specifications for Automatic Sprinkler Systems," with the discussion of the extinguishing effects of water as utilized in automatic sprinkler systems, also apply to water spray systems. The achievement of fire suppression and extinguishment with water spray as with automatic sprinklers is a multifactor phenomenon resulting from a number of extinguishing mechanisms.

Controlling a fire situation is often achieved with the identical mechanisms which may achieve extinguishment and complete suppression. However, due to different characteristics of the fuel with various fuel configurations and geometry, complete extinguishment may not be possible. These fire control situations often occur with flammable liquid fuels having low flash points—including materials like gasoline—involved in fire situations with the fuel bed exceeding depths beyond 6 to 8 inches, which is common in petroleum storage tank situations. In these fire control situations, the water spray system will control the spread of the fire and limit the thermal output and exposure to the exposed vessels or equipment, but will not usually achieve total extinguishment of the fire occurrence.

Water spray systems are typically utilized for the exposure protection of steel structures, vessels, tanks, and containers. This protection prevents thermal damage to the tank supports, the tank, or the tank shell, and averts a rupture or explosion of the tank which would ignite the contents. Such use of water spray systems primarily involves flammable liquid or gas storage containers, and exposed unprotected steel as often found with stills or reactors in petrochemical complexes such as refineries. Water spray systems are also utilized in industry to dissolve, dilute, disperse or cool flammable liquids, vapors, or other materials before the actual ignition of the fuel occurs.

The discharge pattern of the spray from water spray nozzles is determined by the nozzle design, while the flow and reach of the stream is principally determined by the nozzle pressure. Factory Mutual's *Loss Prevention Data* indicated most of the water spray nozzles have a cone or fan type of discharge pattern, with the cone varying from an inclined angle of about 60° to 180°.[4] Figure 11.7 illustrates the various shapes of the water spray nozzle discharge patterns as usually provided by standard nozzles.

From the types of water spray nozzle discharge patterns shown in Figure 11.7, the design of the nozzle will affect the design parameters of the water spray system relative to the area of the equipment or material protected, and

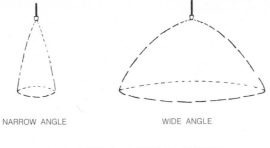

NARROW ANGLE   WIDE ANGLE

SOLID CONE & UMBRELLA SHAPES

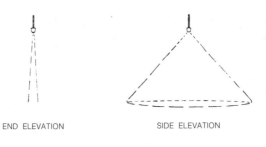

END ELEVATION   SIDE ELEVATION

FLAT FAN SHAPE

Fig. 11.7.   Water spray nozzle discharge patterns. (Factory Mutual System)

the discharge density provided to protect the equipment. The usual size of water spray nozzles found on the water spray systems have $\frac{1}{4}$- to $\frac{1}{2}$-inch discharge orifices, and will discharge from 8 to 50 gpm at 50 psi nozzle pressure. However, as indicated in the electrical transformer diagram in Figure 11.6, the large capacity nozzles will usually discharge 110 to 150 gpm at the 50 psi nozzle pressure.

The various water spray system nozzles produce their water spray discharge through the processes of impingement and dispersion of straight or spiral water streams. The various typical designs of the water spray system nozzles are illustrated in Figure 11.8.

Most water spray nozzles are provided with blow-out plugs when the water spray system is preprimed, or when the nozzles are utilized in conjunction with a preprimed deluge system. In preprimed systems, the use of the blow-out plug in the nozzle provides immediate water discharge when the system is activated. The blow-out plugs are usually designed to retain approximately 10 psi of water pressure, and to blow out with 30 psi of water pressure. The blow-out plugs, as used on either preprimed deluge or preprimed water spray systems, are not listed by Factory Mutual or Underwriters Laboratories. In addition to blow-out plugs, blow-off caps and dust caps are provided to prevent the clogging of water spray system nozzles where they may be exposed to tar, pitch, or dusts.

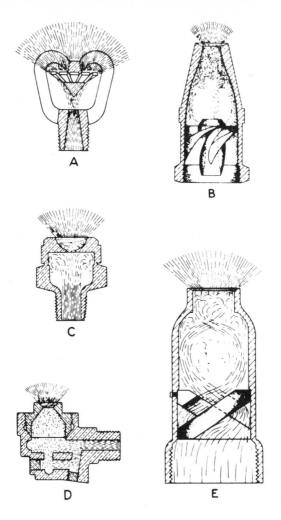

A. Straight stream impinging on an exterior-toothed deflector
B. Straight stream impinging on spiral streams
C. Straight stream discharging through a square orifice, forming a fan-shaped pattern
D. Two straight streams swirled in an upper and lower spiral chamber and discharging through a common orifice
E. Spiral streams impinging on each other and dispersing (large-capacity nozzle)

Fig. 11.8. Water spray system nozzle configuration and design. (Factory Mutual System)

Water spray system nozzles are selected in the design phase of the spray system relative to their area of coverage, discharge range, and nozzle discharge flow characteristics. These nozzle performance characteristics then determine the water spray system design relative to the various pipe sizes needed to provide the required operating pressures at the nozzles in order to achieve the desired discharge rates.

Figure 11.9 illustrates typical water spray system nozzles of the flat type, solid spray, high velocity, directional types.

When very fast operation of the water spray system is desired (such as in explosives manufacturing, hyperbaric chambers, and petrochemical opera-

Fig. 11.9.   Typical water spray system nozzles. (Automatic Sprinkler Corporation of America)

tions), pilot type water spray nozzles may be utilized. This type of nozzle is utilized where the system is preprimed with water pressure above the 10 psi, maximum where blow-out plugs are effective. The pilot nozzles are utilized on wet pipe systems with the piping filled with water under pressure to the nozzle. The nozzle has a pilot valve incorporated into the design of the nozzle; the pilot line with water pressure maintains the nozzle in the set, or closed, position.

When the separate thermal detection system responds to the fire condition, the detection system activates a pilot valve. This activation relieves the pressure on the pilot line to the pilot nozzle, opening the nozzle and allowing immediate high flow discharge through the nozzle. (Any other nozzles connected to the same pilot line would also open and discharge water.) The pilot type of water spray nozzle will automatically close when the detection system is no longer in the actuation state from the fire condition. Once the thermal detection system closes the pilot valve, the pressure is restored in the pilot line and the nozzles return to the set, or closed, position with the increased pressure on the pilot line. Thus, the pilot line water spray nozzle is actually a type of hydraulically operated cycling deluge valve within the nozzle actuated by a separate thermal detection system. The flame type of detectors are often combined with the pilot type of water spray nozzle to achieve ultrafast and efficient fire detection and activation of the water spray system for interior operations such as hyperbaric medical chambers.[5]

Figure 11.10 (left) illustrates the external appearance of the pilot water spray nozzle and (right) an internal view which shows the operational control valve in the nozzle.

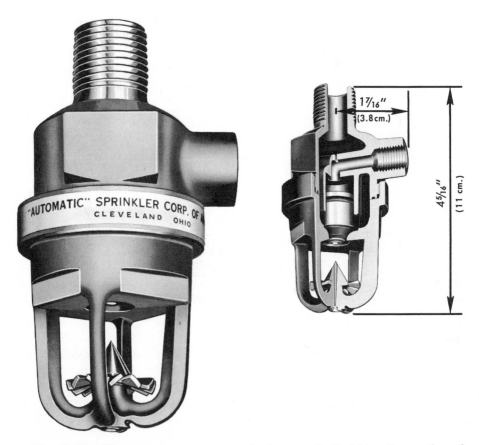

1 7/16″
(3.8 cm.)

4 5/16″
(11 cm.)

Fig. 11.10. Pilot type water spray nozzle. (Automatic Sprinkler Corporation of America)

The pilot type of water spray nozzle was the first effective on-off nozzle, even though it was controlled in the on-off sequence of operations by a separate detection system.

Water Application Rates for Spray Systems

The recommended water application rates for water spray systems will vary according to the design purpose of the water spray system relative to a system for fire extinguishment, fire control, or exposure protection. In design situations involving unusual configurations of containers or flammable liquids for which test data is not available, specific extinguishment experiments may have to be conducted to determine the minimum water discharge density to be achieved from the water spray system.

The *NFPA Water Spray Standard*[3] recommends the following water application rates relative to the design purpose of the water spray system.*

> For the extinguishment of ordinary combustibles or flammable liquids, a discharge density from 0.2 gpm/sq ft to 0.5 gpm/sq ft of protected surface.
>
> For the control of burning, the nozzles shall be designed to impinge on the areas of the source of expected fire, including the areas where flammable liquid spills may accumulate. The water application rate on the surface of the spill or the exposed area of equipment should be a minimum of 0.5 gpm/sq ft of liquid or equipment surface area.

The water spray systems installed for exposure protection must be designed to operate before any deposit of carbon accumulates on the surfaces to be protected and before the possible failure of the container of flammable liquids or gases due to increases in the internal pressure or temperature. Thus, water spray systems designed for exposure protection will, of necessity, have to be fast-response systems. Water should be discharging from all the nozzles on the complete water spray system within 30 seconds from the time of initial operation of the detection system. According to the *NFPA Water Spray Standard,* detection systems to actuate water spray systems should activate the water control valve within 20 seconds under the expected fire conditions, and within 40 seconds under normal test conditions, with a standard heat source.

When the water spray system is providing protection to uninsulated containers or vessels, the water application rates are based on the assumption the container will be equipped with pressure relief devices based upon a maximum allowable heat input of 6,000 Btu per hour per square foot of exposed surface area. Should the container not have this minimum pressure relief afforded by conservation or emergency vents, the water application rate for the protection of the container should be increased to prevent possible rupture. The exposed surface area of the container should have a water application rate of 0.25 gpm/sq ft of uninsulated surface area, and an application rate of 0.10 gpm sq ft of surface area for the skirts of the containers. The structural steel support members for equipment or containers should have minimum water application rates of 0.10 gpm/sq ft for horizontal units; care should be taken to achieve application to the web and the flanges of the beams. For the vertical support beams, the application rate should be a minimum of 0.25 gpm/sq ft of wetted surface area on the structural member. The wetted surface area for structural members is defined as consisting of one side of the web and the inside of the flanges on the same side with the web. Figure 11.11 illustrates a

*From NFPA 15, *Standard for Water Spray Fixed Systems for Fire Protection,* 1973, NFPA, Boston.

specially designed water spray sprinkler head with a fusible element for individual operation, designed to discharge water into the rectangular pockets formed by horizontal structural steel members.

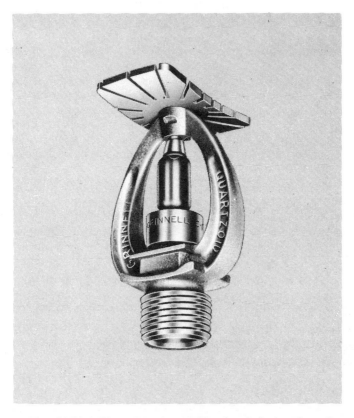

Fig. 11.11. Water spray sprinkler head designed to discharge water into the rectangular pockets formed by horizontal structural steel members. (Grinnell Fire Protection Systems Company, Inc.)

The water application rate recommended for the protection of sections of pipe, tubing, or conduit is a minimum of 0.10 gpm/sq ft of aggregate pipe wall area. Where unprotected steel pipe is used for structural supports, the water application rates should be identical to the steel structural members consisting of a discharge density of 0.10 gpm/sq ft for the horizontal members, and 0.25 gpm/sq ft for the vertical members. Where it is desired to protect runs of nonmetallic cable or tubing, a minimum water application rate of 0.3 gpm/sq ft of projected plane area both over and under each cable tray should be utilized. Where the cable trays are provided with $\frac{1}{16}$-inch metal flame shields, or where other extinguishing systems are utilized to protect the exposing hazard, the minimum water application rate may be reduced 50 percent to a rate of 0.15

gpm per square foot of projected plane area for the cable trays with nonmetallic cable.

Where electrical transformers are to be protected, the recommended NFPA minimum water application rate is a discharge rate of 0.25 gpm/sq ft on the rectangular prism envelope for the transformers and their related equipment, and a discharge density of 0.15 gpm/sq ft for the ground area expected to be involved in any spill fire situation. (A water spray system discharging water onto an electrical transformer installation was shown in Figure 11.5; the diagram of the water spray piping system with nozzle locations for a typical electrical transformer installation was shown in Figure 11.6.)

Figure 11.12 shows a water spray system designed for the exposure protection of spherical LPG (liquified petroleum gas) containers. Note the coverage of the water spray nozzles over all the surfaces of the containers.

Fig. 11.12. Water spray system discharging for exposure protection of spherical LPG containers. (Factory Insurance Association)

When water spray systems are designed for the protection of a number of containers, the water application rates for the total discharge may become rather large (see Figure 11.13). Therefore, provision should be made for adequate drainage of the water discharged from the spray system. Relative to the consideration of large water application rates, Figure 11.13 shows a water spray system on a launching pad at Cape Kennedy; the system was designed for use in case of missile malfunction. The total discharge rate of the water spray system shown in Figure 11.13 is 9,500 gpm.

Fig. 11.13. Water spray system for exposure protection of mobile launcher at Cape Kennedy. (National Aeronautics and Space Administration)

SPECIAL APPLICATIONS OF WATER SPRAY SYSTEMS

Eggleston and Herrera (*Automatic Fire Protection Systems for Manned Hyperbaric Chambers*[5]) have reported on the design and prototype installation of a water spray system for the protection of naval hyperbaric diving chambers. The system was designed with actuation of the water spray system from the operation of one of two infrared flame detectors, or a single ionized products-of-combustion detector. The flame detectors were utilized to detect the expected flaming condition; however, experimental tests indicated that the ionized products-of-combustion detectors were necessary in order to detect the smoldering and preflame conditions. The system was designed for a water application rate of 2.0 to 3.0 gpm/sq ft of floor area. The water spray nozzles selected for the system operated at a nozzle pressure of 60 psi, and the six nozzles utilized five varying discharge angles for the water patterns. The total system response time for water discharge varied from 1.3 to 2.6 seconds, including the detection of the fire occurrence.

In *Fire Protection of HEPA Filters by Using Water Spray*, J. R. Gaskill and J. L. Murrow have reported on the utilization of three nozzles flowing a total of 4 gpm to protect high efficiency particulate air filters from possible thermal exposure damage or degradation.[6] The nozzles were effective in reducing the temperatures from 1500°F to approximately 300°F at the filters, with an air flow rate of 1000 cfm. The studies on the duct filters indicated that water spray nozzles of the pin-type hollow cone discharge were more effective in the protection of the filter than solid-cone nozzles. In the low flow configuration needed, the orifice of the solid-cone nozzle tended to be susceptible to the clogging of the nozzle orifice.

With the development of missile systems in the late 1950s and the early 1960s came the need for a water spray system having unique design and operating characteristics. Thus, the missile booster water quench nozzle was designed to be installed on 1½-inch water pipes for the suppression and control of the burning of an accidentally or inadvertently ignited missile booster. The detection system for the nozzle was incorporated into the design of the nozzle, and consisted of a Pitot tube assembly connected to a metal diaphragm and a series of linked levers. The operation of the missile booster water quench nozzle was actuated by the initial rise in air pressure at approximately 4.3 psig to 4.9 psig. The nozzle was designed to be installed to withstand an operating water pressure of 200 psi maintained on the system piping at the nozzle. The nozzle was also designed to operate at nozzle pressures varying from 70 to 200 psi. The nozzle was actuated by an increase in air pressure, and was actually operated by the mechanical linkage arrangement that opened the nozzle discharge area; the nozzle discharge area consisted of seven discharge outlets. The center and largest discharge outlet was surrounded by six smaller discharge outlets.

The operation of the missile quench nozzle was caused by the flow of air or gas into the Pitot tube assembly. The flow of air or gas into the Pitot tube assembly caused the movement of the diaphragm which then caused the movement of the levers, with the cam lever revolving clockwise to release the nozzle latch; thus, water was discharged from the nozzle. Despite its unique design and actuation features, this nozzle is no longer available and is no longer being manufactured. However, some of these units may still be found in use at some industrial and military installations.

It should be noted the missile water quench nozzle was also adaptable for use with flame detector, thermal rate-of-rise, or other thermal types of detection systems if desired in different applications. The nozzle was also adapted to conventional water spray or deluge type sprinkler systems where fast operation of a large capacity water spray nozzle was needed and desired to control extremely flammable materials.

In an article titled "Fixed Extinguishing Systems—An Overview," Charles W. Bahme has reported the first installation of a water spray system was completed in 1935 to protect electrical transformers at the Hell Gate station of the Consolidated Edison Company in New York City.[7]

An analysis of the advantages and the comparative features of water spray systems for the protection of equipment (including containers, vessels, or tanks for the storage of flammable liquids or gases) and operational areas in petrochemical plants was conducted by fire protection engineers of the Union Carbide Corporation.[8] The advantages, as presented in a Union Carbide Corporation article titled "Water Spray Protection in Hazardous Areas of Petrochemical Plants," are as follows:*

> 1. Water spray replaces manpower that would otherwise be required for firefighting during emergencies (a point of special importance to a plant having few men available).
> 2. Affords time and safety for orderly control and shutdown of processes.
> 3. Gives prompt, efficient, and adequate coverage.
> 4. Prevents secondary failures of process parts and consequent spread of fire.
> 5. Costs less than suitable insulation.
> 6. Sustains cooling (removes heat) continuously throughout a fire.
> 7. Greatly reduces the losses and down time of a general fire.
> 8. Can prevent fire when used during process emergencies.

According to the Union Carbide Corporation article, the application of a water spray system discharge with a flow rate of 0.24 gpm/sq ft to uninsulated containers or equipment reduces the heat absorption rate at the surface of the container to 30 percent of what it would be without the water spray protection. With the insulated containers or vessels, a water spray system discharge application rate of 0.15 gpm/sq ft of surface area reduces the heat absorption rate lower than 15 percent of the unprotected rate on the insulated surface.

Relative to the value of insulation as a means of protecting containers or vessels of flammable liquids, it should be kept in mind that insulation only

---

*From "Water Spray Protection in Hazardous Areas of Petrochemical Plants," Union Carbide Corporation, *Fire Journal,* Vol. 9, No. 5, Sept. 1966, pp. 29–31.

retards the rate of heat transfer, thus extending the time required for the tank contents to reach dangerous temperatures and pressures. However, the water spray system discharge transfers and removes the heat input from the shell of the tank and the tank contents. The Union Carbide Corporation's analysis indicates that water spray system installation costs were less than the costs of providing insulation on containers or vessels: in 1966 their figures indicated that the cost for complete installation of water spray systems for exposure protection to containers or vessels ranged from $50 to $90 per water spray nozzle.

The Union Carbide Corporation's analysis also examined the advantages of water spray system installations as compared to large water streams from monitors. Cited are the following disadvantages of large water stream operations for the protection of vessels or containers at petrochemical plants:*

1. Water streams require manpower during emergencies when personnel are urgently needed for process purposes.

2. Require monitors when there are not enough people to manage hose lines.

3. Demand more water at higher heads than fixed protection—also increase the necessary drainage facilities.

4. Afford lower over-all coverage and wastes most of the water discharged.

5. Monitors subject the operators to danger.

Therefore, the Union Carbide Corporation's analysis indicated a preference for the exposure protection of vessels and containers with fixed water spray systems rather than large water streams from hose lines or monitor nozzles.

SUMMARY

Exposure protection for the combustible portions of buildings is usually provided by exposure sprinkler systems. Exposure sprinkler systems may be either manually or automatically operated, and nozzles are installed to protect openings such as windows, doors, and areas of combustible construction, including cornices. In their studies, Heselden and Hinkley ("Measurement of the Transmission of Radiation through Water Sprays"[2]) found the higher the nozzle pressure, the lower the amount of radiation transmitted through the water spray. Given water pressures above 70 psi at the nozzles, a flow of 4 to 5 gal/ft/min would absorb approximately 60 to 70 percent of the radiation from the fire exposure.

Water spray systems are very similar in operation to the deluge sprinkler systems previously discussed in detail in Chapter 9, "Deluge and Preaction Automatic Sprinkler Systems." Many water spray systems with automatic

---

*From "Water Spray Protection in Hazardous Areas of Petrochemical Plants," Union Carbide Corporation, *Fire Journal*, Vol. 9, No. 5, Sept. 1966, pp. 29–31.

operation utilize deluge valves for the activation of the system, with a thermal or flame type detection system as the releasing device system for the deluge valve. Water spray systems are primarily found in industrial situations, and are usually concentrated in the chemical, petrochemical, and electrical industries. Water spray systems are used to: (1) extinguish fire situations, (2) control the propagation of a fire situation, (3) provide exposure protection to containers, tanks, or vessels—primarily containers utilized for the storage of flammable liquids or gases, and (4) control spills in hazardous materials, and prevent the ignition of such materials.

The water spray systems found in most industrial situations were installed to help extinguish and control fires that could expose containers or equipment. The most common and numerous installations probably involve the protection of electrical transformers and flammable liquid or gas containers. Water spray systems have been found to be one of the most effective and efficient means of preventing the explosive rupture of flammable liquid and gas containers when the containers are exposed to fire conditions.

SI Units

The following conversion factors are given as a convenience in converting to SI units the English units used in this chapter:

$$1 \text{ square foot } = \ 0.0929 \text{ m}^2$$
$$1 \text{ Btu/hr/sq ft} = \ 3.155 \text{ W/m}^2$$
$$1 \text{ inch } \quad\quad = 25.400 \text{ mm}$$
$$1 \text{ psi } \quad\quad = \ 6.895 \text{ kPa}$$
$$\tfrac{5}{9}(°F - 32) \quad = °C$$
$$1 \text{ gpm } \quad\quad = \ 3.785 \text{ litres/min}$$
$$1 \text{ gpm/sq ft } \quad = 40.746 \text{ litres/min}^2$$

ACTIVITIES

1. Describe the most common types of automatic operation for exposure sprinkler systems. What type is usually preferred? Explain why.
2. List several locations that you are familiar with that have exposure sprinkler systems; also list the type of operation relative to automatic or manual control, with the location of the water control valves.
3. When should the water supply mains to exposure sprinkler systems be flow tested and flushed?
4. In your own words, describe the design and function of the pilot type of water spray nozzle.
5. Explain why water spray systems designed for exposure protection are, necessarily, fast-response systems.
6. List four primary uses of water spray systems.
7. List several locations that you are familiar with that have water spray systems; indicate the purpose of the systems, the manual or automatic mode of operation, and the last fire related operation of the system.

8. Determine how many and what size fire department hose lines would be required to protect an LPG storage tank consisting of 2,000 square feet of surface area in a manner equal to a water spray system.

SUGGESTED READINGS

NFPA 15, *Standard for Water Sprays Fixed Systems for Fire Protection,* 1973, NFPA, Boston.

Heselden, A. J. M. and Hinkley, P. L., "Measurement of the Transmission of Radiation Through Water Sprays," *Fire Technology*, Vol. 1, No. 2, May 1965, pp. 130–137.

BIBLIOGRAPHY

[1]NFPA 13, *Standard for the Installation of Sprinkler Systems,* 1975, NFPA, Boston.

[2]Heselden, A. J. M. and Hinkley, P. L., "Measurement of the Transmission of Radiation through Water Sprays," *Fire Technology*, Vol. 1, No. 2, May 1965, pp. 130–137.

[3]NFPA 15, *Standard for Water Spray Fixed Systems for Fire Protection,* 1973, NFPA, Boston.

[4]*Loss Prevention Data,* Factory Mutual System, Norwood, Mass., 1975.

[5]Eggleston, Lester A. and Herrera, William R., *Automatic Fire Protection System for Manned Hyperbaric Chambers*, Southwest Research Institute, San Antonio, Texas, 1971.

[6]Gaskill, J. R. and Murrow, J. L., *Fire Protection of HEPA Filters by Using Water Sprays,* Lawrence Livermore Laboratory, University of California, UCRL-73800, Livermore, Cal., July 6, 1972.

[7]Bahme, Charles W., "Fixed Extinguishing Systems—An Overview," *Firemen*, Vol. 35, No. 3, Mar. 1968, pp. 30–33.

[8]"Water Spray Protection in Hazardous Areas of Petrochemical Plants," Union Carbide Corporation, *Fire Journal,* Vol. 9, No. 5, Sept. 1966, pp. 29–31.

Chapter 12

Supervision of
Automatic Sprinkler Systems

THE NEED FOR SUPERVISION

As indicated throughout this text, the most significant single cause of failure of automatic sprinkler systems has been the result of human action: such action involves the closing of water supply control valves prior to the fire occurrence, or before the fire is under control or completely extinguished. Thus, the thermal rate of heat transmission from the diffusion flame combustion fuses too many sprinkler heads before the water is again turned on. This could be prevented and proper corrective actions initiated before a fire occurrence if responsible personnel received adequate notification whenever water control valves to sprinkler systems were closed.

The electrical contact switch supervision of water control valves to sprinkler systems, standpipes, and private fire hydrant systems is one of the primary service functions available relative to the supervision of the critical components of the automatic sprinkler system. The concept of "supervision," as discussed in this text, involves the automatic monitoring of the operational status of critical components of the automatic sprinkler system, with the reporting of this status to a responsible office, agency, or official. Thus, in many occupancies, the supervision of the water control valve for the sprinkler system is usually provided on a leased or contract basis from a central station agency. Most large industrial or manufacturing facilities have such supervisory services monitored at the appropriate plant response agency—usually the security office, fire station, or safety office at the facility. The receiving of the supervisory service signals at the plant would, of course, be characteristic of a proprietary system.

362

The types of supervisory service provided for sprinkler systems may be classified as: (1) the monitoring functions related to the various water supply components for the sprinkler system, and (2) the monitoring of the operational features of the specific sprinkler system, including the status of the principal water control valve located on the main riser of the system and any P.I.V. or P.I.V.A. valves which control the water flow to the specific sprinkler system.

SUPERVISION OF WATER SUPPLY COMPONENTS

The components of the water supply system, with the exception of the specific water control valves on the riser of the sprinkler system, will be considered as the water supply components. Thus, the monitoring of the essential components of the water supply primarily involves the functional operational status of gravity tanks, pressure tanks, and fire pumps.

Gravity Tank Supervision

The principal causes of sprinkler system failure in gravity tanks usually involves the tank not being filled with water to a level adequate enough to provide the necessary pressure or required capacity of water for the automatic sprinkler system. However, with electrical supervision of the water level in the gravity tank, positive notification is transmitted by low water level signal when the water level is 12-inches lower than the required level. Thus, corrective action may be taken to provide adequate water level and pressure for an effective supply to the automatic sprinkler system.

In temperate climates, another gravity tank water supply problem involves the failure or malfunction of tank heating systems. Such failure or malfunction can subject the water in the tanks to freezing temperatures: even a skim of ice

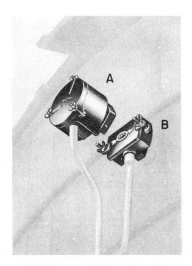

Fig. 12.1. Water level (A) and water temperature (B) indicators on a gravity tank. (A.D.T. Security Systems)

on the water surface can hamper or completely stop the flow of water from a gravity tank. When a gravity tank is being filled during the winter in an area that is subjected to freezing temperatures, the high water level signal will often prevent the tank from overflowing. The overflowing of gravity tanks has caused severe ice conditions. Such conditions can impose an excessive weight load on the tank and, in some cases, can even cause the structural failure of the tank. Figure 12.1 shows one type of supervision detector installed to monitor the temperature and the level of the water in the gravity tank by a central station agency. NFPA 71, *Standard for the Maintenance, Installation, and Use of Central Station Signaling Systems*, hereinafter referred to in this text as the *NFPA Central Station Signaling Systems Standard*, recommends that the water temperature detector activate a signal at a minimum temperature of 40°F, with a restoration signal at the proper temperature.[1]

Pressure Tank Supervision

The supervision of the air or nitrogen gas pressure is critical in pressure tanks that provide a primary or secondary water supply to a sprinkler system. This air or gas pressure supplies the necessary pressure to expel the water from the tank to the sprinkler heads that have fused in the fire area. When installed by central station agencies, the air or gas pressure supervision usually consists of a signal which indicates when the air or gas pressure has exceeded a set limit above or below the recommended pressure (usually approximately 10 psi).

Another supervisory device often found installed on pressure tanks is the water level indicator. This indicator initiates a signal for both high and low water level. The low water level condition is of critical concern because the pressure tank must have sufficient water to extinguish or control the fire situation that requires the tank as the initial or secondary supply of water to the sprinkler system. If undetected, the low water level situation can cause inadequate water pressure for the sprinkler system discharge conditions. If the water level is lowered, a larger volume of the tank will need to be filled with air or gas; thus, the pressure tank has a lower air or gas expellent pressure.

Conversely, a high water level situation in the pressure tank can be of critical concern because such a level will result in a smaller volume of the tank being available for the air or gas pressure. Obviously, such a condition could cause inadequate gas or air discharge pressure for the water from the pressure tank to adequately discharge from all the sprinklers on the topmost branch lines.

Figure 12.2 shows the installation of both the water level supervisory device and the air pressure indicator as installed on a pressure tank by a central station agency. The water level indicator (C) shown in Figure 12.2 is usually installed to indicate a water level change exceeding 3 inches above or below the normal setting. The air pressure supervisory device (D) is operated by a pressure-actuated capsule usually adjusted to initiate a signal when the air or gas pressure varies 10 psi above or below normal.

Fig. 12.2. Water level supervisory device (C) and pressure supervisory device (D) as installed on a pressure tank for a sprinkler system. (A.D.T. Security Systems)

Supervision of Fire Pumps

Fire pumps are usually installed on sprinkler and standpipe systems supplied by public or private water mains when the pressure is inadequate to meet the flow requirements from the top outlet, or the topmost line of automatic sprinklers. Fire pumps are also installed to supply both adequate capacity or flow and pressure at industrial manufacturing or storage occupancies. In addition to supplying sprinkler and standpipe systems, fire pumps also supply complete private water main and hydrant systems, as previously discussed in Chapter 4, "Water Specifications for Automatic Sprinkler Systems." When fire pumps are arranged for automatic starting by fire pump controllers (previously shown in Figure 1.5 of Chapter 1, "Standpipe and Fire Hose Systems"), it is essential to constantly monitor the operational status of the fire pump relative to the availability of electric power for the motor driven pumps. Thus, loss of the electric power needed to operate electrically motor driven fire pumps would immediately result in the activation of a supervisory signal. Also available are supervisory devices that indicate when the steam pressure for steam fire pumps reaches a minimum level of 110 percent of the pressure required to operate the pump at its rated capacity.

Various methods of transmission are used to transmit supervisory signals to many agencies. The type of alarm system used for the transmission of supervisory signals is usually determined by two factors: (1) the means of transmission, and (2) the agency receiving the signal. The various types of alarm systems for the transmission of the supervisory signals are discussed later in this chapter. It should be kept in mind, as previously indicated, supervisory signals do not have to be transmitted beyond the premises of origin. The supervisory system may be installed on a local system type of design with the supervisory signal panel consisting of both visual and audible alarm devices that are installed at a suitable location with on-premise personnel available 24 hours a day.

In many fire department communication or alarm centers, supervisory signals will not be received and the supervisory receiving equipment may not be installed. Usually, the public fire department will reserve the communication and signaling facilities in the alarm center for receiving alarms and signals that indicate an emergency condition or a fire occurrence. Therefore, although most fire departments allow the installation of equipment to receive waterflow alarms from sprinkler systems, they will not allow the installation of equipment for nonfire indicating supervisory signals. The supervisory devices and indicators usually provided to monitor the operation status of the automatic sprinkler system are discussed in the remainder of this chapter.

SUPERVISION OF SPRINKLER SYSTEM COMPONENTS

As indicated throughout this text, and as specifically tabulated in Table 2.2 of Chapter 2, "Fire Department Procedures for Automatic Sprinkler Systems," the greatest single cause of automatic sprinkler system failure is the closed water control valve. As previously noted, for years the Factory Mutual Companies have recommended the water control valves for sprinkler systems be locked in the open position. The *NFPA Sprinkler System Standard*[2] recommends the following procedures for maintaining the water control valves on sprinkler systems in an operational position:*

> Valves controlling sprinkler systems, except underground gate valves with roadway boxes, shall be supervised open by one of the following methods:
> 1. Central station, proprietary, or remote station alarm service.
> 2. Local alarm service which will cause the sounding of an audible signal at a constantly attended point.
> 3. Locking valves open.
> 4. Sealing of valves and approved weekly recorded inspection when valves are located within fenced enclosures under the control of the owner.

---

*From NFPA 13, *Standard for the Installation of Sprinkler Systems,* 1975, NFPA, Boston, p. 43.

According to the *NFPA Sprinkler System Standard*, the option of supervising the position of the water control valve with a central station, proprietary, remote station, or local alarm service is a recommended means of assuring the monitoring of the operational status of water control valves. The supervisory devices to indicate the condition of the water control valve should be installed on all of the O.S. & Y., P.I.V., or P.I.V.A. installations that control water supply mains or risers to the sprinkler systems. There is probably no other single procedure that can provide a greater increase in the reliability and operation of the automatic sprinkler system. Figure 12.3 shows a supervisory device installed on an O.S. & Y. control valve to a sprinkler system riser. This device, as installed by a central station agency, will provide a supervisory signal when the wheel of the O.S. & Y. valve is moved to close the water supply valve. The supervisory device will also provide a signal when the O.S. & Y. valve is reopened. (A supervised O.S. & Y. valve was previously shown in Figure 2.4 of Chapter 2, "Fire Department Procedures for Automatic Sprinkler Systems.")

Fig. 12.3.   Supervisory device installed on O.S. & Y. valve on sprinkler system riser.
(Fire Protection Curriculum, University of Maryland)

Similar supervisory devices are often found installed on P.I.V. and P.I.V.A. at many industrial and mercantile properties. Figure 12.4 shows a supervisory valve device installed on the side of a P.I.V.; the tubing is the metal conduit which protects the electrical wiring.

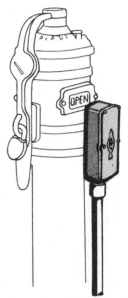

Fig. 12.4. Supervisory device installed on P.I.V.
(A.D.T. Security Systems)

Figure 7.1 of Chapter 7, "The Wet Pipe Automatic Sprinkler System," shows a horizontal P.I.V. on the outside of a building, which is also provided with valve supervision. The supervisory device can be identified by the electrical conduit leading to the device mounted on the side of the valve housing.

Supervision of the Air or Gas Pressure in Dry Pipe Sprinkler Systems

As discussed in Chapter 8, "The Dry Pipe Automatic Sprinkler System," most dry pipe valves are differential type valves. The air or gas pressure in the piping system is, therefore, able to restrain approximately 5 to 6 psi of water pressure for every psi of air or gas pressure. Because of this, low air or gas pressure, if allowed to approach the trip point of the valve, may result in complete or partial tripping of the dry pipe valve clapper. When the dry pipe valve clapper is tripped, water enters the system; should the system be exposed to freezing temperatures, physical damage and water leakage may result. Thus, low air or gas pressure detection for dry pipe automatic sprinkler systems is one type of automatic supervision which is provided by central station agencies.

Also critical is the condition caused by high air pressure on a dry pipe automatic sprinkler system. Such a condition may result in the delayed operation of the dry pipe valve and, in some situations, possible damage to components of the sprinkler system. Thus, most of the supervisory devices provided to monitor the air or gas pressure on dry pipe sprinkler systems will initiate a high air or gas pressure signal as well as a low air or gas pressure signal.

When sprinkler systems of the dry pipe type are provided with supervision of the air or gas pressure, the waterflow alarm signal which indicates the valve has tripped and that water is flowing into the system will usually be preceded by a low air or gas pressure supervisory signal from the valve. Obviously, when the sprinkler head on a dry pipe sprinkler system fuses, the air or gas is exhausted through the opening, in some cases with the assistance of a Q.O.D., and the low air or gas pressure supervisory signal is initiated. This supervisory signal is initiated prior to the pressure reaching the trip point of the valve when the valve trips and water flows into the system initiating a water-flow alarm signal. Figure 12.5 illustrates an air pressure supervisory device on a dry pipe sprinkler valve assembly. It should be noted that the supervisory devices may be utilized with either air or nitrogen gas in the dry pipe system.

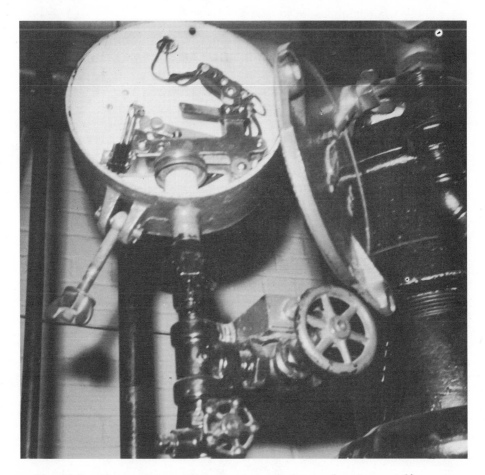

Fig. 12.5. Air or gas pressure supervisory device on dry pipe sprinkler system. (Fire Protection Curriculum, University of Maryland)

In Chapter 9, "Deluge and Preaction Sprinkler Systems," procedures for the supervision of detection devices and circuits were considered according to NFPA recommendations for the supervision of detection systems on deluge and preaction sprinkler systems having 20 or more sprinkler heads. As discussed in that chapter, one of the principal operational features of the preaction sprinkler system involved the supervision of the sprinkler piping system with a low air or nitrogen gas pressure.

Supervisory systems do not initiate corrective action relative to the conditions detected and for which a signal is transmitted. There should be an established procedure for the receipt of supervisory signals that indicate a functional deviation in the operational status of the sprinkler system or the water supply components to the sprinkler system. Beyond the receipt of the supervisory signals, there should be an established procedure for the immediate correction of the indicated malfunction as soon as possible. These corrective procedures should function at any time of day and night, including weekends, holidays, and other nonoperational times for most occupancies.

SYSTEMS FOR THE TRANSMISSION AND RECEIPT OF SUPERVISORY SIGNALS

Central Station Systems

Many occupancies that do not have their own security or fire protection personnel must rely on the services of central station agencies for supervisory service. Central station agencies are private concerns and are primarily located in the larger cities. These private concerns provide burglary, fire alarm, waterflow alarm, and supervisory services for sprinkler systems, watchmen, or other suppression systems. The agencies usually lease or rent the equipment which they install and maintain, charging periodically for services provided.

In the *Fire Protection Equipment List* published annually, Underwriters Laboratories provides a listing service for central station offices.[3] To qualify for the listing, the central station must meet certain standards relative to the number and training of the alarm operators, the location of the central station office, and the installation and servicing of the supervisory and alarm devices. The *NFPA Central Station Signaling Systems Standard* defines a central station system as follows:*

> A system, or group of systems, in which the operation of circuits and devices are signaled automatically to, record in, maintained, and supervised from an approved central station having competent and experienced observers and operators who shall upon receipt of a signal, take such action as shall be required by this standard. Such systems shall be controlled and operated by a person, firm, or corporation whose principal business is the furnishing and maintaining of supervised signaling service.

*From NFPA 71, *Standard for the Installation, Maintenance, and Use of Central Station Signaling Systems*, 1974, NFPA, Boston, p. 6.

In their 1975 listing, Underwriters Laboratories indicated there were 147 central station agencies operating with 339 offices in various cities throughout the United States. Figure 12.6 shows a typical receiving apparatus as commonly found installed in the offices of a central station agency for receiving fire alarm, waterflow alarm, and supervisory signals. On this type of receiving panel, telegraph coded signals are received with visual light and tape indication, accompanied by an audible bell for the fire signals consisting of manual fire alarm, fire detection system, or waterflow alarm. On the other side of the panel, the supervisory signals are received with a visible light and tape indication, accompanied by an audible buzzer signal. The fire signals are punched on the upper side of the paper tape, and the supervisory signals are punched on the lower side of the tape.

Central station office operators receive the coded signal which indicates the type of supervisory device sending the signal, such as low dry pipe system pressure, the identification of the dry pipe sprinkler system, and the code number for the subscriber (this indicates the name and the location of the occupancy where the dry pipe system is installed). The central station personnel then follow a prescribed procedure for notifying responsible persons at the occupancy so the condition may be corrected and a sprinkler system malfunction averted.

Fig. 12.6.   Typical central station type receiving panel. (Fire Protection Curriculum, University of Maryland)

The types of automatic supervision service available for the critical components of both the water supply and the sprinkler system have been examined. It should be noted, however, that central station agencies also provide waterflow alarm devices for all types of sprinkler systems that provide for the transmission of a waterflow fire alarm directly to the central station office when the sprinkler system valve trips. Upon receipt of a waterflow alarm, the central station office immediately notifies the fire department. The fire department then dispatches an appropriate response to the location of the sprinkler system which has tripped. Figure 12.7 is a diagram of the various types of supervisory services, devices, and waterflow alarms available. Also shown are the locations where they may usually be installed on a system with a gravity tank, pressure tank, and fire pump.

Local Systems

In occupancies with personnel in attendance 24 hours a day, such as medical or health care facilities, the supervisory signals for the sprinkler system are often transmitted to a visual or audible device at an attended location on the premises. NFPA 72A, *Standard for the Installation, Maintenance, and Use of Local Protective Signaling Systems for Watchman, Fire Alarm, and Supervisory Service,*[4] appropriately defines a local system as: "A local system is one which produces a signal at the premises protected."*

As indicated, local systems provide a signal only on the premises protected by the automatic sprinkler system. In facilities such as nursing homes and residences for the elderly, water control valves of the O.S. & Y. type at the sprinkler system riser are often supervised by means of a local system operation. Where local systems are utilized for the supervision of automatic sprinkler systems, it is essential for operating personnel to be familiar with the visual and audible alarm signals from the supervisory devices. Personnel must also be trained in the immediate and appropriate response action necessary to restore the sprinkler system to an operable condition, including the appropriate reporting action to be taken.

Remote Station Systems

The remote station system is usually installed where on-premise personnel are not available on a continuing around-the-clock basis, and where the community does not have central station system service available. The remote station system usually involves the transmission of the supervisory signals to the fire or police department by means of leased private telephone lines. When fire alarm and waterflow signals are transmitted on the system in conjunction with the supervisory signals, the system should always be connected directly to the fire department alarm communication headquarters. The system should not be connected to the most convenient fire station, since personnel at such sta-

---

*From NFPA 72A, *Standard for the Installation, Maintenance, and Use of Local Protective Signaling Systems for Watchman, Fire Alarm, and Supervisory Service,* 1974, NFPA, Boston, p. 6.

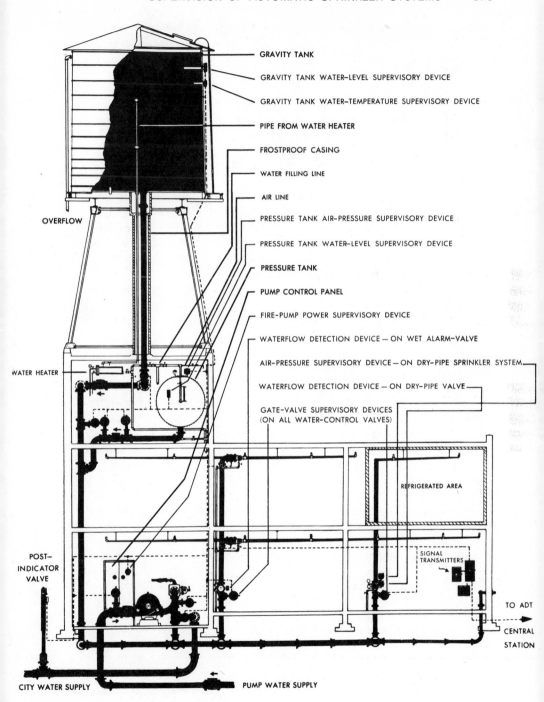

GRAVITY TANK

GRAVITY TANK WATER–LEVEL SUPERVISORY DEVICE

GRAVITY TANK WATER–TEMPERATURE SUPERVISORY DEVICE

PIPE FROM WATER HEATER

FROSTPROOF CASING

WATER FILLING LINE

AIR LINE

PRESSURE TANK AIR–PRESSURE SUPERVISORY DEVICE

PRESSURE TANK WATER–LEVEL SUPERVISORY DEVICE

PRESSURE TANK

PUMP CONTROL PANEL

FIRE–PUMP POWER SUPERVISORY DEVICE

WATERFLOW DETECTION DEVICE — ON WET ALARM–VALVE

AIR–PRESSURE SUPERVISORY DEVICE — ON DRY–PIPE SPRINKLER SYSTEM

WATERFLOW DETECTION DEVICE — ON DRY–PIPE VALVE

GATE–VALVE SUPERVISORY DEVICES
(ON ALL WATER–CONTROL VALVES)

OVERFLOW

WATER HEATER

REFRIGERATED AREA

POST–
INDICATOR
VALVE

SIGNAL
TRANSMITTERS

TO ADT

CENTRAL

STATION

CITY WATER SUPPLY

PUMP WATER SUPPLY

*Fig. 12.7.   Supervisory services and devices as they would be installed on a complete
automatic sprinkler system.* (A.D.T. Security Systems)

tions may not be available to receive signals when the company is out of quarters. NFPA 72C, *Standard for the Installation, Maintenance, and Use of Remote Station Protective Signaling Systems*[5] indicates the following characteristics of remote station systems:*

> The provision of this standard contemplates a system of electrically supervised circuits employing a direct circuit connection between signaling devices at the protected premises and signal receiving equipment in a remote station, such as a municipal fire alarm headquarters, a fire station, or other location acceptable to the authority having jurisdiction.

There are some fire departments and other municipal or governmental authorities that allow the fire alarm and waterflow alarm signals to be transmitted to the fire department by means of a remote station system: however, they do not allow the supervisory signals to be transmitted to the fire department. The supervisory signals are not considered to be emergency signals requiring immediate fire department response or action.

Thus, in some locations, there may be situations where the supervisory signals have to be provided on a local system: otherwise, supervisory service—although desirable—may not be available due to the unavailability of a central station agency.

Proprietary System

Proprietary systems are found at government installations or facilities, and at large industrial complexes. The proprietary system is utilized when the facility has organized and trained security or fire protection personnel. NFPA 72D, *Standard for the Installation, Maintenance, and Use of Proprietary Protective Signaling Systems for Watchman, Fire Alarm, and Supervisory Service*[6] defines a proprietary system as follows:**

> A protective signaling system under constant supervision by competent and experienced personnel in a central supervising station at the property protected. The system includes equipment and other facilities required to permit the operators to test and operate the system, and upon receipt of a signal, to take such action as shall be required under the rules established for their guidance by the authority having jurisdiction. The system shall be maintained and tested by owner personnel or an organization satisfactory to the authority having jurisdiction.
>
> Noncontiguous properties under a single ownership may be considered as "the property" and be connected to a single central supervising station.[6]

Proprietary systems are generally similar to central station systems or remote station systems, the exception being that proprietary systems are limited to the

---

*NFPA 72C, *Standard for the Installation, Maintenance, and Use of Remote Station Protective Signaling Systems*, 1974, NFPA, Boston, p. 5.
**NFPA 72D, *Standard for the Installation, Maintenance, and Use of Proprietary Protective Signaling Systems for Watchman, Fire Alarm, and Supervisory Service*, 1974, NFPA, Boston, p. 7.

premises of the property protected and the signals are received by private fire department or security forces employed by the facility.

Most large industrial complexes and governmental bases utilize proprietary signaling systems for the transmission of fire alarms, waterflow alarms, and the supervisory signals from the sprinkler system and other components of the water supply system.

SUMMARY

The provision of supervisory systems to monitor and detect any malfunction in the operational status of the automatic sprinkler system is one of the most effective means of reducing the sprinkler system failure rate and increasing the operating effectiveness of the automatic sprinkler system. Supervisory systems consist of detection and monitoring devices, the circuits to transmit the signals from the monitoring devices, and the receiving of the supervisory signals at a location having personnel in attendance at all times. However, the initiation of prompt remedial action requires a detailed and predetermined procedure involving key personnel to provide prompt restoration of the automatic sprinkler system to an operational status. The most important type of supervisory service that may be provided would appear to be the supervision of all the water control valves for the automatic sprinkler and any other fire control systems such as water-spray or standpipe systems.

Supervisory signals may be transmitted and processed by any one of the four recommended signaling systems recognized by the NFPA, including a central station system, a local system, a remote station system, or a proprietary system. It should be emphasized that supervisory signals are not emergency signals; thus, supervisory signals will not be handled by many fire departments on a remote station system type of operation. Where industrial or private fire departments are the receiving agencies for proprietary systems, there is usually no problem because of the fire department's responsibility to the principal industrial or governmental agency. It should be recognized that private facility fire departments usually provide most of the supervisory maintenance functions for automatic sprinkler systems as installed in the industrial property. Where central station systems are utilized for supervisory services, the maintenance of the systems and devices are assumed by the central station agency.

SI Units

The following conversion factors are given as a convenience in converting to SI units the English units used in this chapter:

$$1 \text{ inch} = 25.400 \text{ mm}$$
$$1 \text{ psi} = 6.895 \text{ kPa}$$

ACTIVITIES

1. In your own words, write an explanation that describes why you believe there is a need for the supervision of automatic sprinkler systems. Include in your explanation the classification of the principal types of supervisory services provided.
2. (a) What are the usual causes of sprinkler system failure in gravity tanks and in pressure tanks?
   (b) Describe the corrective actions that can be taken to help avoid such failures.
3. Explain the operation of the following supervisory signaling systems. Where is each predominantly used?
   (a) central station system
   (b) local system
   (c) proprietary system
   (d) remote station system
4. Why is it necessary to constantly monitor fire pumps that are arranged for automatic starting by fire pump controllers?
5. Explain the advantages and disadvantages of having supervisory service signals transmitted by a local system, a central station system, a remote station system, or a proprietary system, including in your explanation where each is predominately used.
6. List the names and addresses of the central station offices serving your community. If no central station office serves your community, explain why you think such service has not been provided.
7. (a) Where central station systems are not available in an area, do you think the fire department should provide such service by means of a remote station system?
   (b) List the advantages and disadvantages to the fire department and the community.
8. Make a list of the locations in your area that have some type of supervisory service on sprinkler, water spray, standpipe, or water supply systems.

SUGGESTED READINGS

Alarm Accessories for Automatic Water Supply Control Valves for Fire Protection Service, U.L. 753, 1971, Underwriters Laboratories, Inc., Northbrook, Ill.

Central Stations for Watchman, Fire Alarm, and Supervisory Services, U.L. 827, 1973, Underwriters Laboratories, Inc., Northbrook, Ill.

NFPA 72D, *Standard for the Installation, Maintenance, and Use of Proprietary Protective Signaling Systems for Guard, Fire Alarm, and Supervisory Service*, 1974, NFPA, Boston.

"Protective Signaling Services—Central Station," *Fire Protection Equipment List,* Underwriters Laboratories, Inc., Northbrook, Ill., January 1975, pp. 157–165.

"Waterflow Alarms and Sprinkler System Supervision," *Fire Protection Handbook*, 14th ed., 1976, Chapter 14–5, National Fire Protection Association, Boston.

BIBLIOGRAPHY

[1]NFPA 71, *Standard for the Installation, Maintenance, and Use of Central Station Signaling Systems*, 1974, NFPA, Boston.

[2]NFPA 13, *Standard for the Installation of Sprinkler Systems*, 1975, NFPA, Boston.

[3]*Fire Protection Equipment List*, Underwriters Laboratories, Inc., Northbrook, Ill., January 1975.

[4]NFPA 72A, *Standard for the Installation, Maintenance, and Use of Local Protective Signaling Systems for Watchman, Fire Alarm, and Supervisory Service*, 1974, NFPA, Boston.

[5]NFPA 72C, *Standard for the Installation, Maintenance, and Use of Remote Station Protective Signaling Systems*, 1974, NFPA, Boston.

[6]NFPA 72D, *Standard for the Installation, Maintenance, and Use of Proprietary Protective Signaling Systems for Guard, Fire Alarm, and Supervisory Service*, 1974, NFPA, Boston.

SUBJECT INDEX

X Y Z

5M—9/76—BG—VH (5M)